Handbook of Price Impact Modeling

Handbook of Price Impact Modeling provides practitioners and students with a mathematical framework grounded in academic references to apply price impact models to quantitative trading and portfolio management. Automated trading is now the dominant form of trading across all frequencies. Furthermore, trading algorithms' rise introduces new questions professionals must answer, for instance:

- How do stock prices react to a trading strategy?
- How to scale a portfolio considering its trading costs and liquidity risk?
- How to measure and improve trading algorithms while avoiding biases?

Price impact models answer these novel questions at the forefront of quantitative finance. Hence, practitioners and students use the Handbook as a comprehensive, modern view of systematic trading.

For financial institutions, the Handbook's framework aims to minimize the firm's price impact, measure market liquidity risk, and provide a unified, succinct view of the firm's trading activity to the C-suite via analytics and tactical research.

The Handbook's focus on applications and everyday skillsets makes it an ideal textbook for a master's in finance class and students joining quantitative trading desks. Using price impact models, the reader learns how to:

- Build a market simulator to back test trading algorithms
- Implement closed-form strategies that optimize trading signals
- Measure liquidity risk and stress test portfolios for fire sales
- Analyze algorithms' performance controlling for common trading biases
- Estimate price impact models using the public trading tape

Finally, the reader finds a primer on the database kdb+ and its programming language q, which are standard tools for analyzing high-frequency trading data at banks and hedge funds.

Authored by a finance professional, this book is a valuable resource for quantitative researchers and traders.

Chapman & Hall/CRC Financial Mathematics Series

Recently Published Titles

Machine Learning for Factor Investing: Python Version
Guillaume Coqueret and Tony Guida

Introduction to Stochastic Finance with Market Examples, Second Edition
Nicolas Privault

Commodities: Fundamental Theory of Futures, Forwards, and Derivatives Pricing, Second Edition
Edited by M.A.H. Dempster and Ke Tang

Introducing Financial Mathematics: Theory, Binomial Models, and Applications
Mladen Victor Wickerhauser

Financial Mathematics: From Discrete to Continuous Time
Kevin J. Hastings

Financial Mathematics: A Comprehensive Treatment in Discrete Time
Giuseppe Campolieti and Roman N. Makarov

Introduction to Financial Derivatives with Python
Elisa Alòs and Raúl Merino

Handbook of Price Impact Modeling
Kevin T. Webster

Sustainable Life Insurance: Managing Risk Appetite for Insurance Savings & Retirement Products
Aymeric Kalife with Saad Mouti, Ludovic Goudenege, Xiaolu Tan, and Mounir Bellmane

Geometry of Derivation with Applications
Norman L. Johnson

For more information about this series please visit: https://www.crcpress.com/Chapman-and-HallCRC-Financial-Mathematics-Series/book series/CHFINANCMTH

Handbook of Price Impact Modeling

Kevin T. Webster

CRC Press is an imprint of the
Taylor & Francis Group, an **informa** business
A CHAPMAN & HALL BOOK

First edition published 2023
by CRC Press
6000 Broken Sound Parkway NW, Suite 300, Boca Raton, FL 33487-2742

and by CRC Press
4 Park Square, Milton Park, Abingdon, Oxon, OX14 4RN

CRC Press is an imprint of Taylor & Francis Group, LLC

Library of Congress Cataloging-in-Publication Data

Names: Webster, Kevin Thomas, author.
Title: Handbook of price impact modeling / Dr. Kevin Thomas Webster.
Description: First edition. | Boca Raton : CRC Press, 2023. | Includes bibliographical references and index.
Identifiers: LCCN 2022053015 (print) | LCCN 2022053016 (ebook) | ISBN 9781032328225 (hardback) | ISBN 9781032328232 (paperback) | ISBN 9781003316923 (ebook)
Subjects: LCSH: Stocks--Prices--Mathematical models. | Investments--Mathematical models.
Classification: LCC HG4636 .W43 2023 (print) | LCC HG4636 (ebook) | DDC 332.63/222--dc23/eng/20230104
LC record available at https://lccn.loc.gov/2022053015
LC ebook record available at https://lccn.loc.gov/2022053016

ISBN: 978-1-032-32822-5 (hbk)
ISBN: 978-1-032-32823-2 (pbk)
ISBN: 978-1-003-31692-3 (ebk)

DOI: 10.1201/9781003316923

Typeset in font CMR10
by KnowledgeWorks Global Ltd.

Publisher's note: This book has been prepared from camera-ready copy provided by the authors.

Contents

Preface

This Handbook describes algorithmic trading through the lens of *price impact.* Price impact plays a core role in finance: it models how trades cause price moves. Concretely, price impact quantifies trading costs and is essential to executing transactions, scaling portfolios, and building trading strategies.

Two main parts structure the Handbook to reflect how practitioners use price impact models:

- Acting on price impact
- Measuring price impact

Acting on price impact takes a price impact model as given: the focus is on deriving, testing, and implementing trading strategies from signals using a price impact model. The primary tool is *stochastic control theory,* an extension of stochastic calculus.

Measuring price impact estimates a price impact model from market data, typically in conjunction with trading signals: price predictors called *alpha signals* and liquidity estimates, such as volume curves. The core issue is the presence of common trading biases. Statisticians solve such biases using *causal inference*, an extension of Bayesian probability theory.

Instructors can use the Handbook as a graduate course on algorithmic trading. Furthermore, it is a reference for practitioners at the forefront of quantitative finance.

Preface to Practitioners

The Handbook's framework relates three essential trading variables:

- alpha signals,
- market activity, and
- price impact

using closed-form formulas for real-life trading problems. Practitioners measure, estimate, or predict these variables in real time to implement trading strategies, scale portfolios, and remove trading biases.

Traders can study chapters individually: each chapter begins with a straightforward example to describe its core issue. Furthermore, Chapters 3, 4, 6, and 7 summarize practical formulas for trading desks at the end. Therefore, these summaries serve as references for practitioners. In addition, the Handbook implements two concrete types of trading experiments. First, Section 4.2.3 from Chapter 4 modifies historical prices to respond to new trades and simulates transaction costs in back tests. Second, Section 5.3 from Chapter 5 outlines best practices for A-B tests to confront trading biases. Indeed, simulation and live trading experiments are crucial to TCA, as per Section 3.3 of Chapter 3.

Finally, most quantitative traders and researchers interact with kdb+ and q daily but may find current documentation aimed at developers. Therefore, Appendix A provides a researcher- and trader-friendly introduction to kdb+ to analyze execution data and build trading models.

Preface to Instructors

The Handbook covers a one-semester course:

I The non-technical introduction in Chapter 1 equips the students with the essential language and concepts of modern trading.

II *Acting on price impact* builds out mathematical tools in Chapter 2 before delving into finance applications in Chapters 3 and 4.

III *Measuring price impact* introduces the theory of causal inference in Chapter 5. Then, Chapter 6 solves ubiquitous trading biases. Finally, Chapter 7 estimates price impact models on public trading data emphasizing reproducibility.

The mathematical results are purposefully general: they reflect quantitative trading's current focus on non-parametric trading signals, such as those estimated through machine learning models. Instructors should emphasize this practice but can prove simplified results to their students. For instance, each chapter begins with pedagogical examples.

Chapters are self-standing. Therefore, instructors can reuse parts or chapters within other courses.

- Chapter 1 introduces students to standard trading terms: one includes it in any trading class.

- A general course on stochastic control can delve into Chapter 2 for applications.

- Chapter 3 solves real-life order execution problems and fits in a market microstructure class.
- Chapter 4 applies price impact to portfolio construction, a broad topic in finance courses.
- One can study the trading biases in Chapters 5 and 6 in a probability theory course for current or future finance professionals. Simpson's paradox, covered in the pedagogical example of Section 5.1, is of particular interest in the growing field of Transaction Cost Analysis (TCA) at banks and funds.
- Instructors with access to public trading data can reproduce the empirical price impact study in Chapter 7 as student coursework. The study familiarizes students with real-life trading data and models.

Chapters 3, 4, 6, and 7 end with summaries of practical results. Instructors can expect their students to apply those results to real-life trading problems.

Preface to Students

Students should view each chapter as four distinct blocks:

(a) The pedagogical example at the chapter's start.

(b) The summary of results at the chapter's end.

(c) The chapter's exercises.

(d) The chapter's core (everything else).

The ordering reflects each block's difficulty. Therefore, students need not worry about an advanced block if a previous one remains unclear. For instance, one tackles the exercises after understanding the chapter's pedagogical example and the chapter's summary of results. In addition, students' first read of the chapter's core can be cursory to learn how practitioners formulate trading problems.

Finally, newcomers to trading teams will likely encounter the database kdb+ and its coding language q. Therefore, Appendix A introduces students to analyzing sizable trading data in kdb+.

Acknowledgments

The author is indebted to the fantastic academic community, particularly the Department of Mathematics at Imperial College London, where the author is a visiting associate professor (reader). The author also has the privilege of teaching classes on price impact at the Mathematics in Finance (MAFN) Masters at Columbia University and the Mathematics and Finance MSc at Imperial College London. Special thanks to Johannes Muhle-Karbe, Nick Westray, and Zexin Wang for in-depth discussions on academic price impact models. To name but a few, thank you to D. Nehren, P. Kolm, R. Carmona, J.P. Bouchaud, S. Stoikov, C.A. Lehalle, D. Itkin, M. Nutz, R. Cont, E. Neuman, U. Horst, M. Voss, J. Jacquier, J. Gatheral, and I. Mastromatteo for their valuable time. The author also thanks LOBSTER for providing their data to Imperial College, London. Finally, special thanks to Bob Ross and Beth Hawkins for the editorial work on the book.

Part I

Introduction

Part I motivates the Handbook. The reader learns:

(a) The concept of price impact and its significance to algorithmic trading.

(b) The language traders use to formulate strategies.

(c) Four essential real-life applications of price impact.

Chapter 1 is non-technical and gently introduces trading terms used throughout the Handbook.

1

Introduction to Modeling Price Impact

"A common theme echoed by nearly all market professionals, academic researchers, and other students of the securities markets is that algorithmic trading, in one form or another, is an integral and permanent part of our modern capital markets."
– The U.S. Securities and Exchange Commission (2020)[213]

1.1 The Handbook's Scope

1.1.1 Introduction

This Handbook deals with developing and applying price impact models for algorithmic trading, the predominant form of trading of stocks. The *Staff Report on Algorithmic Trading in U.S. Capital Markets* (2020)[213] by the U.S. Securities and Exchange Commission (SEC) reflects on the material role algorithmic trading has taken in modern capital markets.

> "The use of algorithms in trading is pervasive in today's markets" (p. 5).

Trading algorithms require three essential ingredients: *data, technology,* and *models.* Of the models needed for algorithmic trading, price impact models have a unique, direct effect on the profitability of trading in the real world: price impact models establish the actual cost of trading at scale.

One premise of the Handbook is that trading teams develop and use price impact models in an ad-hoc fashion, for example, after observing inconsistencies between their simulated and live trading data. Unfortunately, such reactive research leads to biases, overlooked assumptions, and inconsistencies in approach. Therefore, it is essential to provide a theoretical and practical framework for price impact to develop models for algorithmic trading systematically. A trader building a real-life algorithm must understand their trading's

DOI: 10.1201/9781003316923-1

influence on stock prices. Price impact models quantify such effects and enable traders to deploy strategies at scale while controlling their execution costs.

The Handbook's focus is the algorithmic trading of *stocks*. However, the reader can adapt the price impact models to other electronic markets, including the fixed income and foreign exchange markets. Section 7.5.3 of Chapter 7 reviews methodologies employed in the literature to model other assets' price impact.

The Handbook refers to trading data and technology as *the trading infrastructure* and assumes that the trading team's infrastructure is appropriately set up and scaled for live trading and research. For instance, Velu, Hardy, and Nehren provide a detailed guide to setting up one's trading and research infrastructure in their book *Algorithmic Trading and Quantitative Strategies* (2020)[227] in Chapter 12, "The Research Stack" (pp. 401–410). Velu, Hardy, and Nehren conclude on trading models and infrastructure:

> "Done correctly though, this should become a real engine for innovation and performance improvement, and a real competitive advantage in a highly fierce field" (p. 410).

The modeling principles in Section 1.1.4 apply to any robust trading infrastructure. Therefore, the core issue is the models used to deal with price impact.

Price impact models must have a solid theoretical and mathematical basis. However, one must be able to translate that mathematical basis into a practical algorithm that is simple enough to be verifiable, back-tested, and deployed to live trading. This Handbook reviews the theoretical basis of price impact models and applies these models to bring research to production in live trading.

1.1.2 What is price impact? Why do traders care about it?

Despite the widespread use of price impact in trading, the term has no uniform definition. For instance, traders may refer to price impact as market impact. Three example definitions follow:

- In their book *Trades, Quotes and Prices* (2018)[37], Bouchaud et al. answer the question, "What is price impact?"

  > "Price impact is a reaction to order-flow imbalance" (p. 209).

- Velu, Hardy, and Nehren (2020)[227] answer the question, "What is market impact?"

  > "Market impact is the cumulative market-wide response of the arrival of new order flow" (p. 314).

- Bouchaud et al. introduce the propagator model of price impact in their seminal paper *Fluctuations and response in financial markets: the subtle nature of "random" price changes* (2004)[34].

  > "We define and study a model where the price, at any instant, is the result of the impact of all past trades, mediated by a non constant 'propagator' in time that describes the response of the market to a single trade" (p. 1).

In summary, price impact's underlying economic idea is

Trading of stock cause price moves for the stock that otherwise would not have happened.

Price impact analyzes trading's effect on stock prices. The study of price impact is distinct from the study of the stock's underlying value: economic theory models the long-term behavior of stock prices using rational expectations and fundamental efficiency. On trading timescales, there is a large body of evidence that price impact and other feedback loops dominate stock price dynamics. For example, Bouchaud et al. (2018, p. 35)[37] estimate it takes years for an analyst to measure fundamental value's contribution to stock prices. Feedback models of trading, such as price impact, explain why fundamental value contributes little to stock prices in the short to medium term.

> For example, "all quantitative volatility/activity feedback models suggest that 80% of the price variance is induced by self-referential effects. This adds credence to the idea that the lion's share of the short- to medium-term activity of financial markets is unrelated to any fundamental information or economic effects" (p. 36, [37]).

The economic literature names this phenomenon *the volatility puzzle*. The reader studies Campbell, Lo, and MacKinlay (1997)[54].

Price impact is a fundamental issue for traders, determining when and how they should intervene in the market. Price impact also influences how long-term portfolio managers trade their stock positions. For example, capital.com [55] describes how a portfolio manager considers price impact when investing.

> "Market impact costs are typically used and considered by large financial institutions when determining the viability of a security purchase."

If practitioners understand issues concerning price impact, they know trading's effect on stock prices. Traders have built careers predicting stock prices considering the interaction between three types of data:

(a) Publicly available trading tapes, such as NYSE Trade And Quote (TAQ) or Nasdaq Equities Market Data[1]

(b) The trader's internal order flow, such as orders submitted by portfolio managers

(c) Publicly available *trading catalysts*, such as announcements of a merger or takeover between companies[2]

Traders act on observed price dislocations based on their understanding of how trades impact prices. For example, Bechler and Ludkovski [21] model how traders adapt and capture better prices when considering the observed order flow imbalance in the market.

> "Incorporating order flow imbalance leads to the consideration of the current market state and specifically whether one's orders lean with or against the prevailing order flow" (p. 1).

Conversely, traders minimize their orders' price impact to reduce their execution costs. Finally, catalysts change how trades impact prices: traders react by updating their models and algorithms.

Of course, traders have always sought to understand and act on price dislocations using the available data and their intuition. However, with algorithmic trading, the focus has shifted away from individual traders' intuition and now demands a systematic approach to trading. Price impact models quantify how trades cause price dislocations and provide reproducible actions for trading strategies.

1.1.3 The causality challenge for price impact models

> "[O]ne would ideally like to assess the impact of a market order by somehow measuring the difference between the mid-price in a world where the order is executed and the mid-price in a world where all else is equal but where the given order is not executed [...] this definition

[1]NYSE and Nasdaq provide customers with public trading data, and the historical datasets are helpful for building price impact models. For example, NYSE describes TAQ data on its website [90]:

> "Daily TAQ (Trade and Quote) provides users access to all trades and quotes for all issues traded on NYSE, Nasdaq and the regional exchanges for the previous trading day. It's a comprehensive history of daily activity from NYSE markets and the U.S. Consolidated Tape covering all U.S. Equities instruments."

In addition, Nasdaq describes the Nasdaq Equities Market Data on their website [172]:

> "Nasdaq Equities Market Data powering trading and investment strategies for a full range of exchange-listed equities in the US, Nordics and Canada."

[2]For example, the paper "Signal Processing: Systematic M&A Arbitrage" by Wang et al. (2016)[233] builds trading signals from merger announcements.

cannot be implemented in a real financial system, because the two situations (i.e. the market order arriving or not arriving) are mutually exclusive, and history cannot be replayed to repeat the experiment in the very same conditions" (pp. 210–211, Bouchaud et al. (2018)[37])

A trading team's mandate is to change and improve trading strategies. But, as illustrated by the thought experiment of Bouchaud et al. (2018)[37], practitioners face a striking hurdle:

A trader cannot directly observe two strategies' outcomes for a single trade.

Because trades cause price impact, traders' actions cause a feedback loop on the data they observe. The Handbook terms this hurdle *the causality challenge* and describes its consequences mathematically in Part III: Measuring Price Impact.

The causality challenge brings to attention three hazards:

- *Fundamental measurement uncertainty*

  Assume a trader submits an order considering a return prediction, also called an alpha signal. It is impossible to conclude whether the trade *predicted* or *caused* the price move. Bouchaud et al. (2018)[37] call this phenomenon "prediction bias" (p. 238). Section 6.3 of Chapter 6 applies causal inference as a practical solution to prediction bias.

- *Price manipulation paradoxes*

  Unfortunately, many models that statistically describe trades' price impact present price manipulation paradoxes. These paradoxes incorrectly imply arbitrary profits via complex, counter-intuitive trading patterns.[3,4] A price manipulation paradox prevents an algorithm from trading or causes the algorithm to submit counter-intuitive trades. Section 2.7 of Chapter 2 connects price manipulation paradoxes and liquidity dynamics.

- *Hidden assumptions*

  Trading infrastructures are technologically complex and can harbor hidden assumptions. The measurement uncertainty caused by the *causality challenge* accentuates trading strategies' sensitivity to such confounding.

  For example, a standard assumption in trading is that orders do not overlap in their execution and price impact. This assumption holds if the trader

---

[3] In his seminal paper *No-Dynamic-Arbitrage and Market Impact* (2010)[104] on price impact models with price manipulation, Gatheral defines "*price manipulation* to be a round-trip trade Π whose expected cost $C[\Pi]$ is negative" (p. 7).

[4] The SEC published a set of case studies titled *Market Manipulation* (2019)[212] describing trading behaviors considered cause for concern and investigation. The SEC defines a common thread among manipulative trading patterns the agency looks out for as 'Bizarre Trading'—is there any sense in the trading you see?" (p. 24). The SEC's concern provides an additional incentive for trading algorithms to avoid counter-intuitive trading patterns.

spaces orders on the same stock away from each other. However, the assumption fails when the trader submits orders on the same stock simultaneously or sequentially. Trading systems may enforce this premise at the source by allowing only a single active order per stock. However, such a *single active order* assumption is arduous to implement across multiple, potentially competing, internal trading teams, leading to what Bouchaud et al. (2018)[37] call "issuer bias" (p. 239).

> "[A] bias may occur if a trader submits several dependent metaorders successively" (p. 239).

Section 3.3.4 of Chapter 3 and Section 6.6 of Chapter 6 provide solutions to issuer bias.

1.1.4 Four core modeling principles

The Handbook proposes four core modeling principles for price impact to remove the obstacles that arise from the *causality challenge*. The four core modeling principles demand that a price impact model be:

(a) *Intuitive*

An *intuitive* price impact model is a model for which a trader can quickly draw conclusions when changing a trading strategy's basic parameters: *size*, *speed*, *market correlation*, and *market conditions*.

(b) *Actionable*

An *actionable* price impact model is a model that helps a trading team establish which strategy performs best. The model should enable researchers to *simulate* how the market would react to different trading strategies. In addition, the model should rule out price manipulation paradoxes.

(c) *Testable*

A *testable* price impact model is a model that a statistician can identify with high confidence *using the available trading data*. The model must outline premises in the data, and the fitting procedure must account for these assumptions.

(d) *Robust*

A *robust* price impact model is a model for which a trader can measure their strategy's sensitivity to underlying data assumptions.

All four core modeling principles refer to a team's trading strategy, infrastructure, or both. Each team sets up its trading strategies and infrastructure differently. The Handbook does not feature trading strategies or infrastructure but focuses on mathematical tools, concrete examples, and applications. The resulting framework enables readers to build trading strategies based on *their* needs and infrastructure.

1.1.5 A brief history of price impact models

The Handbook builds upon a long history of modeling price impact for trading. It spans decades and builds upon the expertise of four communities of academics and practitioners. Throughout the Handbook, extensive footnotes tie its framework to the broader literature. These footnotes serve as a *Rosetta stone* for the rich and deep literature on price impact.

Each wave of price impact models captures some of the Handbook's four core modeling principles, and the following chapters reflect the insights from these research strands. In addition, this section introduces each wave's seminal works.

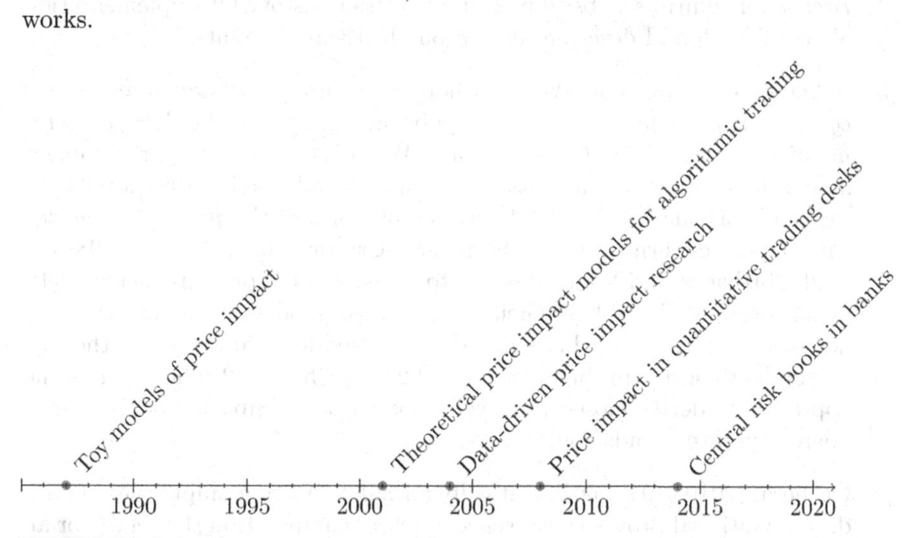

FIGURE 1.1
Illustrative chronology of price impact models.

Economists have built so-called toy models of trade-induced price dislocations since the 1980s. The seminal work in this sub-field is the paper *Continuous auctions and insider trading* (1985)[143] by Kyle. In this literature, price impact arises from idealized market makers updating their prices in reaction to a perfectly knowledgeable insider trader. In addition, Bertsimas and Lo (1998)[26] present a theoretical stochastic control problem to model trading considering price impact. These toy models are illustrative and build intuition for manual trading, also called high-touch trading.

Models with practical actions for algorithms began in the Mathematical Finance community in the early 2000s. Practitioners implemented these price impact models directly in algorithmic trading strategies. The focus during this wave was on developing control problems for use in live trading, proving the absence of price manipulation paradoxes, and deriving closed-form formulas for simple trading strategies.

(a) To the author's best knowledge, the paper *Optimal Execution of Portfolio Transactions* (2001)[10] by Almgren and Chriss is the first paper to articulate an algorithmic execution strategy as a control problem for live trading. Almgren and Chriss (2001)[10] led the widespread use of *implementation shortfall algorithms* that solve a control problem trading off a trading signal, risk aversion, and price impact. This control problem moved price impact models away from illustrative toy models. From that point onward, live trading algorithms modeled price impact to automate execution. Robert Almgren and Christian Hauff founded the firm *Quantitative Brokers*. The firm's website provides "A Brief History Of Implementation Shortfall" [38] and describes execution algorithms' evolution.

(b) In 2005, Obizhaeva and Wang authored the paper *Optimal trading strategy and supply/demand dynamics* (published in 2013 in the *Journal of Financial Markets*)[180]. Obizhaeva and Wang introduced the price impact model the Handbook builds its framework around: the Handbook refers to the model as the *OW model.* It is a simple form of the *propagator model.* The papers by Fruth, Schöneborn, and Urusov (2013, 2019)[97, 98] extend Obizhaeva and Wang's results to a vast class of price impact models. Obizhaeva and Wang base their reduced-form model on a microstructure description of the order book and derive closed-form formulas for the corresponding control problem. Section 2.2.2 of Chapter 2 follows the same approach to derive closed-form solutions to price impact models from a microstructure foundation.

(c) Gatheral (2010)[104] mathematically formalizes price manipulation. In addition, Gatheral proves the absence of price manipulation strategies for an extensive class of propagator models.

(d) The paper *Optimal execution strategies in limit order books with general shape functions* (2010)[7] by Alfonsi, Fruth, and Schied introduces globally concave propagator models. The Handbook refers to this price impact model as the *AFS model.* Gatheral, Schied, and Slynko (2011)[106] study control problems under the AFS model.

(e) Both Bechler and Ludkovski (2015)[21] and Cartea and Jaimungal (2015)[63] extend the Almgren and Chriss (2001)[10] control problem to consider the trades from the rest of the market.

(f) The paper *Optimal execution with non-linear transient market impact* (2017)[80] by Curato, Gatheral, and Lillo solves a control problem with a locally concave propagator model using numerical methods. Local concavity is a crucial but technical feature of price impact models. The Handbook implements an alternative approach by Carmona and Webster (2019)[61] in Section 2.6.3 of Chapter 2 to solve the locally concave OW model in closed form.

The Econophysics community tested price impact models using high-frequency data and led a series of empirical discoveries and practical implications for trading. This data-driven push developed in the early 2000s but accelerated in the late 2000s, and a consensus class of models emerged for price impact: propagator models. Bouchaud et al. (2018)[37] cover this empirical literature.

(a) The paper *Master curve for price-impact function* (2003)[152] by Lillo, Farmer, and Mantegna introduces the *response function* of price impact and analyzes four years of data for 1,000 stocks. With the proper normalizations, Lillo, Farmer, and Mantegna show that price impact is a universal phenomenon that a trader can measure with high confidence. Consequently, the response function moved price impact from a theoretical model to a concrete variable that practitioners measure and confront daily against real-life data. Chapter 7 highlights how the reader concretely observes and quantifies price impact on public trading data.

(b) Bouchaud et al. (2004)[34] analyze Paris stock data and introduce core concepts of the empirical price impact literature.

- Trades on the public tape exhibit long-term autocorrelation and cause price moves as measured by the *response function*.
- Prices are approximately diffusive and exhibit no autocorrelation.
- Bouchaud et al. empirically reconcile the apparent contradiction between predictable trades and unpredictable returns with a price impact model: the *propagator model*.

(c) The paper *Direct Estimation of Equity Market Impact* by Almgren et al. (2005)[11] confronts price impact models with institutional order data. The data is internal to the Citigroup US equity trading desk. Using proprietary trading data with explicit trading intentions, including order start and end times, is a regime shift compared to using public trading data alone.[5]

(d) Cont, Kukanov, and Stoikov (2014)[74] extend the empirical study of price impact to include all order book events.[6] Cont, Kukanov, and Stoikov

- extend empirical results on price impact to limit orders and liquidity-providing strategies and

---

[5]Bershova and Rhakhlin (2013)[25] use data internal to *AllianceBernstein LP* in their paper *The Non-Linear Market Impact of Large Trades: Evidence from Buy-Side Order Flow*. Bouchaud et al. (2015)[35] "use an heterogeneous dataset provided by ANcerno, a leading transaction cost analysis provider. The structure of this dataset allows the identification of metaorders relative to the trading activity of a diversified pool of (anonymized) institutional investors" (p. 3).

[6]Eisler, Bouchaud, and Kockelkoren (2018)[87] also distinguish between different types of trading when fitting their price impact model. By looking at microstructure events on the limit order book, Eisler, Bouchaud, and Kockelkoren estimate the price impact of limit orders and liquidity-providing strategies.

- couple empirically observable price impact parameters to the adverse selection of limit orders.

(e) The paper *Price response in correlated financial markets: empirical results* (2015)[232] by Wang, Schäfer, and Guhr introduces the notion of *cross-impact.* Cross-impact captures the impact of trading stock A on the price of a connected stock B. For example, supply chains provide causal links between companies: if A is a supplier to B, investors assume trades on A reflect information on B.[7]

Quantitative trading desks became prominent at banks and hedge funds in the late 2000s, and detecting price dislocations based on high-frequency trading data has always been essential to quantitative desks. Books on the topic followed a couple of years later.

(a) *High Frequency Trading* (2010)[6] by Aldridge is a specialized book describing high-frequency data and trading strategies. The book covers

- institutional knowledge around market participants, order types, and tick data needed to set up a high-frequency trading business,
- standard econometric approaches to building models for high-frequency trading, and
- business and technology considerations to set up a high-frequency trading system.

(b) *Algorithmic and High-Frequency Trading* (2015)[64] by Cartea, Jaimungal, and Penalva describes high- to medium-frequency trading strategies. The book covers

- institutional and empirical knowledge around high-frequency tick data,
- advanced mathematical tools around stochastic control from a PDE perspective, and
- applications of the PDE approach to control problems in the style of Almgren and Chriss (2001)[10] for high- to medium-frequency trading.

(c) *The Financial Mathematics of Market Liquidity: From Optimal Execution to Market Making* (2016)[111] by Guéant describes high- to medium-frequency trading models. The book covers

- institutional and empirical knowledge of electronic markets,
- optimal liquidation under the Almgren and Chriss impact model,

[7] The paper "Signal Processing: The Logistics of Supply Chain Alpha" by Wu et al. (2015)[239] builds trading signals using supply chain data.

- option pricing under transaction costs, and
- market-making models in Mathematical Finance.

(d) *Quantitative Trading* (2017)[114] by Guo et al. is a generalist book spanning all trading frequencies. The book covers

- portfolio management and low-frequency trading based on economic theory,
- econometric models of high-frequency data,
- high- to medium-frequency trading strategies based on limit order book models and control problems, and
- implementation considerations such as technology, regulation, and risk management.

(e) *Market Microstructure in Practice* (2018)[147] by Lehalle and Laruelle covers microstructure rules and their implications for trading strategies. Lehalle and Laruelle have a unique, in-depth focus on market regulation, such as Reg NMS and MiFID. Therefore, the book describes how practitioners trade within these rules and how regulation shapes algorithmic trading, from market fragmentation to systemic risks.

(f) Velu, Hardy, and Nehren (2020)[227] is another generalist book that spans all trading frequencies. Compared to Guo et al. (2017)[114], Velu, Hardy, and Nehren (2020)[227] place a lighter emphasis on economic theory and focus instead on trading and technology.

The creation of *Central Risk Books* (CRBs) at banks in the 2010s further strengthened the use of price impact models to improve algorithmic trading. CRBs moved quantitative trading teams from disjointed pods into a central trading function to understand and minimize price impact at the firm level and reduce internal competition.[8] Bergault, Drissi, and Guéant (2021)[23] mention central risk books alongside statistical arbitrage in *Multi-asset optimal execution and statistical arbitrage strategies under Ornstein-Uhlenbeck dynamics* when listing applications of price impact models.

> "In practice, operators routinely face the problem of having to execute simultaneously large orders regarding various assets, such as in block trading for funds facing large subscriptions or withdrawals, or when considering multi-asset trades in statistical arbitrage trading strategies. More generally, banks and market makers manage their (il)liquidity and market risk, when it comes to executing trades, in the context of a central risk book" (p. 2).

[8]This Handbook deals with various aspects that relate to CRBs or that may be used in CRBs and with respect to their theoretical basis. However, it does not seek to build price impact models as such for CRBs.

Remark 1.1.1 (Fundamental Review of the Trading Book (FRTB) regulation)**.** *Bergault, Drissi, and Guéant (2021)[23] associate the rise in CRBs to the Fundamental Review of the Trading Book regulation by the BASEL committee:*

> *"the new FRTB (Fundamental Review of the Trading Book) regulation will lead practitioners to assess liquidity risk within a centralized risk book" (p. 26).*

The BASEL committee motivates the FRTB in its Explanatory note on the minimum capital requirements for market risk *(2019)[182] considering the 2008 financial crisis:*

> *"the Committee acknowledged that a number of structural shortcomings that came to light during the crisis remained unaddressed. It therefore conducted a 'fundamental review of the trading book' (FRTB). The objective of the project was to develop a new, more robust framework to establish minimum capital requirements for market risk, drawing on the experience of 'what went wrong' in the build-up to the crisis" (p. 5).*

The BASEL committee emphasizes the inability to measure the liquidity risk and price impact of a firm's overall trading as a critical shortcoming during past crises.

> "Inability to capture the risk of market illiquidity. *The Basel 2.5 framework assumed that individual banks would be able to exit or hedge their trading book exposures over a 10-day period without affecting market prices. However, in times of stress, the market is likely to become illiquid rapidly when the banking system as a whole holds similar exposures. This happened at the height of the crisis as banks were unable to exit or hedge positions in a short time frame, resulting to substantial mark-to-market losses" (p. 7).*

Section 4.3 of Chapter 4 proposes practical solutions to measure and act on liquidity risk.

Finance articles describe CRBs.

(a) efinancialcareers.co.uk (2018)[48] mentions that CRBs "first came to our attention in 2016" and are "the places that (in theory) centralize the execution of trades for the whole bank." Consequently, "central risk desks are always highly quantitative" and are poised to "become increasingly integrated and increasingly powerful."

(b) finextra.com (2016)[93] also "came across the term 'central risk book' [...] a couple of weeks back" in 2016.

> "The purpose of the CRB in all this is to act as a huge repository for all the firm's positions. Then, someone with a brain the size of a small planet sifts through all this and works out what the net exposure of the bank is."

(c) Not all news on CRBs is positive: Bloomberg (2018)[110] reported a $60 million loss at Deutsche Bank's CRB in 2018.

(d) Marketsmedia.com (2017)[236] states that "liquidity is at a premium", which motivates the creation of CRBs at various financial institutions.

> "While it is not a trivial undertaking, a CRB that handles the full portfolio of risk taken by all the individual trading desks is the optimal method for minimizing trading costs, managing risk, and, therefore, providing liquidity."

In conclusion, four communities of academics and practitioners detail price impact models and their applications to algorithmic trading. Therefore, the Handbook synthesizes this knowledge into a concise, practical framework. First, the framework defines essential terms in Section 1.2.

1.2 Trading Terminology

This section describes essential terms used throughout the Handbook, which the reader should study with the mathematical definitions in Chapter 2. The terminology used within the Handbook matches the standard usage in trading teams. Footnotes translate the standard usage to the language from the broader literature. Practitioners may skip this section during their first reading.

1.2.1 Trading strategies

The Handbook defines *trading strategies* as automated processes taking input *signals* and producing *trades* as outputs. What practitioners consider a *signal* and a *trade* differ depending on context.[9]

[9] This definition echoes the definition of a quantitative fund by Guo et al. (2017)[114]:

> The quantitative fund investment process has "three key components: (a) an input system that provides all necessary inputs including rules and data (b) a forecasting engine that generates estimates of model parameters to forecast prices and returns and evaluate risk (c) a portfolio construction engine that uses optimization and heuristics to generate recommended portfolios" (p. 8).

In a trading context, (a) corresponds to the trading infrastructure and (b) to trading signals. For a trading strategy, (c) outputs intended trades instead of recommended portfolios.

After the execution, a trader can compare the *prediction* from a signal and the *intended* trades to their respective *realized* counterparts. Four *basic trading parameters* are universal across strategies: size, speed, market correlation, and market conditions.

Traders evaluate trading strategies using *order slippage.* Traders improve and act on their strategies by distinguishing between *alpha slippage* and *slippage due to price impact.* Practitioners use *trading experiments*, such as *A-B tests* and *back tests,* to assess trading strategies.

The following subsections define the above concepts.

1.2.2 Trading data: fills, orders, and binned data

Trading data depends on the internal systems of financial institutions and the messaging format of trading venues. Therefore, dealing with high-frequency tick data requires a specific technological skill set and a fine-grained knowledge of individual trading venues. For example, the book sections by Aldridge (2010)[6] "Working with Tick Data" (p. 115), Cartea, Jaimungal, and Penalva (2015)[64] "The Data" (p. 39), Bouchaud et al. (2018)[37] "Description of the NASDAQ Data" (p. 422), and Velu, Hardy, and Nehren (2020)[227] "Data Infrastructure" (p. 401) handle high-frequency tick data. In addition, Velu, Hardy, and Nehren (2020)[227] detail the technology used to analyze high-frequency tick data:

> "[Kdb+] has been the dominant player in this space, despite the steep cost and terse and cryptic 'q' language that is hard to master and thus is a very much sought-after skill" (p. 402).

Appendix A provides a primer to kdb+ and q aimed at traders and researchers.

The Handbook does not focus on analyzing the particularities of high-frequency tick data. Instead, the Handbook assumes the reader has access to trading data in one of the multiple standard formats for research.

What all representations of trades have in common is a traded volume. The Handbook understands trade volumes to have a sign: buyer-initiated executions are positive trades; seller-initiated executions are negative trades. However, the term *trade notional* refers to a trade's size regardless of its sign.

Data can represent trades as:

- *Individual fills* on specific trading venues. Figure 1.2 illustrates fill data.

  Individual fills are the most granular information and contain, at minimum, a timestamp, volume, and price. In addition, individual fills can provide context specific to the trading venue that executed the fill, such as which order type triggered the fill. Fill data does not take a cross-sectional format. Instead, traders store fill data on a stock-by-stock basis. Section 1.2.3 defines stock-by-stock and cross-sectional data.

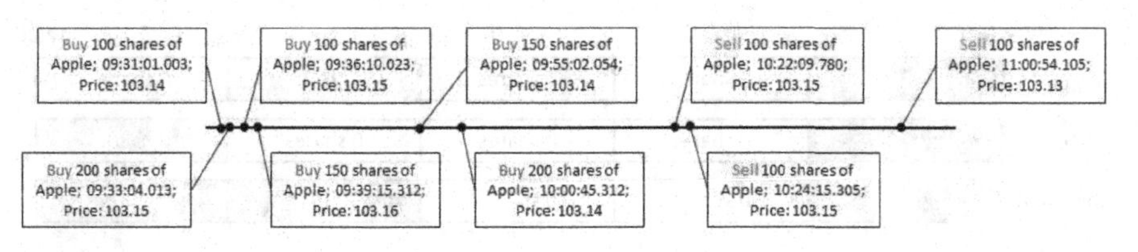

FIGURE 1.2
Illustration of fill data.

- *Orders* that provide a start-time, an end-time, and a traded volume. Figure 1.3 illustrates order data. One also refers to orders as meta-orders or parent-orders.

  Orders aggregate individual fills into compact, more interpretable data. The cost is a loss in data granularity, for example, which venue an individual fill executed on. Orders can contain additional events such as order cancellation or amending times and volumes. Order data does not take a cross-sectional format. Instead, traders store order data on a stock-by-stock basis.

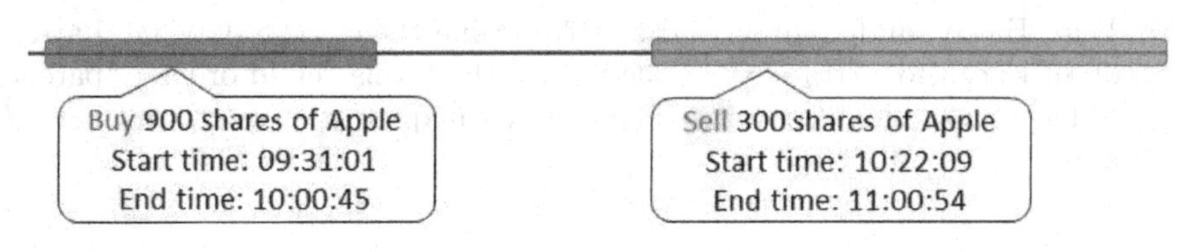

FIGURE 1.3
Illustration of order data.

- *Binned data* on a regular time grid, for example, every thirty minutes, that aggregates the fills traded in that data bin for a given stock universe. Figure 1.4 illustrates binned data.

  Traders stack together the time series of a specific stock universe, for example, the S&P 500 constituents, in a two-dimensional (stock, time) matrix. This type of data, called *cross-sectional data*, facilitates computing statistics across stocks at a given time.

The Handbook refers to these three representations as *fill, order,* and *binned* data. Two other books describe their trading data.

	09:30–10:00	10:00–10:30	10:30–11:00	11:00–11:30	11:30–12:00
Apple	+700 shares	0 shares	0 shares	-100 shares	0 shares
IBM	+100 shares	0 shares	0 shares	0 shares	-300 shares
Microsoft	-500 shares	0 shares	0 shares	300 shares	200 shares

FIGURE 1.4
Illustration of binned data.

(a) Bouchaud et al. (2018)[37] outline "The Ideal Data Set" (p. 231):

> "one would ideally have access to some form of proprietary data or detailed broker data that lists:
>
> - which child orders belong to which metaorders;
> - the values of t_i, p_i and v_i for each child order; and
> - whether each child order was executed via a limit order or a market order" (p. 231)

Child orders map to *fill* data with timestamp, price, and volume. The metaorders map to *order* data.

(b) Velu, Hardy, and Nehren (2020)[227] outline their trade data in "Layers of an Execution Strategy" (p. 348) with the terms "child order", "parent order", and "bins" (p. 348). These terms map to *fill*, *order*, and *binned* data.

1.2.3 Trading signals, alpha signals

Statistical models generate signals in the form of predictions. The most frequent signal type is a so-called *alpha signal*, which predicts future stock price returns in the absence of a strategy's own trading. Velu, Hardy, and Nehren (2020)[227] mention alpha signals as an advanced topic in optimal execution:

> "A final refinement to this approach, is the usage of predictive signals (a.k.a. alpha)". (p. 351)

Other signals predict future volatility, the volume traded on the market, or specific events, such as mergers between stocks.

How one creates trading signals based on public or internal data is out of scope. The Handbook instead focuses on how to act on signals by trading using a price impact model. For this purpose, one answers five questions about a signal:

(a) *Is the signal directional or non-directional?*

A directional signal is a signal to buy or sell.[10] For example, predicting whether the market is buying or selling the stock leads to a directional signal. However, a non-directional signal serves as a scaling factor for models. For example, volatility and volume estimators explain how models scale across different stocks or market regimes.[11] Lee, Fok, and Liu (2001)[146] build a model for trading volumes' intraday patterns using data from the Taiwan stock market.[12]

> "The findings show that intraday trading volume as well as the real orders from both types of investors are J-shaped" (p. 1).

Chapters 2 and 3 describe how a trading strategy reacts to such liquidity signals.

(b) *What time horizon does the signal predict?*

Most trading signals predict changes in variables. For example, returns (price changes) require start and end times.[13] Even a volatility prediction holds over a particular period: an intraday volatility prediction differs from a monthly volatility prediction.

---

[10] Aldridge (2010)[6] points out when "Back-Testing Trading Models" (p. 219) that testing "the accuracy of directional systems presents a greater challenge" (p. 222).

[11] Velu, Hardy, and Nehren (2020)[227] call these "Extraneous Signals: Trading Volume, Volatility, etc." (p. 189) and outline models for such predictions and practical uses to complement directional return signals.

[12] Cartea, Jaimungal, and Penalva (2015)[64] cover non-directional signals based on "seasonal patterns" (p. 61) in Chapter 4 "Empirical and Statistical Evidence: Activity and Market Quality" (p. 61).

[13] This distinction is vital for *dynamic* trading strategies and portfolio optimization.

Velu, Hardy, and Nehren (2020)[227] state that the "static Markowitz framework, as discussed in the last section, is not realistic for practical applications" (p. 234) and replace the standard economics toolbox with "Dynamic Portfolio Selection" (p. 234). Dynamic portfolios require advanced signals, "not simply using the mean and variance of past returns" (p. 235), a focus on "Transaction Costs, Shorting and Liquidity Constraints" (p. 239), and mathematical and computational methods to deal with an entire forward-looking trading trajectory.

Guo et al. (2017)[114] are equally critical of the static approach. An "argument is that after a period of inactivity, the eventual rebalancing concentrates the required trading into a particular moment in time, resulting in market impact and higher transaction costs" (p. 69). Guo et al. improve the static approach with "dynamic optimization principles and methods" (p. 69), which require price impact models and multiple prediction horizons to be realistic.

Finally, Gârleanu and Pedersen (2013)[101] characterize dynamic trading strategies by two principles:

> "1) aim in front of the target and 2) trade partially towards the current aim" (p. 1).

(c) *At what frequency does the signal trigger?*

What information triggers a signal change? Conversely, when no new data arrives, how should the algorithm update the signal? The answer to these questions determines how quickly the signal turns around, affecting a portfolio's trading speed, turnover, and transaction costs. Trigger frequencies can be regular, for example, every second, or event-based, for example, every time the algorithm observes a market trade.[14]

(d) *Is the signal exogenous or endogenous?*

Exogenous signals are simple signals that hold regardless of a trader's actions. Unfortunately, there is a frequent need to use endogenous signals in trading. Traders also refer to endogeneity as *path dependence*: an endogenous signal depends on the path the trading strategy took before the prediction. Price impact models are, because of the causality challenge, always endogenous. Chapter 5 introduces the mathematics of causal inference to deal with signal endogeneity.

(e) *Is the signal cross-sectional?*

A signal is cross-sectional if the prediction on one stock depends on what happens in other stocks. Statisticians use cross-sectional models for two reasons: causal effects and statistical robustness. First, there are structural reasons why stock events may cause effects on other stocks. Second, models that normalize variables or predictions across a stock universe tend to be statistically more robust, even without any causal relationships.[15]

The Handbook defines an *alpha signal* as a directional signal predicting returns that are exogenous to a given trading strategy's actions. An alpha signal can span multiple prediction horizons and trigger frequencies. It could be cross-sectional or stock specific.

---

[14]Velu, Hardy, and Nehren (2020)[227] discuss "Automated News Analysis and Market Sentiment" (p. 256) and point out that "trading now mostly occurs at a relatively high frequency level and hence any news related to a stock gets processed almost instantaneously" (p. 256). As a result, algorithms based on regularly spaced data lag event-based algorithms.

Aldridge (2010)[6] defines "Event Arbitrage" (p. 165): "high-frequency strategies, which trade on the market movements surrounding announcements, are collectively referred to as event arbitrage". Aldridge outlines examples and techniques to build signals for event arbitrage trading strategies.

Isichenko (2021)[121] details "Event-based predictors" (p. 15): "Learning on such events is different as they tend to be infrequent but cause a stronger price response. One can use material company news, mergers and acquisitions, earnings surprises, index rebalancing, or even dividends and splits" (p. 15).

[15]For example, Velu, Hardy, and Nehren (2020)[227] introduce "Cross-Sectional Momentum Strategies" (p. 184).

> "Even if price follows a random walk, it has been shown that strategies based on past performance contain a cross-sectional component that can be favorably exploited" (p. 184).

The reader studies Pedersen's book "Efficiently Inefficient" (2015)[188] for further examples.

1.2.4 Intended, predicted, and realized data

It is a standard practice to refer to the *prediction* and *realization* of a variable using the same term. Traders lift this ambiguity by qualifying the term as *predicted* or *realized*. For example, consider an alpha signal that predicts returns on a specific horizon at a particular trigger frequency. Traders refer to the *realized alpha* as the realization of the return the alpha signal predicts and the alpha signal as the *predicted alpha*.

This realized point of view explains Guo et al.'s (2017)[114] definition of *alpha:*

> "The additional return that a portfolio generates relative to the benchmark is commonly known as the *alpha* of the portfolio" (p. 62).

Guo et al. refer not to a prediction but to the portfolios' actual returns. In trader language, Guo's alpha is the portfolio's *realized alpha* in contrast to the *predicted alpha* or *alpha signal* used in the trading decision.

A similar practice takes place for trades: there is a distinction between what a strategy *intends* to trade and how the trade *realizes*. For example, an algorithm may intend to trade ten thousand shares in the next hour but only realize nine thousand due to adverse market conditions. A trading strategy's outputs, also called a schedule, are qualified as *intended trades*.

A practitioner compares trade intentions with the *realized trades* after the fact to uncover potential biases or errors in the algorithm. Bouchaud et al. (2018)[37] refer to a bias between intended and realized trades as "implementation bias" (p. 238). Section 6.5 of Chapter 6 provides solutions to implementation bias.

1.2.5 Basic trading parameters

Trading strategies keep track of trading parameters to guide their decisions. Traders also use such parameters to compare strategies on an equal footing. Velu, Hardy, and Nehren (2020)[227] call these "normalization analytics", or "normalizing features" (p. 271) needed to handle systematic trading's complexity:

> "the practitioner's approach to address this is to parametrize many of the trading decisions using normalizing features" (p. 272).

Four basic parameters are standard across trading teams.

- *Order size* drives a trade's cost, as per Bouchaud et al. (2018)[37]:

  > "At the heart of most empirical and theoretical studies of metaorder impact lies a very simple question: how does the impact of a metaorder depend on its size Q?" (p. 233).

The greater the order, the more likely it will cause a market reaction and increase price impact and trading costs. In sell-side execution teams, the trading strategy does not decide the order size: the upstream portfolio or client instructs the size.

- *Trading speed* is the most common parameter a trading strategy controls.

  Practitioners measure trading speed in proportion to market activity. For example, traders quote *participation of volume* (PoV), indicating which percentage of the market volume the strategy tracks.

  Faster trading is aggressive and leads to considerable price impact. Slower trading is passive and leads to moderate price impact. Passive trading also exposes the trading strategy to *adverse selection* if it provides liquidity to the market, a topic Chapter 2 covers. Bouchaud et al. (2018)[37] illustrate how adverse selection causes liquidity providers to earn less than the bid-ask spread when providing liquidity in Section 1.3.1 "Adverse Selection" (p. 14).

- Trading in the same direction as the market increases one's *market correlation.*

  Market correlation directly affects trading costs: if the correlation is positive, then a trader competes with the rest of the market for the same shares to buy or sell, increasing trading costs. Conversely, if the correlation is negative, the trader provides liquidity to the market, lowering trading costs but exposing the trader to *adverse selection.* Bechler and Ludkovski (2015)[21] introduce market correlation in their optimal execution model, arguing that "it allows the trader to react to changing market conditions, in particular markets changing from being buy-driven to ask-driven and vice versa" (p. 2). Bouchaud et al. (2018)[37] mention the "synchronization bias" (p. 238) introduced when ignoring market correlation:

  > "The impact of a metaorder can change according to whether or not other traders are seeking to execute similar metaorders at the same time" (p. 238).

  Section 6.4 of Chapter 6 proposes a solution to synchronization bias.

- *Market condition* is a loose term for *volatility.*

  Market conditions include any event or variable associated with *volatility. Volatility* directly scales up trading costs.[16] Examples of market condition variables associated with volatility include the time of the day (*volatility*

[16] Velu, Hardy, and Nehren (2020)[227] note in Section 9.8 "Empirical Estimation of Transaction Costs": "a key aspect of market impact modeling is to provide information about expected pre-trade costs." Velu, Hardy, and Nehren identify five "determinants of cost" as "trade size", "market capitalization", "price volatility", "trading activity", and "time of day" (p. 327). *Market conditions* include the latter four cost determinants.

is highest at the start and end of the trading day), earnings announcements, federal reserve announcements, or index inclusion announcements (*volatility* increases during announcements).

1.2.6 Order slippage, arrival price

Section 1.2.1 defines a trading strategy as a process that takes input signals and provides trades as outputs. A trading strategy maximizes the profits and minimizes the losses (P&L) for a portfolio. An equity portfolio's P&L is reasonably well-defined: practitioners do not use models to measure an equity portfolio's P&L. A portfolio's P&L allows stakeholders, such as investors or regulators, to agree on an investment strategy's merit.

The same statement does not hold for trades: on its own, a single trade does not have a well-defined P&L. One assigns a definite P&L to a trade when complementing it with an offsetting trade, called an exit trade. Instead, traders use models or assumptions to give an approximate P&L to a trade without matching it to an exit trade. For example, traders employ order slippage, alpha, and price impact to estimate a trade's value. This section illustrates these metrics on an example of a single trade. The reader finds formal definitions in Chapter 2.

Assume a trader submits an order at the start of the day to buy 1000 shares of Apple in the next hour using a given trading strategy. At the time they submit the order, and before the trader begins executing the order, Apple's price is \$100. The trader has a signal with perfect accuracy. The signal predicts that, by the end of the trading day, the price will reach \$101. Assume this prediction is exogenous: the trader expects the price to reach \$101 regardless of their strategy.

The order would pay \$100 to buy the 1000 shares in a world without trading frictions. By the end of the day, the order accrues a position worth \$101 per share, netting a \$1000 profit on an initial order worth \$100,000. The trade would have made a hypothetical profit, the order's realized alpha, of 1% in return units. The trader refers to the frictionless scenario as the *arrival benchmark* and the initial price \$100 as the *arrival price*.

In practice, the trader cannot trade their entire order in one fill, and the order *slips* the \$100 arrival price. In expectation, the prices after the trader submits the order are larger than the arrival price, shrinking the trader's profits. For example, if the average price the market fills the trader at is \$100.1, they only capture a fraction of the P&L compared to their arrival benchmark. In this scenario, one states that the order *slipped* by ten basis points (bps) compared to the arrival price. A basis point is a common financial unit for small fractions, for example, low interest rates. One converts from percent to basis points: $1\% = 100\,bps$. Traders quote order slippage as a negative return and estimate the trade's P&L by comparing order slippage to the trade's realized alpha.

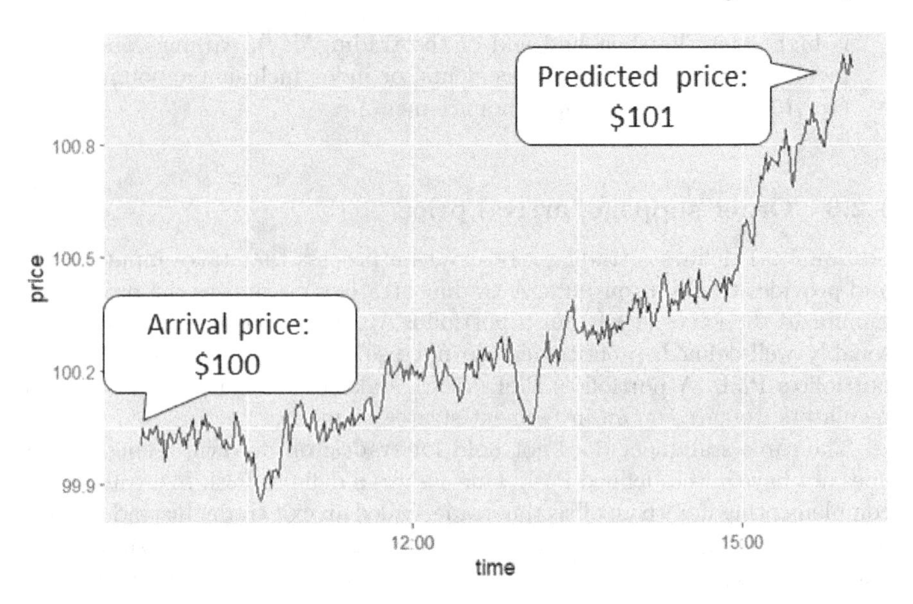

FIGURE 1.5
Illustration of order slippage on a simulated price path.

This single-order example holds constant the order's arrival price and the expected price at the end of the day. Therefore, the order's average fill price is the only quantity the trader's strategy affects. Under these premises, the trade's P&L decomposes into the profit from realized alpha and the order slippage cost.

> Therefore, minimizing order slippage equals maximizing the portfolio's fundamental P&L.

Another term used for order slippage, introduced by Almgren and Chriss (2001)[10], is *implementation shortfall.* Today implementation shortfall frequently refers to the execution algorithm Almgren and Chriss (2001)[10] outline. Section 1.3.1 covers Transaction Cost Analysis (TCA) and discusses order slippage decomposition.

1.2.7 Alpha slippage, slippage due to price impact

Unfortunately, while order slippage yields a concrete objective function for the trader, it does not directionally tell them how to improve their P&L. For example, it does not answer the elementary question:

> *Should the trader have traded faster or slower?*

The trader currently cannot predict how execution parameters affect the order slippage: measuring the order slippage does not tell them *why* the price moved while executing their order. Two opposing forces can drive the price during the order's execution based on the order's trading speed.

(a) The stock price could have moved *regardless of trading*, as one expected the price to reach \$101 by the end of the day. Such price drift would lead the trader to trade *faster* to avoid losing out on capturing their alpha.

(b) The stock price could have moved *in reaction to trading*, as aggressive trading alerts other market participants. For example, liquidity providers may react by pushing up the price, increasing the trader's costs. This price impact would lead the trader to trade *slower* to avoid further pushing the price away from the arrival price.

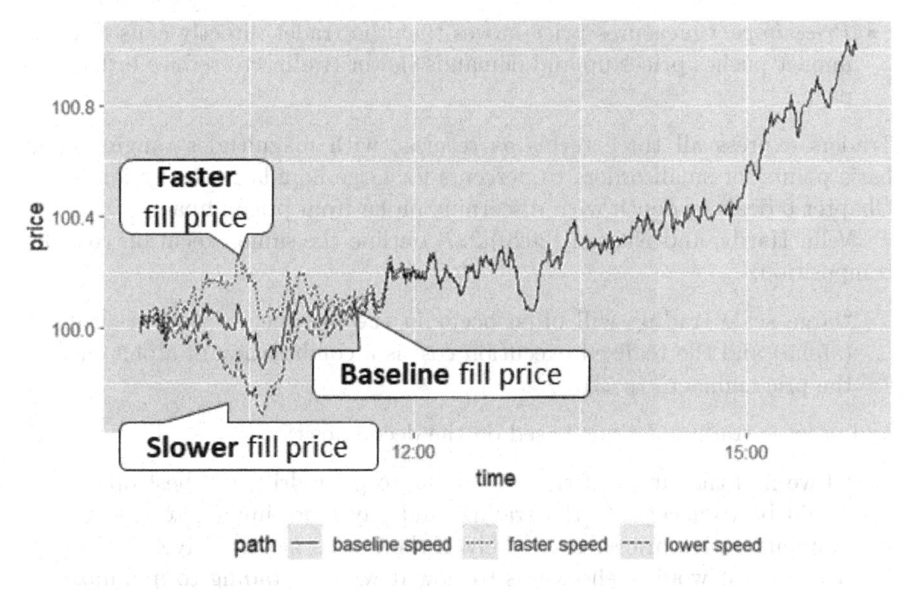

FIGURE 1.6
Illustrated effect of trading speed on prices and order slippage.

Practitioners refer to the former slippage, which occurs regardless of trading, as *alpha slippage.* They refer to the latter slippage, which occurs in reaction to trading, as *slippage due to price impact* or simply *price impact.*[17] Both slippages hurt the trader's profits compared to their unachievable arrival

[17]Bouchaud et al. (2018)[37] introduce a different language in Section 11.2 "Observed Impact, Reaction Impact and Prediction Impact" (p. 210). Because Bouchaud et al. take an outside observer's point of view, they refer to the overall price move as price impact. Bouchaud et al. distinguish between "predictive" price impact, which happens regardless of trading, and "reactive" price impact, which trades cause. These terms correspond to traders' *alpha slippage* and *price impact* language.

benchmark by increasing the price during the order's execution. Without careful separation of the two slippage sources, the trader cannot act to reduce order slippage.

To summarize the terms the single-order example introduces, under the assumptions outlined above:

- *Order slippage* measures the order's trading costs against an unachievable benchmark, the arrival price. Minimizing order slippage equals maximizing the portfolio's fundamental P&L.
- *Alpha slippage* measures price moves that the trader does not cause. Alpha slippage leads to higher prices and demands faster trading to reduce order slippage.
- *Price impact* measures price moves that the trader directly causes. Price impact pushes prices up and demands slower trading to reduce order slippage.

Traders express all three terms as returns, with magnitudes ranging from basis points for small orders to percents for large liquidations. Section 6.3 of Chapter 6 deals in depth with discerning alpha from price impact.

Velu, Hardy, and Nehren (2020)[227] outline the same execution cost decomposition:

> "large scale trading will often occur in the presence of market drift (alpha) and the realized execution cost is a combination of alpha and the price impact" (p. 313).

and provide trading actions based on the decomposition,

> "if we find that most of the cost is due to price drift, the best option would be to accelerate the trading and pay more impact to capture more attractive prices. Conversely, if the cost is mostly driven by impact then it would behoove us to slow down our trading to minimize the impact" (p. 313).

The Handbook employs the term *alpha slippage* over *market drift*, as alpha directly relates to trader terminology. Chapter 3 quantifies the relationship between a strategy's alpha and price impact and implements concrete formulas for trading strategies.

The above example assumes that the stock's start-of-day *and* end-of-day prices are independent of the trader's actions. The first premise holds because the trader did not trade the stock before submitting the order: the order trades in isolation. The second premise holds because the order is small and brief, leaving the end-of-day price unaffected. Chapters 3 and 4 lift both assumptions to cover real-life applications of price impact models.

Section 7.4.5 in Chapter 7 provides orders of magnitude for the price impact of public trades. These estimates align with price impact estimates the literature establishes on proprietary and public trading data.

(a) Almgren et al. (2005)[11] "presents a quantitative analysis of market impact costs based on a large sample of Citigroup US equity brokerage executions" (p. 3). For example, Almgren et al. forecast that the price impact of a purchase of 10% of the day's average volume in IBM is 20bps.

(b) Caccioli, Bouchaud, and Farmer (2012)[53] forecast price impact for orders of 10 times the daily average volume.

> "For stocks we assume $Q = 10V$, which assuming the same participation rate implies a position that would take 50 trading days to unwind. Such positions may seem large, but they do occur for large funds; for instance, Warren Buffet was recently reported to have taken more than eight months to buy a 5.5% share of IBM" (p. 13).

For instance, Caccioli, Bouchaud, and Farmer assess such a sizable order to have a price impact of 2.9% on AAPL.

(c) In "The 2022 Intern's Guide to Trading", Nasdaq (2022, Chart 4)[159] estimates 30bps of trading costs for large-cap stocks, 50bps for mid-cap stocks, and 75bps for small-cap stocks for the average institutional order.

1.2.8 Trading experiments: A-B tests and back tests

Quantitative researchers compare trading strategies using two types of experiments, each with advantages and disadvantages.

(a) *A-B tests*, also-called coin toss experiments or live trading experiments, involve running multiple live algorithms on the market.

Multiple strategies cannot simultaneously run on the same trade, and a team must use a randomization process to segment their trading data. Each trading strategy acts on the segment of the live trades assigned to it. Practitioners refer to trades assigned to a given strategy as a flow.

A team statistically compares each strategy's behavior and merit by probing trading metrics, such as *order slippage,* on each flow. Unfortunately, this comparison can only be statistical: the live trading experiment does not allow a direct comparison between two algorithms on identical data.

The way the team segments trades is essential: a bias in the segmentation process, e.g., systematically assigning sizable, costly orders to one strategy, biases the statistical comparison and challenges the conclusions. Hidden or incorrect assumptions in the trading infrastructure cause common errors and biases when analyzing the results of *A-B tests.* Chapters 5 and 6 provide tools to understand and correct such biases.

(b) *Back tests*, also called simulation experiments, use a data generation model called a market simulator to replay trades under novel strategies.

Unlike in A-B tests, a trading team can run multiple algorithms on the same trades in back tests, with the caveat that a model simulates the results. The simulation allows a direct comparison between two strategies on identical data. Powrie, Zhang, and Zohren (2021)[189] model new trading strategies' price impact in their market simulator.

> "The powerful aspect of having a market simulator is that it effectively enables the re-running of different realities where our interventions with the market either do, or do not, occur" (p. 6).

Simulation errors stem from hidden or incorrect premises in the data generation model, commonly referred to as *model risk.* A trader or portfolio manager should also not *overfit* a back test.

Remark 1.2.1 (Back test overfitting). *Bailey et al. (2014, 2016)[14, 15] define back test overfitting and propose methods to reduce its effect.*

> *"In mathematical finance,* backtest overfitting *means the usage of historical market data (a* backtest*) to develop an investment strategy, where many variations of the strategy are tried on the same dataset. Backtest overfitting is now thought to be a primary reason why quantitative investment models and strategies that look good on paper (based on backtests) often disappoint in practice. Models suffering from this condition target specific idiosyncrasies of a limited dataset, rather than any general behavior, and, as a result, often perform poorly when presented with new data" (p. 1, [15]).*

Bailey et al. introduce the minimal backtest length *relating the number of trials to the confidence a researcher should have in a back test's performance. The authors also quote economist and Nobel Laureate Leontief (1982)[149] in pushing for a better* causal *understanding of observed correlations.*

> *"[E]conometricians fit algebraic functions of all possible shapes to essentially the same sets of data without being able to advance, in any perceptible way, a systematic understanding of the structure and the operations of a real economic system."*

Chapter 5 provides a mathematical theory that extends standard econometric tools to allow for such an analysis. Moreover, the theory enables statisticians to articulate underlying structural (causal) assumptions and prove their identifiability.

Three references outline both the technology and business practices surrounding back tests.

- Waelbroeck et al. (2012)[231] propose the general simulation framework that the Handbook emulates in Section 4.2.3 of Chapter 4.

- Velu, Hardy, and Nehren (2020)[227] outline the technology needed for running simulation experiments in Section 12.3 "Simulation Environment" (p. 404).

  > "This is a topic that deserves an extensive discussion but here we will provide only a cursory overview" (p. 404).

- Aldridge (2010)[6] describes back tests in Chapter 15 "Back-testing Trading Models" (p. 219) and highlights the business case for back tests in trading.

  > "The purpose of back tests is twofold. First, a back test validates the performance of the trading model on large volumes of historical data before being used for trading live capital. Second, the back test shows how accurately the strategies capture available profit opportunities and whether the strategies can be incrementally improved" (p. 219).

The most notable public A-B test in finance is the tick size pilot experiment the SEC ran from October 2016 to October 2018. The SEC (2018)[210] defines the experiment.

> "The Tick Size Pilot Program is a national market system (NMS) plan designed to allow the Commission, market participants, and the public to study and assess the impact of wider minimum quoting and trading increments – or tick sizes – on the liquidity and trading of the common stocks of certain small-capitalization companies."

In its final assessment of the tick size pilot (2018)[211], the SEC outlines the plan's primary attributes.

> "The Plan called for a Pilot to be conducted for two years, beginning October 3, 2016, in which 1,200 securities would be spread across three Test Groups, with the remainder of securities satisfying the Pilot criteria placed into a Control Group" (p. 5).

The A-B test provided real-life trading data for a plan to "widen the quoting and trading increments for certain small-capitalization stocks" (p. 2).

Practitioners can combine A-B tests and back tests; for instance, a trading team simulates every strategy on every live trading flow, leading to a complete matrix of live and simulated performance metrics. If the price impact model is correct and the flow segmentation is unbiased, all results converge to the same conclusion. Otherwise, the trading team investigates discrepancies to uncover hidden assumptions or model weaknesses, as illustrated in Figure 1.7.

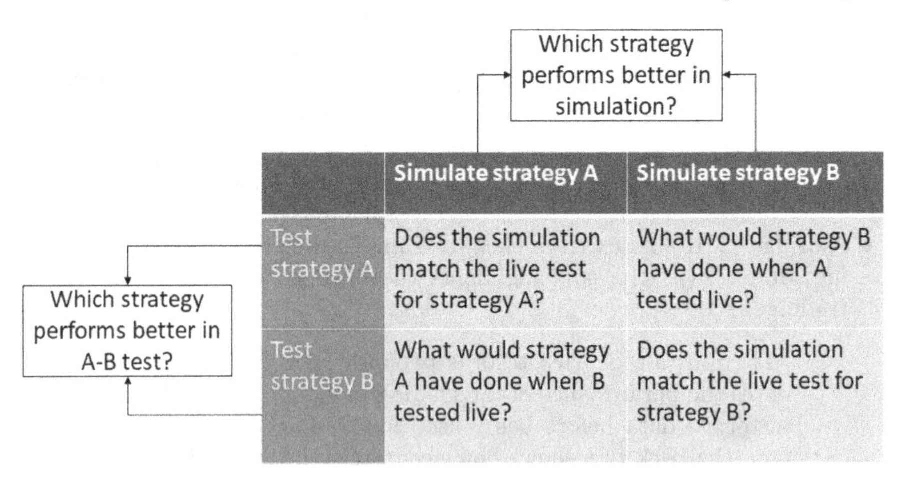

FIGURE 1.7
Matrix of performance evaluations based on both simulated and A-B tested performance metrics.

1.3 Outlining Applications

Section 1.1.2 defines price impact as price moves caused by trading. Practitioners apply price impact models in three broad ways:

(a) To decompose P&L into price impact and alpha slippage and understand which trades add value to a portfolio when considering trading costs.

(b) To simulate new intended trades and test how different actions or trading parameters, such as speeding up or slowing down, affect a portfolio's P&L.

(c) To compute how other participants dislocate prices through their trading. These external price dislocations inform a trader how the market's trades affect their portfolio's performance.

This section presents simplified, non-technical descriptions of price impact model applications for different teams. Chapter 2 covers formal mathematical models, and Chapters 3 and 4 describe these applications in detail. The intended practitioner audience includes *sell-side execution teams, buy-side statistical arbitrage teams, risk management teams, and senior management.*

1.3.1 Transaction cost analysis (TCA) for sell-side execution teams

Transaction Cost Analysis is a standard application of price impact models. TCA is a service sell-side execution teams provide to their clients and takes the form of a report. Bershova and Rahklin (2013)[25] mention the crucial role price impact models play in analyzing transaction costs.

> "Practitioners use impact models as a pre-trade tool to estimate the expected transaction cost of an order and to optimize the execution strategy. On a post-trade basis, impact models serve as performance benchmarks to evaluate trading results" (p. 2).

Guo, Lehalle, and Xu (2022)[113] highlight TCA and price impact's critical role in electronic bond trading.

> "Properly monitoring each individual trade by the appropriate Transaction Cost Analysis (TCA) is the first key step towards this electronic automation. One of the challenges in TCA is to build a benchmark for the expected transaction cost and to characterize the price impact of each individual trade, with given bond characteristics and market conditions" (p. 1).

TCA connects to European trading regulations, most notably to the *Markets in financial instruments directive II* (MiFID II). MiFID II imposed best execution obligations on investment and trading firms in 2018, and TCA is the most common tool firms implement to prove they satisfy these obligations. As a result, MiFID II contributed to an explosion in the use and sophistication of TCA to reduce trading costs for market participants.

- Kolm and Westray (2021)[132] use price impact models to remove a common bias in TCA reports. Kolm and Westray associate the increased focus on TCA with regulatory changes.

  > "With the increasing focus of global regulators on *best execution* it is more important than ever for both buy and sell side firms to demonstrate robust and reproducible TCA" (p. 2).

- Bloomberg [28] highlights its TCA offerings:

  > "Bloomberg's Transaction Cost Analysis (BTCA) solutions seamlessly integrate MiFID II requirements, across multiple asset classes, including OTC derivatives, to set, demonstrate and monitor your order execution policy to help ensure your firm is compliant [...] [and] obtain the best possible result for the end client when executing orders" (p. 1).

- Johnson (2019)[127] describes TCA as a key novelty on the buy-side and connects its rapid adoption to MiFID II.

  > "For the buy side, TCA was once seen as a "tick-the-box" exercise but has become a differentiator, with many large buy-side trading desks now employing dedicated TCA professionals to help them eke out cost savings and improve performance for their institutional clients" (p. 1).

- Basar (2019)[19] estimates that "95% of institutional trading desks in Europe are using transaction cost analysis" (p. 1) and attributes this usage to MiFID II, with the US equities market following closely behind. Consequently,

  > "Fund managers need to invest in technology to manage increasing amounts of data and the trading desk will need to learn new skills" (p. 1).

  Therefore, price impact modeling is becoming a required skill on trading desks.

Remark 1.3.1 (MiFID II requirements). *The European Union [71] documents MiFID II requirements for best execution in Article 27: "Obligation to execute orders on terms most favourable to the client" (p. 48). The regulators*

> *"require investment firms who execute client orders to monitor the effectiveness of their order execution arrangements and execution policy in order to identify and, where appropriate, correct any deficiencies" (p. 49).*

> *"require investment firms to be able to demonstrate to their clients, at their request, that they have executed their orders in accordance with the investment firm's execution policy and to demonstrate to the competent authority, at its request, their compliance with this Article" (p. 50).*

TCA inputs order data and produces statistics explaining the order slippage as outputs. The report establishes how well a trading strategy serves the portfolio for which it is trading. As argued in Velu, Hardy, and Nehren (2020)[227], Section 12.4 "TCA Environment" (p. 408), trading teams require high-performance, robust, and flexible trading infrastructure to produce these reports in a timely and customizable manner. A TCA implementation contains:

(a) a well-defined order set that captures and standardizes the team's trades,

(b) a method to enrich these orders with returns, market data, and internal signals over various horizons, allowing the decomposition of every trade's order slippage,

(c) a method to group orders by basic trading parameters, and

(d) model predictions that lead to actions for the trading strategy.

TCA is a systematic, large-scale version of the example in Section 1.2.6 decomposing order slippage into alpha slippage and slippage due to price impact. The Handbook walks through three sections of a mock TCA report.

(a) Figure 1.8 summarizes the orders, their slippage, alpha, and impact. Next, clients compare a broker's trading metrics to competitors or past performance. Each metric answers a specific question, which the trader refines by splitting and regrouping orders along trading parameters or experiments.

Grouping	Order group	Nb of orders	Notional traded	Order slippage	Realized alpha	Predicted alpha	Price impact	Alpha slippage	Spread & fees
All	All	100k	11bil	32bps (±10bps)	45bps	50bps	20bps	10bps	2bps
Questions to answer		What is the sample significance?		**How did trading perform?**	Was the order size correct?		Was the trade speed correct?		How did venues perform ?

FIGURE 1.8
Mock TCA report, top level.

(b) The first way to group orders is by trading parameters, such as volatility and size. For example, Figure 1.9 concludes that small orders traded too slowly and large orders traded too fast.

Grouping	Order group	Nb of orders	Notional traded	Order slippage	Realized alpha	Predicted alpha	Price impact	Alpha slippage	Spread & fees
By volatility	Low	70k	6bil	21bps (±7bps)	35bps	40bps	[illegible]	[illegible]	1bp
	Medium	20k	4bil	37bps (±15bps)	40bps	55bps	[illegible]	[illegible]	2bps
	High	10k	1bil	73bps (±25bps)	90bps	80bps	[illegible]	[illegible]	3bps
By Size	Small	50k	3bil	11bps (±5bps)	20bps	10bps	1bp	9bps	1bp
	Medium	35k	4bil	27bps (±10bps)	30bps	40bps	15bps	10bps	2bps
	Large	15k	4bil	82bps (±10bps)	80bps	90bps	65bps	15bps	2bps

Impact low, alpha slippage high: trade faster

Alpha slippage low, impact high: trade slower

FIGURE 1.9
Mock TCA report, breakdown by volatility and size.

(c) Another way to split the orders is by trading strategy or experiment. For example, Figure 1.10 shows that the data distributed across trading strategies is biased: the trader allocated orders with more potent alpha to the aggressive algorithm. In addition, the A-B test indicates that trading slower benefits the execution.

1.3.2 Portfolio optimization for buy-side statistical arbitrage teams

Trading costs and price impact play a prominent role in statistical arbitrage portfolios. Statistical arbitrage teams rely on back tests to simulate

Grouping	Order group	Nb of orders	Notional traded	Order slippage	Realized alpha	Predicted alpha	Price impact	Alpha slippage	Spread & fees
By trading strategy	Aggressive strategy	70k	7bil	39bps (±15bps)	50bps			bps	4bps
	Passive strategy	30k	4bil	25bps (±10bps)	40bps			bps	0bps
A-B test (experiment)	A: keep current speed	50k	5.5bil	38bps (±10bps)	45bps	50bps	20bps	15bps	3bps
	B: trade slower	50k	5.5bil	26bps (±10bps)	45bps	50bps	8bps	17bps	1bps

Bias: aggressive strategy was given higher alpha trades

Unbiased A/B test shows slower outperforming faster trading

FIGURE 1.10
Mock TCA report, breakdown by trading strategy and experiment.

their portfolios' performance when introducing new alpha signals. Isichenko (2021)[121] highlights trading simulations' core role in quantitative portfolios.

> "A historical trading simulation, or *backtest,* seeks to model the impossible – trading using current ideas on the past market data – in the hope a quant strategy in such a *Gedankenexperiment* would be successful in the future. [...] A realistic trading simulator is a critical part of quantitative portfolio management" (p. 227).

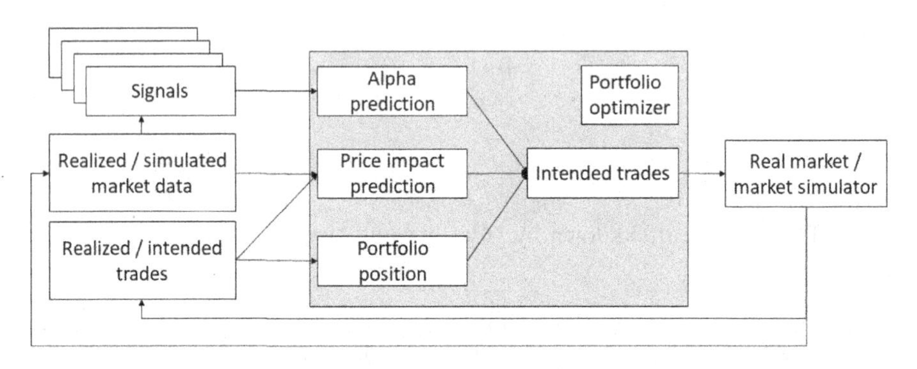

FIGURE 1.11
High-level schema for portfolio optimization.

Figure 1.11 outlines in a simple schema how a researcher can set up a portfolio optimization to compute the trade-off between alpha and impact in a systematic, simulated environment.[18] Statistical arbitrage teams can use this

[18]Guo et al. (2017)[114] and Velu, Hardy, and Nehren (2020)[227] present general techniques for portfolio optimization in a statistical arbitrage setting in Section 3.3 "Multiperiod portfolio management" ([114], p. 69) and Section 6.5 "Transaction Costs, Shorting and Liquidity Constraints" ([227], p. 239).

simulated environment for back tests to show the simulated performance of a new strategy and deploy the portfolio optimization in production to output the strategy's trades. Both Velu, Hardy, and Nehren (2020)[227] and Isichenko (2021)[121] mention this simulation/production duality and outline crucial synchronization difficulties between simulation and production. The reader studies Section 12.3 "Simulation Environment" (p. 404, [227]) and Section 7.1 "Simulation vs production" (p. 228, [121]). Section 4.2 of Chapter 4 covers applications of price impact to statistical arbitrage.

A portfolio optimizer is a simple extension of a trading strategy, taking input signals and producing as outputs trades. In addition, a portfolio optimizer keeps track of intended and realized trades and market data to provide a reproducible portfolio description. The portfolio optimizer works in two separate modes.

- In *real-time*, the portfolio optimizer updates its position, trades, and market data based on realized data. Isichenko (2021)[121] describes the benefits of sharing code between real-time and simulated portfolio trading.

  > "The production trading process has its specifics and could be driven by a code independent of the historical simulator; however, such a design would be more prone to unwanted deviation of production from simulation" (p. 227).

- In the *simulation*, the portfolio optimizer updates its position, trades, and market data based on simulated and intended data. The causality challenge requires a data-generating model to simulate the market data for this new strategy.

With a portfolio optimizer and market simulator, a trading team compares realized and simulated trading and portfolio strategies. Vyetrenko and Xu (2021)[229] leverage such a setup and use reinforcement learning to

> "train a trading agent in a market simulator, which emulates multi-agent interaction by synthesizing market response to our agent's execution decisions from historical data. Due to market impact, executing high volume orders can incur significant cost. We learn trading signals from market microstructure in presence of a simulated market response" (p. 1).

Section 4.2.3 in Chapter 4 re-implements the simulation environment of Waelbroeck et al. (2012)[231].

Before live trading, a report like a TCA report compares the performance of multiple proposed portfolio strategies. For example, one can measure the portfolios' excess returns or Sharpe ratio. Velu, Hardy, and Nehren (2020)[227] outline more performance metrics for portfolios in Section 5.2 "Evaluation of Strategies: Various Measures" (p. 160).

1.3.3 Liquidity reports for risk management teams

While the author was a first-year Ph.D. student at Princeton University deciding on a thesis topic, he attended Bernanke (2010)[24]. In his speech at Princeton on the financial crisis, Ben Bernanke stressed the importance of modeling price dislocations away from fundamental values when liquidity dissipates.

> "The notion that financial assets can always be sold at prices close to their fundamental values is built into most economic analysis, and before the crisis, the liquidity of major markets was often taken for granted by financial market participants and regulators alike" (p. 16).

These price dislocations away from fundamental values are due to price impact driving down prices:

> "To obtain critically needed liquidity, firms were forced to sell assets quickly, but these 'fire sales' drove down asset prices" (p. 16).

Ben Bernanke was the Federal Reserve's chair during the 2008 financial crisis and won the 2022 Nobel Prize in Economics for his work on banks and liquidity crises.

The literature on fire sales and the associated market liquidity risk is still young: there is no consensus methodology. However, any modern approach models price impact.

- The European Securities and Markets Authority (ESMA) published in 2020 its guidelines on liquidity stress testing [209].

  > "Managers using RST should simulate assets being liquidated in a way that reflects how the manager would liquidate assets during a period of exceptional market stress. RST should take into account [...] the role of transaction costs and whether or not fire sale prices would be accepted" (p. 13).

  An RST is a Reverse Stress Test, a stress test that aims to find the minimal shock that causes a fund to shut down. Accurately simulating transaction costs during a fire sale liquidation requires a robust price impact model.

- Caccioli, Bouchaud, and Farmer (2012)[53] accentuate the need to move away from fundamental economic models during liquidity crises. Emphasis was added.

  > "We take advantage of what has been learned recently about market impact to propose a method for impact-adjusted valuation that results in better risk control than mark-to-market valuation. This is in line with other recent proposals that valuation should be based on liquidation prices. **Estimating liquidation prices requires a good understanding of market impact**" (p. 2).

- Cont and Wagalath (2013, 2016)[76, 77] link price impact and endogenous asset correlations: correlations increase during fire sales due to price impact. Therefore, liquidity risk is essential to risk management.

  > "More generally, our study shows that "liquidity risk" and "correlation risk", often treated as separate sources of risk, may be difficult to disentangle in practice" (p. 22, [76]).

- Cont and Schaanning (2017)[75] model "market impact and price-mediated contagion" (p. 2) in their paper "fire sales, indirect contagion and systemic stress testing". Cont and Schaanning focus on price impact and fire sales' effects on systemic risk across multiple financial institutions and apply the methodology to data of EU bank portfolios.

- Bucci et al. (2020)[44] stress the importance of "Crowding Effects in Institutional Trading Activity" (p. 1).

  > "This paper is devoted to the important yet unexplored subject of *crowding* effects on market impact, that we call 'co-impact'."

  The Handbook employs the term *external price impact.* Chapter 2 introduces external impact, and Section 4.3.3 in Chapter 4 applies it to a fire sale model. Crowding effects and external impact drive fire sales and liquidity crises. Section 6.4 of Chapter 6 presents a causal model for crowding.

- Roncalli et al. (2021)[197] is an "article focused on asset liquidity risk modeling" where "we propose a market impact model to estimate transaction costs" for stress testing purposes.

Risk management teams anticipate future risk situations and prepare for them with three steps. Price impact models play a role in all three steps.[19, 20]

(a) Identify a risk situation.

  A frequent source of risk is a fire sale. In a fire sale, market participants sell a stock in a correlated manner. The ensuing price impact adversely affects the portfolio, even if the portfolio is not actively trading: the fire sale's price impact directly dislocates the portfolio's mark-to-market value.

[19] Bechler and Ludkovski (2015)[21] state that "order flow feedback [...] has been questioned in connection with liquidity crises" (p. 2) and propose a model that uses price impact to identify the contribution of trade correlation with the market on liquidation costs. Section 4.3.3 of Chapter 4 tackles the effect of trade correlation with the market on liquidation costs.

[20] The *Handbook of Financial Risk Management* by Roncalli (2020)[196] Chapter 2 (Market Risk) "begins with the presentation of the regulatory framework. It will help us to understand how the supervision on market risk is organized and how the capital charge is computed" (p. 37). This section cites the relevant regulatory framework in Remark 1.3.2 and refers to the *Handbook of Financial Risk Management* for a detailed analysis.

(b) Create a stress test.

A stress test simulates the market reaction to the risk situation and the resulting portfolio effect. For example, a price impact model simulates how a fire sale dislocates prices, leading to losses on the stocks held in the portfolio and increasing trading costs. A risk manager estimates these dislocations' magnitude and duration using a price impact model.

(c) Outline actions the portfolio can take before or during a risk situation.

A portfolio's action before or during a risk situation is to trade out of the position. This *hedge* reduces the portfolio's exposure to the downside risk and temporary price dislocation. However, the hedging trade will pay price impact costs due to the dislocation, leading to a trade-off between the portfolio's risk exposure and the hedge's trading costs.

Remark 1.3.2 (Principles and guidelines on stress testing). *Both European and US regulators document stress testing methodologies. Three best practices emphasize the need for actionable, robust liquidation cost models.*

(a) The Stress testing principles *(2018)[181], outlined by the Basel Committee set up after the 2008 financial crisis, highlight the importance of understanding assumptions and model risk when building stress tests.*

> *"When using the results of stress tests, banks and authorities should have a clear understanding of their key assumptions and limitations, for instance in terms of scenario relevance, risk coverage and model risk" (p. 4).*

(b) ESMA's guidelines on liquidity stress testing (2020)[209] highlight the need for different stress test scenarios when determining liquidation costs.

> *"LST should employ hypothetical and historical scenarios and, where appropriate, RST. LST should not overly rely on historical data, particularly as future stresses may differ from previous ones" (p. 13).*

LST stands for Liquidity Stress Testing.

(c) The Basel stress testing principles (2018)[181], Section 3, highlight that stress tests are forward-looking and should lead to actions from decision-makers, such as traders, risk managers, and regulators.

> *"Stress Testing should be used as a risk management tool and to inform business decisions. As a forward-looking risk management tool, stress testing constitutes a key input into banks' and authorities' activities related to risk identification, monitoring and assessment. As such, stress testing should also contribute to formulating and pursuing strategic and policy objectives" (p. 4).*

A liquidity report estimates hedging costs and their sensitivity to basic trading parameters under a stress test. A liquidity report follows a similar template to a TCA report, with the distinction that a trader bases a TCA report on *realized* orders and *historical* market data. In contrast, a risk manager bases a liquidity report on *hypothetical* orders and *simulated* market data. Figure 1.12 shows a simplified mock liquidity report.

Scenario	Target Position	Target risk exposure	Hedging trade	Price dislocation due to market reaction	Price dislocation due to hedging trade	Half-life of total price dislocation
No hedging	100 mil	10 mil	0 mil	40bps	0bps	5 days
Moderate hedging	50mil	5mil	50mil	45bps	30bps	10 days
Full hedging	0mil	0mil	100mil	50bps	60bps	15 days
Scenario	How much risk does the portfolio hedge?			How much does the fire sale dislocate prices?		How long does the dislocation last?

FIGURE 1.12
Mock liquidity report for a hypothetical stress test.

The report outlines three hedging scenarios in the stress test. At stake is a \$100 million position, with a 10% downside risk scenario: the initial risk exposure is \$10 million.

(a) With *no hedging*, the portfolio exposes itself to \$10 million of risk. In addition, while the portfolio submits no hedge, the rest of the market trades, leading to 40bps of price impact caused by the market. This dislocation does not translate into trading costs as the portfolio is not trading. However, the dislocation causes temporary mark-to-market losses of \$400k on the stock position until the price impact reverts. This reversion has a half-life of 5 days.

(b) With *moderate hedging*, the portfolio lowers its risk to \$5 million. The corresponding trade causes 30bps of price impact and triggers a further dislocation of 5bps from the rest of the market. The hedge costs \$375k, with additional temporary mark-to-market losses on the portfolio's positions of \$375k. The temporary losses revert with a half-life of 10 days.

(c) With *full hedging*, the portfolio holds no position in the stock anymore. There is no risk exposure or temporary dislocation effects on the stock position, but the hedge costs 110bps. The portfolio locks in losses of \$1.1 million to avoid the \$10 million risk exposure.

1.3.4 Portfolio consolidation analysis for senior management

One question arising at the senior level of banks and hedge funds is whether to consolidate portfolios. For example, two geographically separate teams in

Toronto and New York overlap in the stocks they trade. Accordingly, senior management requests a cost and benefit analysis of a team merger, answering questions along four dimensions:

- *Organizational strategy*

  How does the consolidation affect the overall organizational structure senior management is trying to achieve? For example, is the geographic split rooted in a deeper local understanding of the market, or does the overlap in stocks traded lead to a lack of clarity between the two teams?

- *People and team synergy*

  How do the two teams interact? Would merging the two teams incur an upfront training or relocation cost? Would a single team communicate more effectively?

- *Portfolio or trading economics*

  Would a combined portfolio achieve a higher P&L than the two portfolios achieve separately? The economics favor consolidation due to two reasons. First, a combined trading team seeks out offsetting orders and positions and reduces overall trading, a strategy called *internalization*. Second, a combined portfolio searches for a global optimum and outperforms two portfolios searching for local optima. Quantifying these economic benefits requires a vision of how the team trades the combined portfolio and a model to estimate the combined trading costs.

- *Infrastructure sharing*

  Do the two teams benefit from sharing their trading infrastructure? Senior management should expect a degree of technological divergence. A considerable divergence leads to upfront costs when consolidating the two trading infrastructures. Conversely, a shared trading infrastructure benefits the firm over time: the two teams share technology innovations and maintenance costs.

This section focuses on the economic dimension's trading and price impact considerations. First, the trading team builds a plan for portfolio consolidation and quantifies the expected economic benefits. Then, senior management evaluates the economic plan along with the consolidation's three other dimensions.

Section 4.4 of Chapter 4 applies price impact models to portfolio consolidation. When two portfolios trade separately within the same organization, they can do so in two ways:

(a) With knowledge of each others' trades

(b) Without knowledge of each others' trades

This knowledge may be explicit, for example, if they share the same data, or implicit, for instance, if their traders talk amongst each other.

In the case of mutual knowledge, there may be a degree of collaboration, but the two teams compete when trading. Price impact will incentivize each team to trade faster than the other team to minimize their own order slippage. Unfortunately, this *arms race* goes against the firm's best interest. Indeed, the firm favors minimizing the combined portfolio's price impact. In game theory terms, the Nash equilibrium and the Pareto equilibrium for two simultaneously trading portfolios are far apart.[21] In the case without mutual knowledge, the game theory element is absent: instead, each portfolio trades optimally based on their information, and consolidation benefits from sharing data. This global optimum outperforms the individual portfolios' local optima.

Chapters 2 and 4 describe the analysis's mathematical tools. This section walks through a mock report summarizing metrics for each portfolio and the combined portfolio.

<table>
<tr><th>Portfolio</th><th>Trade notional</th><th>Spread & fees paid</th><th>Realized alpha</th><th>Impact due to Toronto</th><th>Impact due to New York</th><th>Total P&L</th></tr>
<tr><td>Toronto</td><td>5bil</td><td>2bps</td><td>90bps</td><td>30bps</td><td>10bps</td><td>24mil</td></tr>
<tr><td>New York</td><td>10bil</td><td>2bps</td><td>80bps</td><td>20bps</td><td>25bps</td><td>33mil</td></tr>
<tr><td>Combined</td><td>13bil</td><td>2bps</td><td>100bps</td><td colspan="2">35bps</td><td>82mil</td></tr>
</table>

Netting reduces total trading

Non-netting trades realize higher alpha

Both portfolios impact each other's trades

The combined optimum outperforms two local optima

FIGURE 1.13
Mock portfolio consolidation report.

In Figure 1.13, two economic benefits arise from consolidating the hypothetical New York and Toronto portfolios.

- First, opposite-sided trades net down, leading to *internalization.*

  Internalization reduces trading, leading to less bid-ask spread and fees in total and a higher alpha for the remaining trades.

- Second, same-sided trades have less combined impact.

[21]The application of game theory to price impact is rooted in probabilistic mean-field game theory, for which the reference book is Carmona and Delarue (2016)[58, 59]. Carmona and Delarue describe in Section 1.3.2 "A Price Impact Model":

> "The model in this section is of great importance in modern financial engineering because it is used as input to many optimal execution engines" (p. 33).

A theoretical approach extends the Kyle model (1985)[143] to a finite-player game, as in Holden and Subrahmanyam (1992)[117]. The idealized traders in such models lead to qualitative results relating price impact to general price equilibrium models.

The two portfolios' price impact is substantial: in particular, the Toronto portfolio has an outsized effect on the New York portfolio. As a result, a trading strategy considering the two teams' combined price impact reduces the firm's trading costs materially.

1.4 Roadmap

1.4.1 What to expect from the Handbook

After introducing the basis of the price impact models, the Handbook offers the reader three types of content for their daily work. The first type relates to the theoretical and practical basis of discrete applications that fit into any team and trading infrastructure. The second type is a simple framework viewing algorithmic trading through the lens of price impact. Lastly, the Handbook illustrates the framework through problems and examples.

1.4.2 A brief summary of each chapter

The book presents seven chapters and three appendices. Four parts organize the chapters and appendices. The central parts are part II, *Acting on Price Impact,* and part III, *Measuring Price Impact.* These parts each begin with a technical chapter to root the applications in rigorous mathematics. Many chapter elements are self-standing: the reader is free to skip sections that do not concern their area of work or for which the mathematical analysis is too challenging.

Furthermore, Chapters two to seven each start with a simple pedagogical example and end with a summary of their results. The summaries of Chapters two and five cover mathematical definitions and propositions. The remaining chapters' summaries describe formulas, techniques, and applications for practitioners.

I Introduction

1 *Introduction to modeling price impact*

This first chapter motivates the study of price impact models, provides a simple glossary of related trading terminology, and outlines four concrete applications fleshed out mathematically in Chapters three and four.

II Acting on price impact

2 *Mathematical models of price impact*

Chapter two moves on to the mathematical definition of price impact models. The chapter builds up a large class of models from a microstructure foundation and proves various extensions and results to cover future applications. The results include liquidity bounds that rule out price manipulation considering volume curves. The last section summarizes the mathematical models and results in a concise reference.

3 *Applications of price impact models*
The third Chapter tackles two applications of price impact models: optimal execution and TCA, leveraging the mathematical framework from Chapter two to provide practical solutions to both problems.

4 *Further applications of price impact models*
The fourth Chapter expands applications to portfolio construction problems. These applications extend the use-case for price impact models beyond sell-side execution teams to include buy-side, risk management, and senior management teams.

III Measuring price impact

5 *An introduction to the mathematics of causal inference*
Chapter five provides a technical introduction to the mathematical theory of causal inference and applies the theory to the A-B testing of trading algorithms. The chapter offers a related machine learning technique called *causal regularization*. The last section summarizes the mathematical definitions and theorems in a concise reference.

6 *Dealing with biases when fitting price impact models*
Chapter six identifies four common causal biases in trading and solves them. The chapter emphasizes identifying, communicating, and accounting for causal assumptions connected to price impact.

7 *Empirical analysis of price impact models*
Chapter seven concretely measures price impact, reproducing empirical results from the literature on a single dataset and methodology. The chapter also reviews cross-impact and fitting methods for price impact beyond equities.

IV Appendix

A. *Using kdb+ for trading models*
Kdb+ is a standard programming language used in financial institutions to analyze sizable trading data and build price impact models. Appendix A provides a primer aimed at quantitative traders and researchers.

B. *Functional convergence theorems for microstructure*
For completeness, Appendix B provides mathematical results not typically covered in textbooks.

C. *Solutions to exercises*
Finally, Appendix C provides solutions for exercises from each chapter.

Part II

Acting on Price Impact

Part II takes a price impact model as given and describes how traders and algorithms *act* on it. The reader learns how to

(a) derive optimal strategies from trading signals and prove the absence of price manipulation opportunities.

(b) implement an order's execution schedule considering price impact and trading signals, such as alpha and volume predictions.

(c) diagnose past orders' execution to improve trading algorithms.

(d) build alpha signals based on the price impact of public trades.

(e) measure and anticipate liquidity risk.

(f) optimize two teams' joint trading costs.

Chapter 2 is technical and lays the mathematical foundation for the applications in Chapters 3 and 4. Therefore, the reader can skip Chapter 2 in their first study. Furthermore, Section 2.7 highlights the main takeaway of Chapter 2 for practitioners:

> Volume predictions cannot increase faster than a multiple of how impact decays without introducing price manipulation opportunities.

Finally, Section 2.8 summarizes mathematical results.

Chapter 3 focuses on order-centric applications of price impact. Chapter 4 highlights portfolio-centric applications of price impact. Both chapters conclude by summarizing concrete formulas for practitioners in Sections 3.4 and 4.5.

2

Mathematical Models of Price Impact

This chapter defines a class of price impact models initiated by Obizhaeva and Wang (2013)[180] and derives their mathematical properties. The reader may skip this chapter in their first study and rely on the summary in Section 2.8. This chapter achieves three goals:

(a) Establish *reduced-form models* of price impact as limits of discrete microstructure models.

Not only are reduced-form models intuitive and straightforward, but they also show which empirical details matter and which features *average away.* This simplification empowers the reader's practical work and highlights which features require modeling attention.

(b) Derive closed-form trading strategies based on stochastic control problems.

Modern algorithmic trading strategies implement control problems. However, the control problems deployed in production may not always have closed-form solutions and often require numerical methods. This chapter derives closed-form solutions to match most empirical facts on price impact. The corresponding models provide intuition and simplify the numerical implementation of trading strategies, as applied in Section 3.2.5 of Chapter 3.

(c) Prove no price manipulation bounds for volume predictions.

Price impact connects price manipulation to market activity: if market activity rises too fast, a trader may generate profits from price impact alone. This chapter replicates and extends the no-price manipulation result of Fruth, Schöneborn, and Urusov (2013, 2019)[97, 98]. The result provides concrete bounds on volume predictions ruling out price manipulation strategies.

The chapter proves essential mathematical results in three steps:

(a) The chapter derives *reduced-form trading models* from a discrete microstructure model. Therefore, each reduced form model links to precise microstructure assumptions.

(b) The chapter replicates the results by Obizhaeva and Wang (2013)[180] using a streamlined method by Fruth, Schöneborn, and Urusov (2013,

DOI: 10.1201/9781003316923-2

2019)[97, 98]. The proof introduces the notions of *impact space* and *myopic control problem* and provides intuition on the optimal trading strategy. Section 2.1 presents the argument in its simplest form.

(c) The chapter extends the OW model in three essential dimensions, ending with a *generalized OW price impact model* that captures known price impact facts while maintaining closed-form solutions:

- Extensions to the objective function, including *alpha signals*
- Extensions related to time, including time-varying, stochastic liquidity parameters
- Extensions related to external impact to consider the rest of the market

Throughout the chapter, footnotes highlight the contributions and similarities in the academic literature.

2.1 A Pedagogical Example

For traders, this chapter's essential takeaway is that, for a large class of price impact models, there is a myopic relationship between signals and impact:

> For the optimal strategy, price impact equals a simple formula of trading signals and their decay rate.

The applications in Chapters 3 and 4 leverage this myopia property repeatedly.

This section presents the myopic trading strategy in its most elementary form for teaching purposes. Let Q be a trading position, α an alpha signal, and I an impact model. Assume all quantities to be deterministic, smooth functions. The chapter clarifies and generalizes these assumptions. The trading algorithm maximizes its expected P&L over an interval $[0, T]$. For a given strategy Q_t, its realized P&L at the end of the day is

$$\underbrace{S_T Q_T}_{\text{end of day fundamental price}} - \underbrace{\int_0^T (S_t + I_t) dQ_t}_{\text{impacted transaction price}} = \int_0^T (S_T - S_t - I_t) dQ_t$$

This equation leads to the expected P&L formula

$$\int_0^T (\alpha_t - I_t)\, dQ_t$$

where α_t predicts the price move $S_T - S_t$.

> In expectation, each fill captures alpha and pays impact.

Assume I is proportional to an exponential moving average of past trades:

$$I_t = \lambda \int_0^t e^{-\beta(t-s)} dQ_s$$

for constant liquidity parameters $\lambda, \beta > 0$. In particular, I solves the ordinary differential equation (ODE)

$$I_t' = -\beta I_t + \lambda Q_t'$$

with $I_0 = 0$. The ODE expresses how one computes I given Q.

Maximizing across all possible strategies Q is a daunting task: the present price impact depends on past decisions, and current decisions affect future impact. Mathematically, the trading decisions $dQ_t = Q_t' dt$ link across time, and one cannot solve them independently.

The crucial insight of Fruth, Schöneborn, and Urusov (2013, 2019)[97, 98] is that one can invert the ODE to derive the trading speed Q' given I:

$$Q_t' = \frac{\beta}{\lambda} I_t + \frac{1}{\lambda} I_t'.$$

From a trading perspective, this formula is intuitive:

Given a strategy's price impact, one backs out its trades.

Mathematically, this inversion leads to an equivalent problem where one maximizes over I instead of Q:

$$\begin{aligned}
\int_0^T (\alpha_t - I_t)\, dQ_t &= \frac{1}{\lambda} \int_0^T (\alpha_t - I_t)\,(\beta I_t + I_t')dt \\
&= \frac{1}{\lambda}\left(\int_0^T (\beta\alpha_t I_t - \beta I_t^2)\, dt + \int_0^T \alpha_t I_t' dt - \int_0^T I_t I_t' dt\right) \\
&= \frac{1}{\lambda}\left(\int_0^T (\beta\alpha_t I_t - \beta I_t^2)\, dt + \alpha_T I_T - \int_0^T \alpha_t' I_t dt - \frac{1}{2} I_T^2\right) \\
&= \frac{1}{\lambda}\left(\int_0^T (\beta\alpha_t I_t - \alpha_t' I_t - \beta I_t^2)\, dt + \alpha_T I_T - \frac{1}{2} I_T^2\right).
\end{aligned}$$

One derives the third line through integration by parts:

$$\int_0^T \alpha_t I_t' dt - \int_0^T I_t I_t' dt = \alpha_T I_T - \int_0^T \alpha_t' I_t dt - \frac{1}{2} I_T^2.$$

This final formula exhibits more terms than the original one but is notably easier to solve: one optimizes the I_t independently. An algorithm solves the quadratic objective in I_t separately for each time t, including $t = T$. Therefore, the optimal strategy is such that I_t maximizes

$$\beta\alpha_t I_t - \alpha_t' I_t - \beta I_t^2$$

for all $t \in (0,T)$, and I_T maximizes

$$\alpha_T I_T - \frac{1}{2} I_T^2.$$

Finally, the optimal solution I_t^* is *myopic*

$$I_t^* = \frac{1}{2}\left(\alpha_t - \beta^{-1}\alpha_t'\right)$$

and only depends on the alpha level α, its decay α', and the impact decay parameter β. For slow-decaying alphas, the optimal impact is half the trader's alpha for $t \ll T$. At time T, the optimal impact state equals

$$I_T^* = \alpha_T = 0.$$

The corresponding optimal trading speed is

$$\begin{aligned} Q_t^{*'} &= \frac{\beta}{\lambda} I_t^* + \frac{1}{\lambda} I_t^{*'} \\ &= \frac{\beta}{2\lambda}\left(\alpha_t - \beta^{-1}\alpha_t' + \beta^{-1}\alpha_t' - \beta^{-2}\alpha_t''\right) \\ &= \frac{\beta}{2\lambda}\left(\alpha_t - \beta^{-2}\alpha_t''\right) \end{aligned}$$

for $t \in (0,T)$. In particular, the optimal strategy trades proportionally to the alpha level unless the alpha presents convexity.

This chapter shows that myopic formulas hold for the optimal trading strategy in general settings.

2.2 Mathematical Setup

Let $T > 0$ and $\left(\Omega, \mathcal{F}, (\mathcal{F}_t)_{t\in[0,T]}, \mathbb{P}\right)$ be a filtered probability space. Denote by

- D the space of càd-làg functions with finite variation taking value in $\mathbb{R}$.
- E_t the space of $\mathcal{F}_t$-measurable, bounded random variables for $t \leq T$. E is a shorthand for E_T.
- $\mathcal{S}_n(t)$ the space of bounded semimartingales $(Z_s)_{s\in[0,t]}$ taking value in $\mathbb{R}^n$ for $t \leq T$ and $n > 0$. The boundedness assumption simplifies the presentation. Furthermore, Remark 2.2.4 articulates why boundedness is not restrictive in practice given the modern emphasis on machine learning methods.

Notation 2.2.1 (Shorthands). *$\mathcal{S}_n$ is a shorthand for $\mathcal{S}_n(T)$, and $\mathcal{S}$ is a shorthand for $\mathcal{S}_1$.*

- $\mathcal{S}_{m,n}$ the space of functionals

$$F : [0, T] \times \Omega \times \mathcal{S}_m \to \mathcal{S}_n,$$

for $m, n > 0$, such that for all $t \in [0, T]$ the function

$$(t, \omega, Z) \to F_t(\omega, Z)$$

is $\mathcal{B}([0, S]) \otimes \mathcal{F}_t \otimes \mathcal{B}(\mathcal{S}_m(t)) / \mathcal{B}(\mathcal{S}_n(t))$-measurable. The measurability condition states that the functional does not present look-ahead bias:

The functional does not depend on future outcomes.

Notation 2.2.2 (Causal dependencies). *To make causal relationships explicit, one may write out the dependencies of a functional F explicitly. For example, this is helpful when studying price impact from a causal perspective. Chapters 5 and 6 cover causal inference.*

For instance, $F_t(\omega, L)$ is a process-valued functional with a causal dependence on L. The value of F given (t, ω, L) is $F_t(\omega, L_{\cdot \le t}(\omega))$. One may drop the explicit dependencies to lighten the notation.

Notation 2.2.3 (Stochastic integrals). *The semimartingales in this chapter have jumps. These jumps lead to additional notation for stochastic integrals to differentiate the value before and after the jump. For instance, consider the integration by parts formula*

$$X_T Y_T = X_{0-} Y_{0-} + \int_0^T X_{t-} dY_t + \int_0^T Y_{t-} dX_t + [X, Y]_T$$

for X, Y two semimartingales. This formula is exactly Corollary 2 from Protter (2005, p. 68)[191] and uses the convention

$$\int_0^T X dY = \int_{[0,T]} X dY$$

as per Protter (2005, p. 60)[191].

To lighten the notation, consider the following two conventions throughout the chapter:

(a) *Stochastic integrals with semimartingale integrands use their integrand's left-limit. That is, define*

$$\int X dY = \int X_- dY$$

(b) *Initial conditions at time $t = 0$ refer to a semimartingale's* left limit. *For instance, when stating $X_0 = 0$, this is shorthand for $X_{0-} = 0$, and one singles out the right limit with X_{0+}. This slight inconsistency with the càd-làg semimartingale X allows the reader to use the notation X_0 for a process's initial condition to denote X_{0-} and is* specific *to $t = 0$.*

The simplified notation yields the integration by parts formula over $[0, T]$

$$X_T Y_T = X_0 Y_0 + \int_0^T X_t dY_t + \int_0^T Y_t dX_t + [X, Y]_T$$

for X, Y two semimartingales.

Remark 2.2.4 (Bounded signals and applications to machine learning). *For simplicity, this chapter bounds the variables of interest to focus on deriving optimal trading strategies. The reader extends the results from this chapter to broader variables using, for instance, Chapter 2 "Semimartingales and Stochastic Integrals" (p. 51) by Protter (2005)[191] to lift boundedness assumptions.*

From a practical perspective, the boundedness assumption is compatible with the modern emphasis on non-parametric trading signals. Historically, Mathematical Finance applications specify price dynamics using unbounded parametric models, such as the Bachelier or Ornstein-Uhlenbeck models. Today, practitioners estimate price dynamics, such as alpha signals, using machine learning and derive optimal trading strategies for generic signals. Furthermore, machine learning signals reflect price dynamics and model features non-parametrically: the researcher does not specify the signal with explicit formulas. Therefore, control problems in Mathematical Finance must allow non-parametric signals for trading applications.

One incorporates machine learning into control problems by specifying signals as Bayesian estimators: for instance, consider a generic alpha signal

$$\alpha_t = \mathbb{E}\left[S_T - S_t \middle| \mathcal{F}_t \right],$$

where S is a price and T is the signal's prediction horizon. In trading terms,

$$\mathbb{E}\left[\cdot \middle| \mathcal{F}_t \right]$$

is a model-agnostic description of an algorithm's best predictor given the information $\mathcal{F}_t$ available at time t. One can constrain machine learning methods to implement a bounded signal for α_t. Such a bounded signal satisfies this chapter's assumptions and allows for closed-form optimal trading strategies.

2.2.1 Defining price impact and instantaneous transaction costs

This section introduces two base variables: prices and trades. Because of market microstructure, these variables come in multiple flavors. The different price descriptions lead to formal definitions of price impact and instantaneous transaction costs. The chapter takes a given *trader's* vantage point, also referred to as *the agent.*

Definition 2.2.5 (Trading process). *Define a trading process as a stochastic process $L \in \mathcal{S}_n$ for $n > 0$. Its variations dL describe trades.*

Define $Q \in \mathcal{S}$ as the agent's trading process. If $L = Q$, one does not model any other trades; otherwise, Q is a component of L.

The components of L other than Q model different portfolios' trades. For instance, L may include the public trading tape.

Definition 2.2.6 (Price processes, impact, and instantaneous transaction costs). *Let $n > 0$ and $L \in \mathcal{S}_n$ be a trading process that includes the trader's process $Q \in \mathcal{S}$.*

Define the following prices:

(a) $S_t(\omega)$ is the unperturbed price, and $S \in \mathcal{S}$

S is the stock's price without the trading process L and its price impact.[1]

(b) $P_t(\omega, L)$ is the perturbed price, and $P \in \mathcal{S}_{n,1}$. Define the price impact I that L causes by the difference

$$I = P - S.$$

P is the observable price and differs from the unperturbed price S because L causes price impact.[2]

(c) $\tilde{P}_t(\omega, L, Q)$ is the trade *price for the trading process Q considering L's impact and lives in $\mathcal{S}_{n+1,1}$. In the frequent case $L = Q$, $\tilde{P}$ includes both the price impact of Q and the spread costs paid by Q.*[3]

---

[1] The *unperturbed price* appears in the literature under different names.

- Gatheral, Schied, and Slynko (2011)[106] refer to the *unperturbed price* as "the unaffected price" (p. 2).
- Dang (2017)[82] refers to the *unperturbed price* as the "fair price" (p. 1).
- Becherer, Bilarev, and Fentrup (2018)[20] refer to the *unperturbed price* as the "unaffected fundamental price" (p. 3).

[2] Becherer, Bilarev, and Fentrup (2018)[20] refer to the *perturbed price* as the "actual price" (p. 3).

[3] Both the *trade price* and the *instantaneous transaction costs* appear in the literature under different names.

- Amgren and Chriss (2001)[10] refer to the *trade price* as the "actual price" (p. 9) and *instantaneous transaction costs* as "temporary market impact" (p. 9).
- Busseti and Lillo (2012)[47] refer to the *trade price* as the "effective price" (p. 5).
- Graewe and Horst (2017)[109] refer to the *trade price* as the "execution price" (p. 2) and *instantaneous transaction costs* as "instantaneous impact" (p. 2).

Define the instantaneous transaction costs, or spread costs, incurred by the trading process Q at time t as the difference

$$s_{t-} = \tilde{P}_{t-} - P_{t-}.$$

Notation 2.2.7 (Left limit). *Z_{t-} describes the left limit of Z at time t. In modern electronic markets, a trade at time t resolves using the state of the world at time $t-$. Indeed, when a fill executes at time t, an algorithm submitted it at time $t-$ considering the limit order book at time $t-$. The venue then matches the fill with some infinitesimal latency, leading to the new state of the world at the time t.*

In summary, Q represents the agent's cumulative trades, and the general trading process L includes Q. When the trader executes a fill at time t, they trade at the price

$$\tilde{P}_{t-}(\omega, L, Q) = S_{t-}(\omega) + I_{t-}(\omega, L) + s_{t-}(\omega, L, Q)$$

and observe the price

$$P_t(\omega, L) = S_t(\omega) + I_t(\omega, L).$$

Figure 2.1 illustrates the three price definitions.

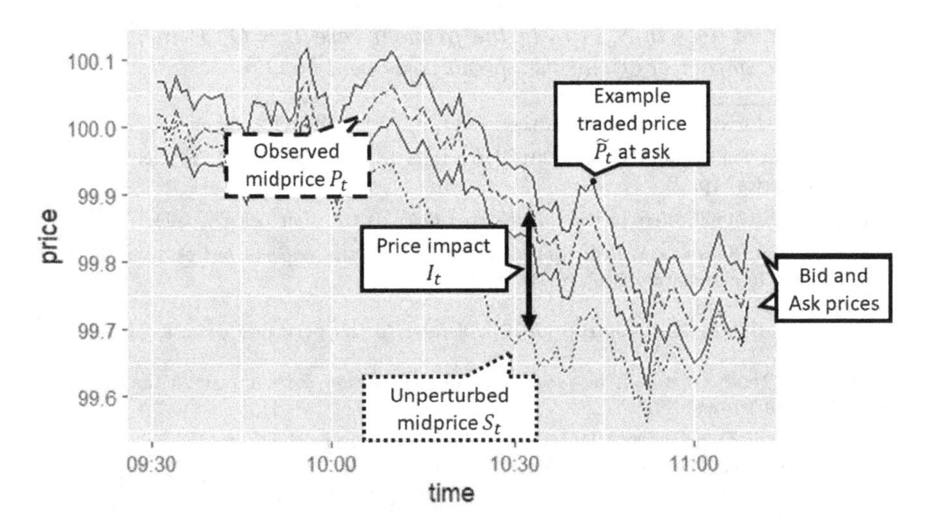

FIGURE 2.1
Illustration of Definition 2.2.6.

These formal definitions allow the reader to make precise assumptions about the market microstructure and single out their effect on trading strategies. Sections 2.2.2 and 2.2.3 provide examples of microstructure models and assumptions. Finally, Section 2.2.4 takes these instances' continuous-time limit to derive *reduced-form models*. Such *reduced-form models* are tractable and ideally suited for applications, such as in Chapters 3 and 4.

2.2.2 Establishing P&L in discrete time

This section describes the trader's P&L. Assume given a trading process L, the agent's trading process Q, and the three prices S, P, and $\tilde{P}$ from Definition 2.2.6. Then, four steps establish a continuous-time formula for P&L.

(a) One discretizes the variables on a regular time grid.

(b) One defines formulas for the trader's cash position and P&L in discrete time.

(c) One describes microstructure assumptions. See Section 2.2.3 for examples.

(d) One proves the continuous-time limit and derives *reduced-form models*. See Section 2.2.4 for examples.

Notation 2.2.8 (Time discretization). *An integer N discretizes time. Let*

$$t_n^N = \frac{n}{N} T = n \Delta t^N$$

and

$$X_n^N = X_{t_n^N}.$$

Define similarly Q^N, S^N, P^N, I^N, $\tilde{P}^N$, and s^N. The notation

$$\Delta_n Z = Z_n - Z_{n-1}$$

describes the increments of variable Z.

To define the trader's P&L, one establishes their cash position and marks their stock with a price. Denote by K_n^N the trader's cash position at time t_n^N and assume an initial position $K_0^N = K_0$. Consider a fill $\Delta_n Q^N$. The fill executes at time t_{n-1}^N and provides the updated position $Q_{t_n^N}$. The fill price is $\tilde{P}_{n-1}^N$: if the trade is a buy, then the fill *pays* that price. If the trade is a sell, then the fill *receives* that price.

Definition 2.2.9 (Cash position). *Define the trader's cash position K_n^N as the stochastic process with initial condition K_0 satisfying the equation*

$$\Delta_n K^N = -\tilde{P}_{n-1}^N \Delta Q_n^N.$$

Depending on how the trader marks their stock position, Definition 2.2.9 leads to two P&L metrics.

(a) The *accounting P&L* marks the trader's stock position with the observed price P.

Because P is the observed market price, this definition matches the trader's accounting systems, also called the *mark-to-market* P&L.

Definition 2.2.10 (Accounting P&L). *Define the trader's accounting P&L X^N as*

$$X_n^N = P_n^N Q_n^N + K_n^N. \tag{2.1}$$

(b) To determine an optimal trading strategy, the trader must mark their position with the unperturbed price S, leading to their *fundamental P&L.*

Marking the position to the unperturbed price removes the trader's incentive to inflate their P&L in the short term. Instead, the fundamental P&L reflects the position's long-term expected value when the price dislocation from price impact has reverted. Section 4.3.1 of Chapter 4 describes how price impact distorts accounting P&L.

Definition 2.2.11 (Fundamental P&L). *Define the trader's fundamental P&L Y^N as*

$$Y_n^N = S_n^N Q_n^N + K_n^N. \tag{2.2}$$

Remark 2.2.12 (When do accounting and fundamental P&L match?). *Under the liquidation constraint $Q_T = 0$, the two P&L definitions coincide at $t_N^N = T$. This constraint is common in execution and market-making applications.*

One can express equations 2.1 and 2.2 as difference equations. The Mathematical Finance literature refers to such equations as *self-financing equations* or *wealth equations.*

Proposition 2.2.13 (Discrete self-financing equations). *The processes X^N and Y^N satisfy*

$$\Delta_n X^N = Q_{n-1}^N \Delta_n P^N - s_{n-1}^N \Delta_n Q^N + \Delta_n P^N \Delta_n Q^N \tag{2.3}$$

and

$$\Delta_n Y^N = Q_{n-1}^N \Delta_n S^N - I_{n-1}^N \Delta_n Q^N - s_{n-1}^N \Delta_n Q^N + \Delta_n S^N \Delta_n Q^N. \tag{2.4}$$

Proof. One derives the formulas by discrete differentiation. The accounting P&L X^N satisfies

$$\begin{aligned}
\Delta_n X^N &= Q_{n-1}^N \Delta_n P^N + P_{n-1}^N \Delta_n Q^N + \Delta_n P^N \Delta_n Q^N + \Delta K_n^N \\
&= Q_{n-1}^N \Delta_n P^N + \left(P_{n-1}^N - \tilde{P}_{n-1}^N\right) \Delta_n Q^N + \Delta_n P^N \Delta_n Q^N \\
&= Q_{n-1}^N \Delta_n P^N - s_{n-1}^N \Delta_n Q^N + \Delta_n P^N \Delta_n Q^N.
\end{aligned}$$

The fundamental P&L Y^N satisfies

$$\begin{aligned}\Delta_n Y^N &= Q^N_{n-1}\Delta_n S^N + S^N_{n-1}\Delta_n Q^N + \Delta_n S^N \Delta_n Q^N + \Delta K^N_n \\ &= Q^N_{n-1}\Delta_n S^N + \left(S^N_{n-1} - \tilde{P}^N_{n-1}\right)\Delta_n Q^N + \Delta_n S^N \Delta_n Q^N \\ &= Q^N_{n-1}\Delta_n S^N - I^N_{n-1}\Delta_n Q^N - s^N_{n-1}\Delta_n Q^N + \Delta_n S^N \Delta_n Q^N.\end{aligned}$$

□

Remark 2.2.14 (Interpretation of the additional terms). *The two self-financing equations add* trading frictions *compared to the* frictionless *equation traditionally used in finance.*

(a) The temporary transaction cost *term*

$$-s^N_{n-1}\Delta_n Q^N$$

captures the spread the trader pays or receives. It is positive when the fill provides liquidity and negative otherwise.

(b) The price impact *term*

$$-I^N_{n-1}\Delta_n Q^N$$

quantifies how the price dislocation I^N_{n-1} affects a fill.

(c) The adverse selection *terms*

$$\Delta_n P^N \Delta_n Q^N; \quad \Delta_n S^N \Delta_n Q^N$$

capture the instantaneous correlation between the fill and the stock price. Microstructure heavily affects adverse selection.[4]

Remark 2.2.15 (Adverse selection). *Liquidity providers frequently observe counterparties with superior short-term information fill their limit orders. This* adverse selection *leads to the inequality*

$$\mathbb{E}\left[\Delta_n S^N \Delta_n Q^N \,\middle|\, \mathcal{F}_{t^N_{n-1}}\right] \leq 0$$

when the trader provides liquidity. Market makers consider adverse selection *when deploying strategies.*

(a) When providing liquidity, the negative adverse selection term partly or completely erases the spread gains.

[4]Carmona and Webster (2019)[61] first derived this term from a microstructure model for market-making applications. Carmona and Webster also empirically measure adverse selection for the representative liquidity provider on NASDAQ data, finding it to drive their P&L. In addition, Webster et al. (2015)[235] measure adverse selection on the Russell 3000 over ten years.

(b) Conversely, a liquidity taker with a high-frequency signal offsets their instantaneous costs by adversely selecting market makers.[5]

2.2.3 Examples of microstructure assumptions

This section first covers the *frictionless* case used in the seminal work by Black and Scholes (1973)[27]. In that setting, the new observed price equals the last traded price. This assumption reflects the trading data of the time: transaction prices were public, but there was no continuously quoted order book providing market information on fills. In the frictionless case, temporary costs and adverse selection cancel each other.

Definition 2.2.16 (Frictionless trading)**.** *Define a frictionless trading model as a model where the assumption $P_n^N = \tilde{P}_{n-1}^N$ holds.*

Remark 2.2.17 (Modern trading frictions)**.** *Modern electronic markets compute the fill price $\tilde{P}_{n-1}^N$ based on the order book at time t_{n-1}^N and the fill quantity $\Delta_n Q^N$. All three friction terms from Remark 2.2.14 become relevant and require a microstructure model.*

Obizhaeva and Wang (2013)[180] provide microstructure assumptions based on a modern limit order book. Obizhaeva and Wang make two microstructure assumptions:

- The trader only buys and executes at the ask. For simplicity, let P denote the ask price and ignore the bid-ask spread altogether.
- The order book is proportional to the Lebesgue measure for price levels above the ask price P.

The assumption on the order book's shape requires a detailed description of what happens when a fill executes to define the three friction terms from Remark 2.2.14. The related material by Obizhaeva and Wang (2013)[180] spans pages six to nine. Figure 2.2 illustrates a fill's execution under the OW microstructure model. Let λ be a positive constant. When the trader fills a quantity $\Delta_n Q^N > 0$ at time t_{n-1}^N:

- They deplete the ask levels until reaching the price
$$P_{n-1}^N + \lambda \Delta_n Q^N.$$
- The fill's average execution price per share is
$$\tilde{P}_{n-1}^N = P_{n-1}^N + \frac{\lambda}{2} \Delta_n Q^N.$$

[5]Muhle-Karbe and Webster (2017)[171] provide a model: high-frequency traders act with superior short-term information over the timescale Δt^N. This informational advantage guarantees that the adverse selection term harms liquidity providers, as observed empirically in Carmona and Webster (2019)[61].

Muhle-Karbe, Wang, and Webster (2022)[169] leverage an adverse selection model to merge delta-hedging and market-making into a cohesive trading strategy.

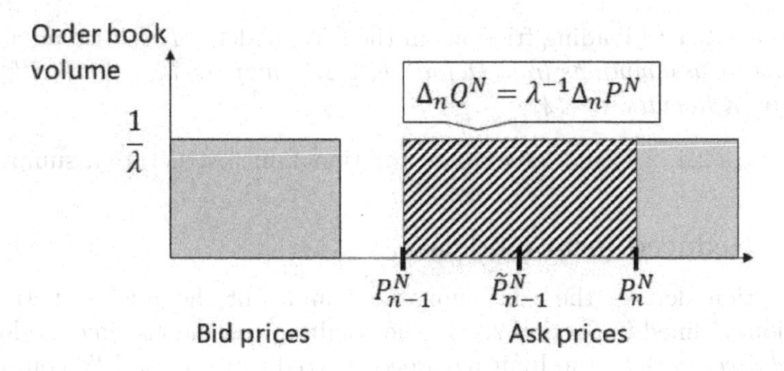

FIGURE 2.2
Illustration of a buy fill under the OW microstructure from Definition 2.2.18.

- The order book remains proportional to the Lebesgue measure for price levels above the new ask

$$P_n^N = P_{n-1}^N + \lambda \Delta_n Q^N.$$

- The fill $\Delta_n Q^N$ does not adversely select the limit order book. Mathematically, assume the convergence $\Delta_n Q^N \Delta_n S^N \to 0$ holds uniformly on compacts in probability (u.c.p.) as $N \to \infty$.[6]

Finally, assume the price dislocation on the ask exponentially reverts over time:

$$\Delta_n I^N = -\beta I_{n-1}^N \Delta t^N + \lambda \Delta_n Q^N.$$

Definition 2.2.18 formalizes the OW microstructure model.

Definition 2.2.18 (OW microstructure model)**.** *Define the following three microstructure assumptions:*

(a) linear instantaneous transaction costs

$$s_n^N = \frac{\lambda}{2} \Delta_n Q^N.$$

(b) exponential decay price impact

$$\Delta_n I^N = -\beta I_{n-1}^N \Delta t^N + \lambda \Delta_n Q^N.$$

(c) no adverse selection

$$\Delta_n Q^N \Delta_n S^N \to 0$$

u.c.p. as $N \to \infty$.

[6] If the trading strategy Q is deterministic and S is an Itô process, as for Obizhaeva and Wang (2013)[180], the no adverse selection assumption holds. Section 2.4.2, covers trading strategies for which the no adverse selection assumption does not hold.

Remark 2.2.19 (Trading frictions in the OW model). *The three named microstructure assumptions from Definition 2.2.18 map the three trading friction terms from Remark 2.2.14.*

Sections 2.4, 2.5, and 2.6 explore additional microstructure assumptions.

2.2.4 Reduced form models

This section derives the continuous-time limits of the microstructure assumptions defined in Section 2.2.3. The resulting continuous-time models are *reduced-form models*: the limiting procedure reduces the model's complexity and tells the reader which features matter.

Proposition 2.2.20 (Black and Scholes self-financing equation). *Consider the frictionless trading model from Definition 2.2.16.*

Then the continuous-time limit of the accounting P&L X^N follows the self-financing equation

$$dX_t = Q_t dP_t,$$

and the convergence happens u.c.p. as $N \to \infty$.

Proof. The choice $\tilde{P}^N_{n-1} = P^N_n$ and equation 2.3 from Proposition 2.2.13 yield the discrete self-financing equation

$$\begin{aligned} \Delta_n X^N &= Q^N_{n-1}\Delta_n P^N + \left(P^N_n - \tilde{P}^N_{n-1}\right)\Delta_n Q^N \\ &= Q^N_{n-1}\Delta_n P^N. \end{aligned}$$

The u.c.p. convergence follows from Theorem 21 of Protter (2005, p. 64)[191]. □

Proposition 2.2.21 (Original OW self-financing equation). *Consider the original OW microstructure from Definition 2.2.18:*

(a) linear temporary transaction costs.

(b) exponential decay price impact.

(c) no adverse selection.

Then the continuous-time limit of the fundamental P&L Y^N follows the self-financing equation

$$dY_t = Q_t dS_t - I_t dQ_t - \frac{\lambda}{2} d[Q,Q]_t,$$

and the impact process I^N converges to the solution to the Stochastic Differential Equation (SDE)

$$dI_t = -\beta I_t dt + \lambda dQ_t$$

with initial condition I_0. The convergence happens u.c.p. as $N \to \infty$.

Proof. The SDE follows from the *exponential decay price impact* assumption

$$\Delta_n I^N = -\beta I_{n-1}^N \Delta t^N + \lambda \Delta_n Q^N.$$

Indeed,

$$I_n^N = \lambda \left(1 - \beta \Delta t^N\right)^n \sum_{m=1}^{n} \left(1 - \beta \Delta t^N\right)^{-m} \Delta_m Q^N$$

solves the discrete equation. First, one controls the gap

$$\left(1 - \beta \Delta t^N\right)^{-m} - e^{\beta t_m^N}$$

between discrete and continuous-time exponentials. Then, one applies Theorem 21 of Protter (2005, p. 64)[191] to obtain convergence.

The self-financing equation 2.4 from Proposition 2.2.13 satisfies

$$\begin{aligned}
\Delta_n Y^N &= Q_{n-1}^N \Delta_n S^N - I_{n-1}^N \Delta_n Q^N - s_{n-1}^N \Delta_n Q^N + \Delta_n S^N \Delta_n Q^N \\
&= Q_{n-1}^N \Delta_n S^N - I_{n-1}^N \Delta_n Q^N - \frac{1}{2}\lambda \left(\Delta_n Q^N\right)^2 + \Delta_n S^N \Delta_n Q^N
\end{aligned}$$

by the *linear instantaneous transaction costs* assumption. The convergence follows from Theorem 21 of Protter (2005, p. 64)[191] and the *no adverse selection* assumption. □

Remark 2.2.22 (Linear temporary transaction costs). *The reader may be surprised that the term $\frac{\lambda}{2} d[Q,Q]_t$ survives in the reduced form model. Two arguments emphasize the term.*

(a) Obizhaeva and Wang (2013)[180], derive the term in the continuous-time limit in equation (29) on page 16. Obizhaeva and Wang highlight that "the execution cost consists of two parts: the costs from continuous trades and discrete trades" (p. 16).

Because Obizhaeva and Wang restrict themselves to deterministic strategies $Q \in D$, $d[Q,Q]_t$ is non-zero only when Q jumps, which corresponds to their "discrete trades" (p. 16). Corollary 2.3.6 from Section 2.3.2 proves that the optimal trading strategy under the OW model does *include jumps, making this term non-negligible.*

(b) When one introduces stochastic liquidity or alpha signals, the set of admissible Q lifts from D to S, and $d[Q,Q]_t$ is non-zero even without jumps. Dynamic signals are crucial extensions to the OW model.

Carmona and Leal (2021)[60] theoretically investigate strategies with non-finite variations. Carmona and Leal leverage the trading model of Cartea and Jaimungal (2016)[63]. They empirically track individual traders using public trading data from the Toronto stock exchange and "extend the theoretical analysis of an existing optimal execution model to accommodate the presence of Itô inventory processes" (p. 1). Carmona and Leal also "compare empirically the optimal behavior of traders in such fitted models, to their actual behavior as inferred from the data" (p. 1).

The observation $\frac{\lambda}{2}d[Q,Q]_t = \frac{1}{2}d[I,Q]_t$ simplifies the self-financing equation

$$dY_t = Q_t dS_t - I_t dQ_t - \frac{1}{2}d[I,Q]_t. \tag{2.5}$$

The self-financing equation 2.5 holds in more general settings.

2.3 The Obizhaeva and Wang (OW) Propagator Model

This section applies the OW price impact model to order scheduling and reproduces the optimal execution strategy of Obizhaeva and Wang (2013)[180].

2.3.1 An optimal execution problem

Optimal execution minimizes an order's arrival slippage, as defined in Section 1.2.6 of Chapter 1. In addition, executing an order of size $\bar{Q}$ is equivalent to liquidating a position $Q_0 = -\bar{Q}$. For simplicity and to trivially satisfy the *no adverse selection* assumption from Definition 2.2.18, this section explores deterministic trading strategies $Q \in D$. Sections 2.4 and 2.5 lift the space of admissible trading processes to $Q \in \mathcal{S}$ to take into account stochastic alpha and liquidity signals.

Definition 2.3.1 (OW optimal execution problem). *Let Q_0 be a given initial position. For $Q \in D$, let I solve the ODE*

$$dI_t = -\beta I_t dt + \lambda dQ_t$$

for $\lambda > 0$ and with initial condition $I_0 = 0$. Furthermore, let $S \in \mathcal{S}$.

Define the optimal execution problem as the control problem

$$\sup_{Q \in D} \mathbb{E}\left[\int_0^T Q_t dS_t - \int_0^T I_t dQ_t - \frac{1}{2}[I,Q]_T\right] \tag{2.6}$$

under the constraint $Q_T = 0$.

Remark 2.3.2 (Objective function). *The objective function corresponds to the fundamental P&L Y_T derived in Proposition 2.2.21. With a zero terminal position, the fundamental and accounting P&L match at time T:*

$$Y_T = X_T.$$

2.3.2 Closed-form optimal trading strategy

Lemma 2.3.3 simplifies the control problem 2.6 from Definition 2.3.1.

Lemma 2.3.3. *Let $S \in \mathcal{S}$ be an unperturbed price and $Q \in \mathcal{S}$ a trading process. Define*

$$\alpha_t = \mathbb{E}\left[S_T - S_t \middle| \mathcal{F}_t \right].$$

Assume $\alpha \in \mathcal{S}$. Then the identity

$$\mathbb{E}\left[\int_0^T Q_t dS_t\right] = \mathbb{E}\left[\int_0^T \alpha_t dQ_t + [\alpha, Q]_T\right]$$

holds.

Proof. Define the martingale

$$\mathbb{E}\left[S_T \middle| \mathcal{F}_t \right] = Z_t.$$

Because $S, \alpha \in \mathcal{S}$, one has $Z \in \mathcal{S}$. Hence, α satisfies

$$d\alpha_t = -dS_t + dZ_t.$$

By Itô, one has

$$\begin{aligned} 0 &= \alpha_T Q_T \\ &= \int_0^T \alpha_t dQ_t + \int_0^T Q_t d\alpha_t + [\alpha, Q]_T \\ &= \int_0^T \alpha_t dQ_t - \int_0^T Q_t dS_t + \int_0^T Q_t dZ_t + [\alpha, Q]_T. \end{aligned}$$

Shifting the term $\int_0^T Q_t dS_t$ to the left-hand side and taking expectations concludes. □

Remark 2.3.4. *Therefore, the control problem simplifies to*

$$\sup_{Q \in D} \mathbb{E}\left[\int_0^T (\alpha_t - I_t)\, dQ_t - \frac{1}{2}[I, Q]_T + [\alpha, Q]_T\right]$$

under the constraint $Q_T = 0$, where $\alpha_t = \mathbb{E}\left[S_T - S_t \middle| \mathcal{F}_t \right]$.

Three parallels emerge with the trading terminology from Section 1.2 of Chapter 1.

- *As per Section 1.2.3, α_t is a signal predicting the trading strategy's realized alpha based on the trader's information $\mathcal{F}_t$.*
- *The order's* alpha slippage, *which Section 1.2.7 defines, is*

$$\int_0^T \alpha_t dQ_t + [\alpha, Q]_T.$$

- *The* slippage due to price impact, *which Section 1.2.7 defines, is*

$$-\int_0^T I_t dQ_t - \frac{1}{2}[I,Q]_T.$$

Proposition 2.3.5 is the chapter's crucial result and introduces the map to *impact space.* Proposition 2.3.5 corresponds to "Lemma 8.6 (Costs rewritten in terms of the price impact process)" (p. 30) from Fruth, Schöneborn, and Urusov (2013)[97].

Proposition 2.3.5 (Translating the problem into impact space)**.** *Consider the OW optimal execution problem from Definition 2.3.1*

$$\sup_{Q\in D} \mathbb{E}\left[\int_0^T Q_t dS_t - \int_0^T I_t dQ_t - \frac{1}{2}[I,Q]_T\right]$$

with initial condition Q_0 and terminal constraint $Q_T = 0$. Define

$$\alpha_t = \mathbb{E}\left[S_T - S_t | \mathcal{F}_t\right]$$

and assume $\alpha \in \mathcal{S}$.

The execution problem is equivalent to

$$\sup_{I\in D} \frac{1}{\lambda}\mathbb{E}\left[\int_0^T \left(\beta\alpha_t I_t - \beta I_t^2\right) dt - \int_0^T I_t d\alpha_t - \frac{1}{2} I_T^2\right]$$

subject to the linear constraint

$$I_T + \int_0^T \beta I_t dt = -\lambda Q_0.$$

The one-to-one map from D to D

$$dQ_t = \frac{1}{\lambda}\left(\beta I_t dt + dI_t\right)$$

recovers the trading process Q from the impact state I.

Proof. The proof hinges on the simple observation that the ODE

$$dI_t = -\beta I_t dt + \lambda dQ_t$$

defines a one-to-one map between the control variable Q and the state variable I. One inverts the relationship and chooses I as the problem's control variable.[7] The initial and terminal constraints on Q are equivalent to

$$\begin{aligned}\lambda\left(Q_T - Q_0\right) &= \int_0^T \beta I_t dt + \int_0^T dI_t \\ &= I_T + \int_0^T \beta I_t dt.\end{aligned}$$

[7]Fruth, Schöneborn, and Urusov (2013)[97] first introduced this method: "we are going to exploit the fact that there is a one-to-one correspondence between Θ and D" (p. 29). In their setting, Θ is the position represented by Q in this chapter, and D is the price impact I.

Lemma 2.3.3 and the one-to-one map simplify the control problem:

$$\sup_{I \in D} \mathbb{E}\left[\int_0^T (\alpha_t - I_t)\frac{1}{\lambda}(dI_t + \beta I_t dt) - \frac{1}{2\lambda}[I, I]_T + \frac{1}{\lambda}[\alpha, I]_T\right].$$

Finally, the term

$$\int_0^T (\alpha_t - I_t)(dI_t + \beta I_t dt) - \frac{1}{2}[I, I]_T + [\alpha, I]_T$$
$$= \int_0^T \alpha_t dI_t + [\alpha, I]_T - \int_0^T I_t dI_t - \frac{1}{2}[I, I]_T + \int_0^T (\beta \alpha_t I_t - \beta I_t^2)\, dt$$

equals

$$\alpha_T I_T - \int_0^T I_t d\alpha_t - \frac{1}{2}I_T^2 + \int_0^T (\beta \alpha_t I_t - \beta I_t^2)\, dt.$$

$\alpha_T = 0$ concludes the proof. □

Section 2.3.3 discusses the intuition behind this result in depth. This section concludes by replicating the result of Obizhaeva and Wang (2013)[180].

Corollary 2.3.6 (Solution to the original OW problem). *Consider the OW optimal execution problem from Definition 2.3.1 and assume the unperturbed price S is a martingale.*

Then the optimal impact I^* satisfies

$$\forall t \in (0, T), \quad I_t^* = -\frac{\lambda}{2 + \beta T}Q_0; \quad I_T^* = -\frac{2\lambda}{2 + \beta T}Q_0,$$

and therefore, the optimal trading strategy is

$$\forall t \in (0, T), \quad dQ_t^* = -\frac{\beta}{2 + \beta T}Q_0 dt.$$

In particular, the strategy jumps at times $t = 0, T$. Both jumps are of size

$$-\frac{1}{2 + \beta T}Q_0.$$

Proof. S is a martingale and $\alpha_t = 0$. The control problem simplifies to

$$\inf_{I \in D} \int_0^T \beta I_t^2 dt + \frac{1}{2}I_T^2$$

with the linear constraint

$$I_T + \int_0^T \beta I_t dt = -\lambda Q_0.$$

Introduce a Lagrange multiplier $\nu \in \mathbb{R}$ for the constraint. For a given Lagrange multiplier ν, the unconstrained optimization problem is

$$\inf_{I \in D} \int_0^T \left(\beta I_t^2 - \beta \nu I_t\right) dt + \frac{1}{2} I_T^2 - \nu I_T.$$

Pointwise optimization solves the problem:

$$\forall t \in (0, T), \quad I_t^* = \frac{\nu}{2}; \quad I_T^* = \nu.$$

To satisfy the linear constraint, set $\nu = -\frac{2\lambda}{2+\beta T} Q_0$. Finally, one obtains Q^* via the one-to-one map between I and Q:

$$\begin{aligned} dQ_t^* &= \frac{1}{\lambda}\left(\beta I_t^* dt + dI_t^*\right) \\ &= -\frac{\beta}{2 + \beta T} Q_0 dt. \end{aligned}$$

□

2.3.3 Intuition behind the optimal trading strategy

The essential insight from Corollary 2.3.6 is that *optimal execution is a pointwise control problem in I.* This *myopia* property explains the effectiveness of Fruth, Schöneborn, and Urusov's (2013)[97] map from Q to I. Paraphrasing the intuition in trading terms, under the OW model,

(a) Trading decisions are *path-dependent* in Q: the trader requires the history of Q to decide on a strategy in *trade space.*

(b) Trading decisions are *myopic* in I: the trader does not require the history of I to decide on a target in *impact space.*

Statement (b) holds for a given value of the Lagrange multiplier due to the terminal position constraint. (b) also holds without caveat in the unconstrained case, for example, when the agent trades considering an alpha signal. For instance, alpha signals drive trading for a statistical arbitrage portfolio, and the algorithm may be unconstrained at time T.

Not every price impact model leads to a myopic problem in I. However, myopia is an intuitive and helpful property to establish for a sizable class of models. The following sections provide standard price impact models and control problems that naturally extend the OW model. Wherever possible, one proves that the problem is myopic in impact space. The academic literature tackles these extensions, and footnotes reference the rich literature on the mathematics of price impact models.

2.4 Extensions Related to the Objective Function

This section maintains the dynamics of I but lifts other assumptions made by Obizhaeva and Wang (2013)[180].

2.4.1 Alpha signal

The first extension introduces an alpha signal by removing the unperturbed martingale price assumption. Therefore, the trader must predict S or, equivalently, model α. Two instances follow: a deterministic and stochastic alpha signal. Corollary 2.4.1 solves the optimal execution problem with a deterministic signal. Proposition 2.4.2 lifts the unwind constraint and solves a trading problem with stochastic alpha signal. The latter problem models applications beyond optimal execution, such as statistical arbitrage, where there is no terminal constraint. For such strategies, alpha signals drive trading.

To the author's best knowledge, Lorenz and Schied (2013)[156] first solved the original OW model with an alpha signal.[8] Corollary 2.4.1 solves a simplified version of the problem in Theorem 1 on page 8 by Lorenz and Schied (2013)[156] using the method by Fruth, Schöneborn, and Urusov (2013)[97].

Corollary 2.4.1 (Solution with deterministic alpha). *Consider the OW optimal execution problem from Definition 2.3.1 and assume the following* deterministic alpha assumption *holds:*

- $\alpha_t = \mathbb{E}\left[S_T - S_t \middle| \mathcal{F}_t\right]$ *is a deterministic* C^2 *function. Denote its first two derivatives by* α'_t *and* α''_t.

Then the optimal impact I^* *satisfies*

$$\forall t \in (0, T), \quad I^*_t = \frac{1}{2}\left(\alpha_t - \beta^{-1}\alpha'_t + \nu\right); \quad I^*_T = \nu$$

[8]Three other papers and one book tackle trading and optimal execution with signals. First, Almgren and Chriss (2001)[10] introduce "The Value of Information" (p. 24) by modeling a drift in the unperturbed price and solving the control problem on page 26.

> "[T]he price process may have a drift. For example, if a trader has a strong directional view, then he will want to incorporate this view into the liquidation strategy" (p. 25).

Gârleanu and Pedersen (2016)[102] introduce "multiple signals predicting returns, and general signal dynamics" (p. 1). Their solution tracks a Markovitz solution through regularization. Neuman and Voss (2022)[177] "study optimal liquidation in the presence of linear temporary and transient price impact along with taking into account a general price predicting finite-variation signal" (p. 1). They provide a closed-form formula for an Ornstein-Uhlenbeck alpha for illustration on page nine.

Finally, Isichenko (2021)[121] solves the original OW model with a generic alpha signal in Section 6.4.2 "Optimization with exponentially decaying impact" (p. 192) in closed form.

where

$$\nu = -\frac{1}{2+\beta T}\left(2\lambda Q_0 + \alpha_0 + \beta \int_0^T \alpha_t dt\right).$$

The optimal trading trajectory is

$$dQ_t^* = \frac{1}{2\lambda}\left(\beta\alpha_t - \beta^{-1}\alpha_t'' + \beta\nu\right) dt.$$

Proof. First, apply Proposition 2.3.5 and map the objective into *impact space*

$$\sup_{I \in D} \frac{1}{\lambda}\mathbb{E}\left[\int_0^T \left(\beta\alpha_t I_t - \alpha_t' I_t - \beta I_t^2\right) dt - \frac{1}{2} I_T^2\right].$$

Introduce a Lagrange multiplier ν for the constraint

$$\sup_{I \in D} \frac{1}{\lambda}\mathbb{E}\left[\int_0^T \left(\beta(\alpha_t + \nu) I_t - \alpha_t' I_t - \beta I_t^2\right) dt + \nu I_T - \frac{1}{2} I_T^2\right].$$

For a given ν, the problem is myopic in I, leading to the pointwise optimum

$$\forall t \in (0,T), \quad I_t^* = \frac{1}{2}\left(\alpha_t - \beta^{-1}\alpha_t' + \nu\right); \quad I_T^* = \nu.$$

The constraint reads

$$-2\lambda Q_0 = \nu(2+\beta T) - \alpha_T + \alpha_0 + \beta \int_0^T \alpha_t dt,$$

which leads to

$$\nu = -\frac{1}{2+\beta T}\left(2\lambda Q_0 + \alpha_0 + \beta \int_0^T \alpha_t dt\right).$$

Apply the one-to-one map to obtain the dynamics of Q^*:

$$\begin{aligned} dQ_t^* &= \frac{1}{\lambda}\left(\beta I_t dt + dI_t\right) \\ &= \frac{1}{2\lambda}\left(\beta\alpha_t - \alpha_t' + \beta\nu + \alpha_t' - \beta^{-1}\alpha_t''\right) \\ &= \frac{1}{2\lambda}\left(\beta\alpha_t - \beta^{-1}\alpha_t'' + \beta\nu\right). \end{aligned}$$

□

Lifting the unwind constraint removes the need for a Lagrange multiplier and simplifies the trading strategy with stochastic signals. Indeed, the unconstrained case also reflects statistical arbitrage problems.

Proposition 2.4.2 (Problem with stochastic alpha signal). *Consider the control problem*

$$\sup_{Q \in \mathcal{S}} \mathbb{E}\left[\int_0^T Q_t dS_t - \int_0^T I_t dQ_t - \frac{1}{2}[I, Q]_T\right]$$

under the following assumptions:

- *S is an Itô process in $\mathcal{S}$*

  $$dS_t = \mu_t dt + \sigma_t dW_t$$

  with $\mu, \sigma \in \mathcal{S}$.

- *I satisfies the SDE*

  $$dI_t = -\beta I_t dt + \lambda dQ_t$$

  for given $\beta, \lambda > 0$.

  Then, the optimal impact I^* satisfies

$$\forall t \in (0, T), \quad I_t^* = \frac{1}{2}\left(\alpha_t + \beta^{-1}\mu_t\right); \quad I_T^* = 0$$

where $\alpha_t = \mathbb{E}\left[S_T - S_t | \mathcal{F}_t\right]$.

Proof. Lemma 2.3.3 still applies, and the map between Q and I remains one-to-one, except that it maps $\mathcal{S}$ instead of D. Therefore, the stochastic control problem maps *into impact space*

$$\sup_{I \in \mathcal{S}} \frac{1}{\lambda}\mathbb{E}\left[\int_0^T \left(\beta\alpha_t I_t - \beta I_t^2\right) dt - \int_0^T I_t d\alpha_t - \frac{1}{2}I_T^2\right]$$

without constraints on I. One has

$$\mathbb{E}\left[\int_0^T I_t d\alpha_t\right] = -\mathbb{E}\left[\int_0^T I_t \mu_t dt\right].$$

The optimal I^* follows from pointwise optimization. □

Remark 2.4.3 (Intuition). *Corollary 2.4.1 and Proposition 2.4.2 clearly describe trading with an alpha signal.*

- *The control problem remains myopic in the impact state I.*
- *The optimal impact state I^* shifts by*

  $$\frac{1}{2}\left(\alpha_t - \beta^{-1}\alpha_t'\right)$$

  or

  $$\frac{1}{2}\left(\alpha_t + \beta^{-1}\mu_t\right),$$

  reflecting both the alpha level *and* decay.

- *The shorter the reversion timescale β^{-1}, the more consequential the alpha level is over its decay.*

Gârleanu and Pedersen (2013)[101] highlight alpha decay's role when determining optimal trading strategies under transient price impact models.

> *"The alpha decay is important because it determines how long time the investor can enjoy high expected returns and, therefore, affects the trade-off between returns and transactions costs" (p. 2).*

For more intuition, refer to exercise 2, which solves the execution with constant drift alpha. Section 4.2.2 of Chapter 4 tackles the implications of Proposition 2.4.2 for alpha research.

Remark 2.4.4 (Consistency with the microstructure model)**.** *Corollary 2.4.1 and Proposition 2.4.2 introduce minor inconsistencies with assumptions made by Obizhaeva and Wang (2013)[180].*

(a) A non-monotone alpha signal may invalidate the assumption that the trader only trades on one side of the limit order book.

(b) The stochastic alpha case may invalidate the no adverse selection *assumption from Definition 2.2.18.*

Section 2.4.2 addresses (a). Inconsistency (b) requires a more refined microstructure model. One either needs to model adverse selection or use a different discretization Q^N of Q. This chapter ignores (b) by specifying the model in reduced form.

In all following sections, remove the unwind constraint and assume the *deterministic alpha assumption* from Corollary 2.4.1 holds.

2.4.2 Two-sided trading

To extend the OW model to allow for both buying and selling of stocks, recall the discrete self-financing equation 2.4 from Proposition 2.2.13

$$\Delta_n Y^N = Q^N_{n-1}\Delta_n S^N - I^N_{n-1}\Delta_n Q^N - s^N_{n-1}\Delta_n Q^N + \Delta_n S^N \Delta_n Q^N.$$

The main difference with the original OW model is that the price does not model the ask anymore but the midprice. The instantaneous transaction cost s^N_{n-1} presents an additional component

$$\pm\frac{1}{2}\bar{s}^N_{n-1}\,sign\left(\Delta_n Q^N\right)$$

where $\pm$ is positive when taking liquidity and negative when providing liquidity. $\bar{s}^N > 0$ is the spread between the best ask and bid prices.

Unfortunately, the new term has *at least* two possible limits. The convergence depends on how $\bar{s}^N$ scales with N and assumptions on Q. Propositions 2.4.5 and 2.4.6 prove two continuous-time limits.

Proposition 2.4.5 (Self-financing equation in the *low-frequency* regime). *Assume*

- *the continuous-time process Q has finite variation.*
- *one replaces the* temporary linear transaction costs *assumption from Definition 2.2.18 (a) by*

$$s_n^N = \frac{\bar{s}}{2} sign\left(\Delta_n Q^N\right) + \frac{\lambda}{2}\Delta_n Q^N$$

for a positive constant $\bar{s}$.

- *the* exponential decay price impact *model from Definition 2.2.18 (b) holds.*
- *the* no adverse selection *assumption from Definition 2.2.18 (c) holds.*

Then the continuous-time limit of Y^N follows the self-financing equation

$$dY_t = Q_t dS_t - \frac{\bar{s}}{2}\left|dQ_t\right| - I_t dQ_t - \frac{1}{2}d[I,Q]_t$$

where I satisfies the SDE

$$dI_t = -\beta I_t dt + \lambda dQ_t.$$

The convergence happens u.c.p. as $N \to \infty$.

Proof. Proposition 2.4.6 extends Proposition 2.2.21. The additional term

$$\sum_n \frac{\bar{s}}{2}\left|\Delta_n Q^N\right|$$

converges to

$$\int \frac{\bar{s}}{2}\left|dQ\right|$$

by Theorem 53 of Protter (2005, p. 41)[191]. □

Proposition 2.4.5 applies when taking liquidity on stocks with large bid-ask spreads. However, when market-making on liquid stocks, different assumptions hold. For instance, one empirically observes that $\Delta_n S^N \Delta_n Q^N$ does not converge to zero and drives market-making P&L on a wide range of stocks.[9]

Therefore, one lifts the assumption that Q has finite variation and replaces it with the assumption that the bid-ask spread s^N shrinks as $N \to \infty$. This section covers the standard case where $s^N \propto \sqrt{\Delta t^N}$, also called the *vanishing* bid-ask spread assumption for liquid stocks. The assumption answers a crucial thought experiment:

[9] Carmona and Webster (2019)[61] empirically observe a negative covariation between trades and prices on 120 stocks over a month on NASDAQ ITCH data. Webster et al. (2015)[235] extend the analysis and follow 3000 stocks over ten years, finding that adverse selection is a universal phenomenon across the US equities market.

> How would the market's trading frequency change if minimum spreads were halved?

The vanishing bid-ask spread assumption states that the market's frequency quadruples when spreads halve, in line with empirical studies. Furthermore, high-frequency traders drive the increase in trading activity when spreads decrease.

Leland (1985)[148] first introduced the vanishing bid-ask spread to model options delta-hedging. Lott (1993)[157] proved the convergence of Leland's delta-hedging strategy. Çetin, Soner, and Touzi (2010)[51] extend the result to the super-replication of general claims. Finally, Muhle-Karbe, Wang, and Webster (2022)[169] add limit orders with adverse selection to Leland's delta-hedging strategy and prove the corresponding limit.

Carmona and Webster (2019)[61] and Muhle-Karbe and Webster (2017)[171] model market making in the presence of adverse selection by high-frequency traders. Carmona and Leal (2021)[60] test model predictions on individual high-frequency traders' actual strategies. The paper leverages public trading data on the Toronto stock exchange and high-frequency econometric methods by Aït-Sahalia and Jacod (2014)[4].

> "In order to illustrate the significance of the consequences of our theoretical results, we compare their implementations to the actual inventory and wealth processes of active traders using real trading data from the Toronto Stock Exchange" ([60], p. 2).

Proposition 2.4.6 (Self-financing equation in the *high-frequency* regime). *Assume*

- *$Q \in \mathcal{S}$ is an Itô process*

$$dQ_t = a_t dt + l_t dW_t$$

for an adapted Wiener process W and $a, l \in \mathcal{S}$ with l bounded away from zero.

- *one replaces the* temporary linear transaction costs *assumption from Definition 2.2.18 (a) by*

$$s_n^N = \pm \frac{\bar{s}\sqrt{\Delta t^N}}{2} sign\left(\Delta_n Q^N\right) \pm \frac{\lambda}{2} \Delta_n Q^N$$

where $\pm$ is positive when taking liquidity and negative when providing liquidity. $\bar{s}$ is a positive constant.

- *the* exponential decay price impact *model from Definition 2.2.18 (b) holds.*

Then the continuous-time limit of Y^N follows the self-financing equation

$$dY_t = Q_t dS_t - I_t dQ_t \mp \frac{\bar{s}}{\sqrt{2\pi}} l_t dt \mp \frac{\lambda}{2} l_t^2 dt + d[Q, S]_t$$

where I satisfies the SDE

$$dI_t = -\beta I_t dt + \lambda dQ_t.$$

The convergence happens u.c.p. as $N \to \infty$.

Proof. Proposition 2.4.6 extends Proposition 2.2.21. For the additional term

$$\frac{\bar{s}\sqrt{\Delta t^N}}{2}\left|\Delta_n Q^N\right|,$$

the u.c.p. convergence is an application of Theorem 7.2.2 from the book *Discretization of Processes* (2012)([122], p. 217) by Jacod and Protter. See Appendix B for further details. □

2.4.2.1 Bid-Ask spread as a regularization term

Propositions 2.4.5 and 2.4.6 allow the reader to consider the bid-ask spread by adding one of two penalty terms:

(a)

$$\frac{\bar{s}}{2}\left|dQ_t\right|$$

(b)

$$\frac{\bar{s}}{\sqrt{2\pi}} l_t$$

where l_t is the volatility of Q.

Both terms lead to non-myopic problems and require dynamic programming and numerical solutions. Cartea, Jaimungal, and Penalva (2015)[64] review numerical methods for optimal execution.

One interprets the two terms as *regularization terms* that track the case without bid-ask spread. While the problem with spread requires numerical methods, the corresponding solution tightly follows the closed-form solution without bid-ask spread.

For instance, when considering $\alpha \in C^2$, the first penalty introduces a no-trade zone around the solution without bid-ask spread. Fruth, Schöneborn, and Urusov (2013)[97] prove the existence of the no-trade area in a general setting and formalize a "wait-region/buy-zone conjecture" (p. 13).

More generally, one models bid-ask spreads as instantaneous costs. Popular models include immediate *quadratic costs*. For instance, Gârleanu and Pedersen (2016)[102] prove that the optimal strategy with instantaneous quadratic costs *tracks* the closed-form solution. In addition, Gârleanu and Pedersen (2016)[102] note that "the case with vanishing transitory costs, the optimal continuous-time portfolio has positive quadratic variation. With transitory costs, however, our optimal continuous-time strategy is smooth and has a finite turnover" (p. 3).

Ekren and Muhle-Karbe (2019)[88] extend the results of Gârleanu and Pedersen (2016)[102] with closed-form formulas using asymptotics for liquid stocks:

> "the corresponding optimal portfolio not only tracks the frictionless optimizer, but also exploits the displacement of the market price from its unaffected level" (p. 1).

2.5 Extensions Related to Time

Practitioners may criticize the original OW model for its assumption that impact decays at a constant exponential rate. Indeed, empirical results point to a dynamic dependence on time. The most notable dependency is on the time of day, as per Section 7.4.2 in Chapter 7. Therefore, this section extends the OW model to consider such observed features.

2.5.1 Time change

First, empirical studies indicate that measuring *time* in a different clock than the regular wall clock improves the performance of price impact models for specific applications. For instance, Busseti and Lillo (2012)[47] compare propagator models under three time measures using data from the London Stock Exchange and NASDAQ. Busseti and Lillo (2012)[47] note that "there are several possible definitions of aggregated time in financial data" (p. 10) and derive the "propagator model under other time measures" (p. 10) from the trade clock initially proposed by Bouchaud et al. (2004)[34]. The other two clocks introduced are the intuitive "real time" (p. 10) and volume time. Busseti and Lillo define volume time as "time advances by one unit when a given volume has been traded in the market" (p. 10).

Besides the regular wall clock, three common ways to measure time are

(a) the *volume clock*, where one measures time by the cumulative notional traded on the public tape.[10]

[10]Mastromatteo, Tóth, and Bouchaud (2014)[164] measure price impact in the volume clock to home in on the concavity of price impact. See Section 2.6 for more details on concave price impact models. Busseti and Lillo (2012)[47] compare the predictive power of trade time against clock time for propagator models. Busseti and Lillo (2012)[47] (table 3, p. 14) show that trade time provides better predictive power than clock time for their tick-by-tick application.

(b) the *trade clock*, where one measures time by the cumulative number of trades on the public tape.[11]

(c) the *tick clock*, where one measures time by the cumulative number of ticks on the limit order book.[12]

All three clocks express the same economic intuition:

Price impact decays faster with market activity.

Mathematically, time measures are straightforward: the standard tool is a *time change*. The reader can study Çinlar (2011)[52].[13]

Definition 2.5.1 (Time change). *Define a time change τ as an increasing stochastic process in $\mathcal{S}$ such that $\tau_0 = 0$. Define the time-changed filtration as $\hat{\mathcal{F}}_s = \mathcal{F}_{\tau_s^{-1}}$. Define τ as smooth if it takes value in C^1 and denote by τ' the derivative of a smooth time change τ.*

Remark 2.5.2 (Empirical time changes). *Time changes are concrete objects that one can observe on the trading tape. For example, when measuring time using the volume clock, the discrete τ^N satisfies*

$$\tau_n^N \propto \sum_{k=0}^{n} |\Delta_k M^N|$$

where M represents trading on the public tape. The next proposition shows that this discrete time change converges to a well-defined, smooth time change.

Proposition 2.5.3 (Time change convergence). *Let $M \in \mathcal{S}$ be an Itô process*

$$dM_t = m_t dt + v_t dW_t$$

for an $\mathcal{F}$-adapted Wiener process W and $m, v \in \mathcal{S}$ with $v > 0$. Then

$$\sqrt{\Delta t^N} \sum_{k=0}^{\lceil Nt \rceil} |\Delta_k M^N| \to \sqrt{\frac{2}{\pi}} \int_0^t v_s ds$$

and the convergence happens u.c.p.

---

[11] Bouchaud et al. (2004)[34] "always reason in terms of trade time, i.e. time advances by one unit every time a new trade (or a series of simultaneous trades) is recorded" (p. 4). With this clock, price impact adjusts to "intra-day seasonalities and also clustering of the trades in time" (p. 4).

[12] A message that changes the order book's state defines a tick. All order book changes trigger a tick by default, but one defines *simple ticks* as events on the best bid or ask.

[13] Çinlar (2011)[52] introduces "Random time changes" (p. 422) to extend classic results about stochastic processes.

> "Many interesting Markov processes are obtained from Wiener processes by random time changes" (p. 422).

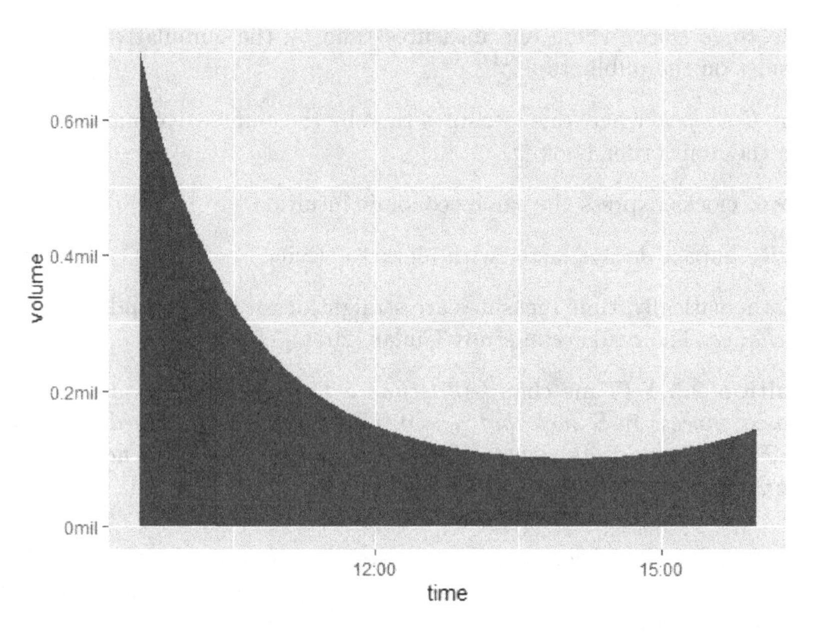

FIGURE 2.3
Example of a τ'_t for a simulated volume clock, as per Definition 2.5.1.

Proof. The u.c.p. convergence is an application of Theorem 7.2.2 from Jacod and Protter (2012)([122], p. 217). See Appendix B for more details. □

Definition 2.5.4 (Time change). *Consider a time change τ and a process $Z \in \mathcal{S}_n$ for $n > 0$. Denote its time-changed value by $\hat{Z}_s = Z_{\tau_s^{-1}}$. Similarly, denote by $\hat{T}$ the random variable such that $\tau_T = \hat{T}$.*

Remark 2.5.5 (Time changed impact model). *A smooth time change's effect on the OW price impact model is straightforward. Assume $\hat{I}$ satisfies*

$$d\hat{I}_s = -\beta \hat{I}_s ds + \lambda d\hat{Q}_s$$

then I satisfies

$$dI_t = -\beta \tau'_t I_t dt + \lambda dQ_t.$$

Therefore, a time change is equivalent to replacing the time decay parameter β with the stochastic process $\beta_t = \beta \tau'_t$.

Remark 2.5.5 motivates lifting the assumption of a constant decay parameter β and leads to Proposition 2.5.6.[14] Crucially, the control problem remains myopic in I.

[14]Graewe and Horst (2017)[109] employ the term "stochastic resilience" (p. 1) to describe a time-changed OW model. Graewe and Horst provide an execution problem that leads to a "fully coupled *three-dimensional* stochastic Ricatti equation (BSDE system)" (p. 3). Graewe and Horst's problem includes a risk term in the spirit of Almgren and Chriss (2001)[10].

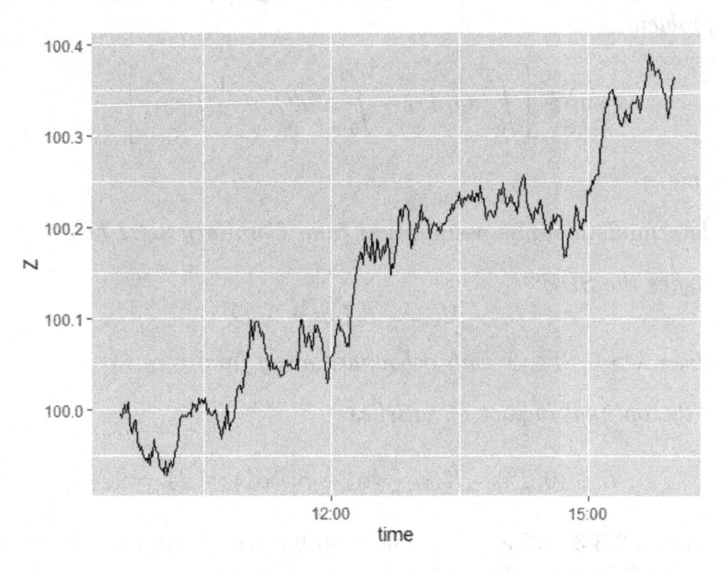

(a) Example of a process Z.

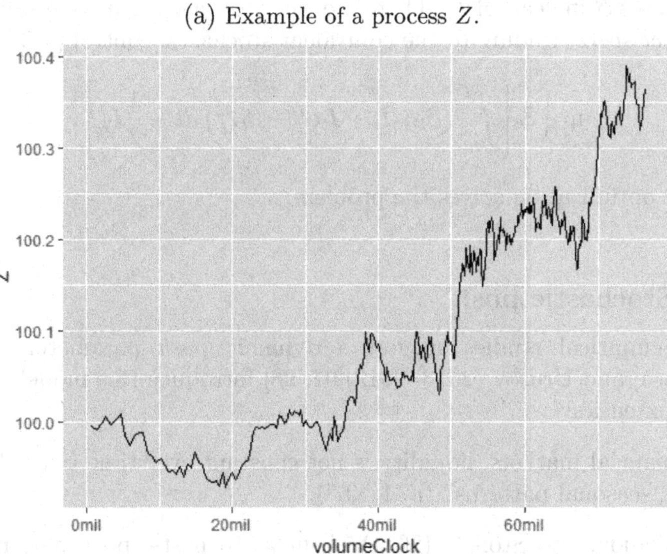

(b) Example of a time-changed process $\hat{Z}$.

FIGURE 2.4
Example of a process Z and its time-changed counterpart $\hat{Z}$ obtained from applying the volume clock from Figure 2.3, as per Definition 2.5.4. Practitioners commonly use time-changes to recast a curve's x-axis into an economically intuitive unit, e.g., trading volumes.

Proposition 2.5.6 (Price impact with time change). *Consider the stochastic control problem*

$$\sup_{Q\in\mathcal{S}} \mathbb{E}\left[\int_0^T Q_t dS_t - \int_0^T I_t dQ_t - \frac{1}{2}[I,Q]_T\right]$$

and assume

- *the* deterministic alpha assumption *from Corollary 2.4.1 holds.*
- *I satisfies the SDE*

$$dI_t = -\beta_t I_t dt + \lambda dQ_t$$

for given $\lambda > 0$, $\beta \in \mathcal{S}$ with β bounded away from zero.

Then the optimal impact I^* satisfies

$$\forall t \in (0,T), \quad I_t^* = \frac{1}{2}\left(\alpha_t - \beta_t^{-1}\alpha_t'\right); \quad I_T^* = 0.$$

Proof. Lemma 2.3.3 still applies. The map between Q and I remains one-to-one but maps $\mathcal{S}$ instead of D. Therefore, the stochastic control problem maps *into impact space*, leading to the equivalent stochastic control problem

$$\sup_{I\in\mathcal{S}} \frac{1}{\lambda}\mathbb{E}\left[\int_0^T \left(\beta_t\alpha_t I_t - I_t\alpha_t' - \beta_t I_t^2\right) dt - \frac{1}{2}I_T^2\right].$$

Pointwise optimization solves the problem. □

2.5.2 Stochastic push

Further, empirical studies suggest a dynamic *push* parameter λ. Fruth, Schöneborn, and Urusov (2013, 2019)[97, 98] introduce this model and motivate the extension:

> "In financial markets, liquidity is not constant over time but exhibits strong seasonal patterns" (p. 1, [97]).

Cont, Kukanov, and Stoikov (2014)[74] fit a stochastic push parameter and motivate dynamic coefficients in Section 3.3 "Intraday patterns" (p. 14).

> "Since the market depth follows a predictable pattern of intraday seasonality, the price impact coefficient must also have a predictable intraday pattern" (p. 14).

Figure 5 on page 14 of [74] displays an intraday curve for λ_t. In addition, Section 7.4.2 of Chapter 7 provides another illustration of an intraday λ_t. Fruth, Schöneborn, and Urusov (2021)[99] distinguish between "daily and weekly patterns" (p. 1) and "stochastic liquidity" (p. 1).

"In addition, there exist random changes in liquidity such as liquidity shocks that superimpose the deterministic evolution. To benefit from times when trading is cheap, institutional investors continuously monitor the available liquidity and schedule their order flow accordingly" (p. 1).

Proposition 2.5.7 proves that the control problem with stochastic λ *remains myopic* in I.

Proposition 2.5.7 (Price impact with stochastic push; mapping to impact space). *Consider the stochastic control problem*

$$\sup_{Q\in\mathcal{S}} \mathbb{E}\left[\int_0^T Q_t dS_t - \int_0^T I_t dQ_t - \frac{1}{2}[I,Q]_T\right]$$

and assume

- *I satisfies the SDE*
$$dI_t = -\beta_t I_t dt + \lambda_t dQ_t$$
for given $\beta, \lambda \in \mathcal{S}$ with β bounded away from zero.

- *λ has the representation*
$$\lambda_t = e^{\gamma_t}$$
where γ has derivative $\gamma' \in \mathcal{S}$.

Then the control problem is equivalent to

$$\sup_{I\in\mathcal{S}} \mathbb{E}\left[\int_0^T \frac{1}{\lambda_t}\left((\beta_t+\gamma_t')\,\alpha_t I_t - \left(\beta_t + \frac{1}{2}\gamma_t'\right) I_t^2\right) dt - \int_0^T \frac{1}{\lambda_t} I_t d\alpha_t - \frac{1}{2\lambda_T} I_T^2\right].$$

Proof. Lemma 2.3.3 still applies. The map between Q and I remains one-to-one and maps $\mathcal{S}$ instead of D. Therefore, the stochastic control problem maps *into impact space.* One maximizes the equivalent objective $\mathbb{E}[Z]$ over $I \in \mathcal{S}$, where

$$Z = \int_0^T \frac{1}{\lambda_t}(\alpha_t - I_t)(\beta_t I_t dt + dI_t) - \int_0^T \frac{1}{2\lambda_t} d[I,I]_t + \int_0^T \frac{1}{\lambda_t} d[\alpha, I]_t$$

Furthermore, one has

$$\begin{aligned} Z = &\int_0^T \beta_t e^{-\gamma_t}\left(\alpha_t I_t - I_t^2\right) dt + \int_0^T e^{-\gamma_t}\alpha_t dI_t - \int_0^T e^{-\gamma_t} I_t dI_t \\ &- \frac{1}{2}\int_0^T e^{-\gamma_t} d[I,I]_t + \int_0^T e^{-\gamma_t} d[\alpha, I]_t. \end{aligned}$$

The equations

$$\int_0^T e^{-\gamma_t}\alpha_t dI_t + \int_0^T e^{-\gamma_t} d[\alpha, I]_t = \int_0^T e^{-\gamma_t}\gamma_t'\alpha_t I_t dt - \int_0^T e^{-\gamma_t} I_t d\alpha_t$$

and

$$\int_0^T e^{-\gamma_t} I_t dI_t + \frac{1}{2}\int_0^T e^{-\gamma_t} d[I,I]_t = \frac{1}{2}e^{-\gamma_T} I_T^2 + \frac{1}{2}\int_0^T \gamma_t' e^{-\gamma_t} I_t^2 dt$$

follow from Itô's identity. They yield

$$Z = \int_0^T \frac{1}{\lambda_t}\left((\beta_t + \gamma_t')\alpha_t I_t - \left(\beta_t + \frac{1}{2}\gamma_t'\right) I_t^2 \right) dt - \int_0^T \frac{1}{\lambda_t} I_t d\alpha_t - \frac{1}{2\lambda_T} I_T^2.$$

□

The optimal impact state I^* is in closed form. Using a simple condition on β, γ', one establishes well-posedness and rules out price manipulation strategies. Section 2.7 extends the no price manipulation result and connects it to volume predictions, such as those used in VWAP (Volume Weighted Average Price) algorithms.

Corollary 2.5.8 (Myopic solution to price impact with stochastic push). *Consider the control problem defined in Proposition 2.5.7. Assume, furthermore, that*

(a) the deterministic alpha assumption *from Corollary 2.4.1 holds.*

(b) $2\beta + \gamma'$ *is bounded away from zero. Refer to this assumption as the* no price manipulation *condition.*

Then the optimal impact I^* *satisfies*

$$\forall t \in (0,T), \quad I_t^* = \frac{\beta_t + \gamma_t'}{2\beta_t + \gamma_t'}\alpha_t - \frac{1}{2\beta_t + \gamma_t'}\alpha_t'; \quad I_T^* = 0.$$

Proof. Under the deterministic alpha assumption, the objective function simplifies

$$\mathbb{E}\left[\int_0^T \frac{1}{\lambda_t}\left((\beta_t + \gamma_t')\alpha_t I_t - \alpha_t' I_t - \left(\beta_t + \frac{1}{2}\gamma_t'\right) I_t^2 \right) dt - \frac{1}{2\lambda_T} I_T^2 \right].$$

The objective function is myopic in I and is strictly concave if and only if $2\beta + \gamma' > 0$. The corollary follows from the objective function's pointwise optimization in impact space. □

The generalized OW model is well-suited for machine learning methods: the solution allows generic signals for β, λ, α and does not preclude non-parametric models. Furthermore, despite being dynamic, the generalized OW model is inherently tractable due to its myopia in impact space. Therefore, the OW strategy reduces the computational burden for trading and research systems compared to other solutions. Section 3.2.5 of Chapter 3 illustrates this tractability with a closed-form implementation and back test of an execution schedule.

2.5.2.1 Sensitivity analysis in impact space

This tractability also simplifies the optimal strategy's sensitivity analysis, per equation 2.7 of Corollary 2.5.9.

Corollary 2.5.9 (Practical formulas). *Consider the control problem from Proposition 2.5.7 and solution I^* from Corollary 2.5.8.*

Let $Q \in \mathcal{S}$ be a candidate trading process and I its impact. Then the following formulas hold:

$$Q_t = \frac{1}{\lambda_t} I_t + \int_0^t \frac{\beta_s + \gamma_s'}{\lambda_s} I_s ds,$$

$$\mathbb{E}\left[Y_T(Q^*)\right] = \mathbb{E}\left[\int_0^T \frac{2\beta_t + \gamma_t'}{2\lambda_t} \left(I_t^*\right)^2 dt\right],$$

and

$$\mathbb{E}\left[Y_T(Q)\right] = \mathbb{E}\left[Y_T(Q^*)\right] - \mathbb{E}\left[\int_0^T \frac{2\beta_t + \gamma_t'}{2\lambda_t} \left(I_t - I_t^*\right)^2 dt + \frac{1}{2\lambda_T} I_T^2\right]. \tag{2.7}$$

Proof. First, λ is bounded away from zero. The one-to-one map yields

$$\begin{aligned} Q_t &= \int_0^t \frac{\beta_s}{\lambda_s} I_s ds + \int_0^t \frac{1}{\lambda_s} dI_s \\ &= \int_0^t \frac{\beta_s}{\lambda_s} I_s ds + \frac{1}{\lambda_t} I_t + \int_0^t \frac{\gamma_s'}{\lambda_s} I_s ds \end{aligned}$$

where the last equality stems from integration by parts. For the second identity, one has

$$\begin{aligned} \mathbb{E}\left[Y_T(Q^*)\right] &= \mathbb{E}\left[\int_0^T \frac{1}{\lambda_t}\left(\left(\beta_t + \gamma_t'\right)\alpha_t I_t^* - \alpha_t' I_t^* - \left(\beta_t + \frac{1}{2}\gamma_t'\right)\left(I_t^*\right)^2\right) dt\right] \\ &= \mathbb{E}\left[\int_0^T \frac{2\beta_t + \gamma_t'}{\lambda_t}\left(\left(\frac{\beta_t + \gamma_t'}{2\beta_t + \gamma_t'}\alpha_t - \frac{1}{2\beta_t + \gamma_t'}\alpha_t'\right) I_t^* - \frac{1}{2}\left(I_t^*\right)^2\right) dt\right] \\ &= \mathbb{E}\left[\int_0^T \frac{2\beta_t + \gamma_t'}{2\lambda_t}\left(I_t^*\right)^2 dt\right]. \end{aligned}$$

The third identity follows from

$$\begin{aligned} \mathbb{E}\left[Y_T(Q)\right] &= \mathbb{E}\left[\int_0^T \frac{2\beta_t + \gamma_t'}{\lambda_t}\left(I_t^* I_t - \frac{1}{2} I_t^2\right) dt - \frac{1}{2\lambda_T} I_T^2\right] \\ &= \mathbb{E}\left[\int_0^T \frac{2\beta_t + \gamma_t'}{2\lambda_t}\left(\left(I_t^*\right)^2 - \left(I_t^* - I_t\right)^2\right) dt - \frac{1}{2\lambda_T} I_T^2\right] \\ &= \mathbb{E}\left[Y_T(Q^*)\right] - \mathbb{E}\left[\int_0^T \frac{2\beta_t + \gamma_t'}{2\lambda_t}\left(I_t^* - I_t\right)^2 dt + \frac{1}{2\lambda_T} I_T^2\right]. \end{aligned}$$

□

Equation 2.7 of Corollary 2.5.9 defines a distance in impact space and proves that this distance measures the P&L difference between two strategies when one is optimal. To the author's best knowledge, this result is originally due to Ackermann, Kruse, and Urusov (2022)[2]: equation (21) on page thirteen. It is a practical result that simplifies auxiliary problems in optimal execution, such as quantifying a sub-optimal strategy's opportunity cost. Chapters 3 and 4 explore further consequences and applications of Proposition 2.5.7 and its corollaries.

2.5.3 Linear propagator models

A different approach replaces the exponential decay with a general time kernel. For instance, empirical results indicate that impact decays with a power-law over long timescales. Bucci et al. (2018)[43] leverage institutional order data to study price impact over multiple days.

> "[T]he decay continues the next days, following a power-law function at short time scales, and apparently converges to a non-zero asymptotic value at long time scales ($\sim$ 50 days) equal to $\approx 1/2$ of the impact at the end of the first day" (p. 1).

In the paper's context, "short time scales" refers to the next day. Moreover, "non-zero asymptotic" impact implies long-term realized alpha. For instance, a realized alpha horizon of fifty days is consistent with the holding period of the associated financial institutions. The literature calls these *linear propagator models*, of which the OW model is the simplest example.

Definition 2.5.10 (Linear propagator model for price impact). *Let $Q \in \mathcal{S}$ be a trading process and define a linear propagator model I by*

$$I_t = \int_0^t k(t-s)dQ_s$$

for a positive, decreasing function $k \in C^2$ called the decay kernel.

Example 2.5.11 (Relation to the OW model). *The original OW model is a linear propagator model with decay kernel $k(s) = \lambda e^{-\beta s}$.*

Unfortunately, the control problem for a general propagator model is not myopic in I. However, one can solve the problem through two means.

(a) Curato, Gatheral, and Lillo (2017)[80] solve the general linear and non-linear propagator model with "brute force numerical optimization of the cost functional" (p. 1) and "an approach based on homotopy analysis" (p. 1). Furthermore, Dang (2017)[82] uses a fixed-point solution to a "Fredholm integral equation of the second kind" (p. 1) to solve the problem numerically.

(b) Abi Jaber and Neuman (2022)[1] derive an analytical solution for arbitrary alpha signal in a general linear propagator model. Their method relies on an "operator-valued Ricatti equation" (p. 1), lifting the problem's dimension to regain tractability. Crucially, Abi Jaber and Neuman's method applies to decay kernels with possible singularities, such as the power-law kernel studied by Bucci et al. (2018)[43].

2.6 Extensions Related to External Impact

> "Execution costs have a very significant impact on the profitability of trading strategies. Crowded trade flow increases such costs, and substantially diminishes the net gains of a portfolio" (p. 5, Capital Fund Management (2019)[161]).

In the impact models so far, one only models the agent's trading process Q and captures the rest of the market through the dynamics of S. This section explicitly models the market flow. Embed the trader's Q in the trading process $L = (Q, M) \in \mathcal{S}_2$. M describes the liquidity-taking trades of the rest of the market. One informally refers to the impact caused by flow M as the *external price impact* and the impact caused by Q as *the trader's price impact.*[15] $\bar{I}$ denotes the total price impact generated by M and Q, leaving I for the trader's price impact, where relevant.

2.6.1 Microstructure assumptions

Because the agent trades alongside another liquidity-taking process, one must establish *who executes first* when modeling concurrent fills.[16] At the microstructure level, the trading order does not influence the price impact $\bar{I}^N$ but affects the instantaneous transaction costs s_n^N. Therefore, Definition 2.6.1 extends Definition 2.2.18 to include external impact.

[15] Bucci et al. (2020)[44] names external impact 'co-impact'.

> "This paper is devoted to the important yet unexplored subject of *crowding* effects on market impact, that we call 'co-impact'."

Section 6.4 in Chapter 6 covers crowding and distinguishes it from the related topic of leakage.

[16] Establishing who trades first is vital when applying game theory to price impact. Different assumptions lead to contrasting Nash equilibria. For example, under conditions, Schied, Strehle, and Zhang (2017)[205] numerically observe and mathematically prove that "the equilibrium strategies of both agents typically oscillate between buy and sell trades, a behavior that is reminiscent of the hot-potato game during the 2010 flash crash" (p. 1). Section 4.4.2 of Chapter 4 studies game theory with price impact.

Definition 2.6.1 (Microstructure extension with external impact). *Let $Q \in \mathcal{S}$ be the agent's and $M \in \mathcal{S}$ be the market's trading process. Define the following three microstructure assumptions:*

(a) linear instantaneous transaction costs with external impact

$$s_n^N = \frac{\lambda}{2}\Delta_n Q^N + \lambda \Delta_n M^N.$$

(b) exponential decay price impact with external impact

$$\Delta_n \bar{I}^N = -\beta I_{n-1}^N \Delta t^N + \lambda \left(\Delta_n Q^N + \Delta_n M^N\right).$$

(c) no adverse selection

$$\Delta_n Q^N \Delta_n S^N \to 0$$

u.c.p. as $N \to \infty$.

Remark 2.6.2 (Interpretation). *Recall Definition 2.2.18 grounding the OW model's microstructure.*

- *Because of the order book's shape, a fill depletes the ask level by $\lambda \Delta_n Q^N$ for the trader's fill. For a market fill, the same formula holds, and the ask level depletes by $\lambda \Delta_n M^N$.*
- *The situation is subtle for the execution price. When trading alone, the agent attains the execution price*

  $$\tilde{P}_{n-1}^N = P_{n-1}^N + \frac{\lambda}{2}\Delta_n Q^N.$$

  When trading side-by-side with the market, consider the following two scenarios.

  (a) The fill $\Delta_n Q$ takes place before *$\Delta_n M$: the trader pays the price observed* before *the market fill $\Delta_n M$*

  $$\tilde{P}_{n-1}^N = P_{n-1}^N + \frac{\lambda}{2}\Delta_n Q^N.$$

  (b) The fill $\Delta_n Q$ takes place after *$\Delta_n M$: the trader pays the price observed* after *the market fill $\Delta_n M$*

  $$\tilde{P}_{n-1}^N = P_{n-1}^N + \frac{\lambda}{2}\Delta_n Q^N + \lambda \Delta_n M^N.$$

  Because the trader does not systematically execute fills immediately before the rest of the market, scenario (b) models their trading costs accurately. Indeed, from a trade latency perspective, (b) rules out the agent consistently trading right before or simultaneously with other market participants: it is a no-front-running *assumption. Therefore, the* linear instantaneous transaction costs *with external impact from Definition 2.6.1 hold.*

Proposition 2.6.3 is a straightforward extension of Proposition 2.2.21.

Proposition 2.6.3 (Original OW model with external impact)**.** *Consider the original OW microstructure assumptions with external impact from Definition 2.6.1. Then the fundamental P&L* Y^N *converges to*

$$dY_t = Q_t dS_t - \bar{I}_t dQ_t - \frac{\lambda}{2} d[Q,Q]_t - \lambda d[M,Q]_t,$$

and the impact process $\bar{I}^N$ *converges to the solution to the SDE*

$$d\bar{I}_t = -\beta \bar{I}_t dt + \lambda \left(dQ_t + dM_t\right).$$

The convergence happens u.c.p. as $N \to \infty$.

2.6.2 Optimal trading strategy with external impact

Proposition 2.6.4 (External impact as an alpha signal)**.** *Let* $Q \in \mathcal{S}$ *be the agent's,* $M \in \mathcal{S}$ *the market's trading process, and* $S \in \mathcal{S}$ *the unperturbed price. Assume the reduced form model derived in Proposition 2.6.3 holds. Furthermore, for* $\beta, \lambda > 0$, *define*

$$dI_t = -\beta I_t dt + \lambda dQ_t$$

with $I_0 = 0$ *and*

$$d\bar{\alpha}_t = -\beta \bar{\alpha}_t dt - \lambda dM_t$$

with $\bar{\alpha}_0 = -\bar{I}_0$. *Then the following two identities hold:*

(a)

$$\bar{I} = I - \bar{\alpha}.$$

(b)

$$\mathbb{E}\left[Y_T\right] = \frac{1}{\lambda}\mathbb{E}\left[\bar{\alpha}_T I_T - \frac{1}{2} I_T^2 + \beta \int_0^T \left(\left(\alpha_t + \bar{\alpha}_t\right) I_t - I_t^2\right) dt - \int_0^T I_t d\left(\alpha_t + \bar{\alpha}_t\right)\right].$$

Proof. (a) follows from the exponential kernel's linearity. By inspection

$$\begin{aligned} d\left(I_t - \bar{\alpha}_t\right) &= -\beta I_t dt + \lambda dQ_t + \beta \bar{\alpha}_t dt + \lambda dM_t \\ &= -\beta \left(I_t - \bar{\alpha}_t\right) dt + \lambda \left(dQ_t + dM_t\right) \end{aligned}$$

and $I_0 - \bar{\alpha}_0 = \bar{I}_0$.

(b) follows from Lemma 2.3.3 and the objective function in impact space

$$\begin{aligned}
\mathbb{E}[Y_T] &= \mathbb{E}\left[\int_0^T (\alpha_t - \bar{I}_t)\, dQ_t - \frac{\lambda}{2}[Q,Q]_T - \lambda[M,Q]_T + [\alpha,Q]_T\right] \\
&= \mathbb{E}\left[\int_0^T (\alpha_t + \bar{\alpha}_t - I_t)\, dQ_t - \frac{1}{2}[I,Q]_T + [\alpha+\bar{\alpha},Q]_T\right] \\
&= \frac{1}{\lambda}\mathbb{E}\left[\int_0^T (\alpha_t + \bar{\alpha}_t) dI_t + [\alpha+\bar{\alpha}, I]_T\right] \\
&\quad + \frac{1}{\lambda}\mathbb{E}\left[\int_0^T \left(\beta(\alpha_t+\bar{\alpha}_t)I_t - \beta I_t^2\right) dt - I_T^2\right] \\
&= \frac{1}{\lambda}\mathbb{E}\left[\bar{\alpha}_T I_T - \frac{1}{2} I_T^2 + \int_0^T \left(\beta\left(\alpha_t+\bar{\alpha}_t\right) I_t - \beta I_t^2\right) dt\right] \\
&\quad - \frac{1}{\lambda}\mathbb{E}\left[\int_0^T I_t d\left(\alpha_t + \bar{\alpha}_t\right)\right].
\end{aligned}$$

□

Remark 2.6.5 (Interpretation). *Proposition 2.6.4 proves:*

One trader's impact is another trader's alpha.

Mathematically, the impact M causes behaves like an alpha signal that persists after the terminal time T.[17]

Proposition 2.6.6 (Generalized OW model with external flow). *Let $Q \in \mathcal{S}$ be the agent's, $M \in \mathcal{S}$ the market's trading process, and $S \in \mathcal{S}$ the unperturbed price. For β, λ as in Proposition 2.5.7, assume:*

- *The trading process M is an Itô process*

$$dM_t = m_t dt + v_t dW_t.$$

  with $m, v \in \mathcal{S}$.

- *The* trader's price impact *I satisfies*

$$dI_t = -\beta_t I_t dt + \lambda_t dQ_t$$

  with $I_0 = 0$.

- *The alpha signal $\bar{\alpha}$ implied from* external impact *satisfies*

$$d\bar{\alpha}_t = -\beta_t \bar{\alpha}_t dt - \lambda_t dM_t.$$

[17] One can also interpret a trader's own past impact as an alpha going forward. Exercise 3 formalizes this point of view.

- *The self-financing equation*

$$dY_t = Q_t dS_t + (\bar{\alpha}_t - I_t)\, dQ_t - \frac{1}{2} d[I, Q]_t + d[\bar{\alpha}, Q]_t$$

holds.

- *S satisfies the* deterministic alpha assumption *from Corollary 2.4.1.*

Then, for the control problem

$$\sup_{Q \in \mathcal{S}} \mathbb{E}\left[Y_T\right],$$

the optimal price impact is

$$\forall t \in (0,T), \quad I_t^* = \bar{\alpha}_t + \frac{\lambda_t}{2\beta_t + \gamma_t'} m_t + \frac{\beta_t + \gamma_t'}{2\beta_t + \gamma_t'} \alpha_t - \frac{1}{2\beta_t + \gamma_t'} \alpha_t'; \quad I_T^* = \bar{\alpha}_T.$$

Proof. The additional external impact term maps to impact space:

$$\mathbb{E}\left[\int_0^T \frac{\bar{\alpha}_t}{\lambda_t} \left(\beta_t I_t dt + dI_t\right) + \int_0^T \frac{1}{\lambda_t} d[\bar{\alpha}_t, I_t]\right].$$

Integration by parts yields

$$\mathbb{E}\left[\int_0^T \frac{\bar{\alpha}_t}{\lambda_t} \left(\beta_t + \gamma_t'\right) I_t dt - \int_0^T \frac{1}{\lambda_t} I_t d\bar{\alpha}_t + \frac{\bar{\alpha}_T}{\lambda_T} I_T\right].$$

The control problem remains myopic in I:

$$\forall t \in (0,T),\ I_t^* = \frac{\beta_t + \gamma_t'}{2\beta_t + \gamma_t'} \left(\alpha_t + \bar{\alpha}_t\right) - \frac{1}{2\beta_t + \gamma_t'} \left(\alpha_t' - \beta_t \bar{\alpha}_t - \lambda_t m_t\right); \ I_T^* = \bar{\alpha}_T.$$

□

2.6.3 Local concavity

So far, the models split the price impact caused by the trader and the market linearly. However, empirical studies propose non-linear effects. Fill-level concavity is the first instance of non-linearity. One refers to *local* concavity in contrast to *global* concavity, which takes place at the order level. Section 2.6.4 covers global concavity.

Definition 2.6.7 (Locally concave propagator model)**.** *Assume the following dynamics for impact at the microstructure level:*

$$\Delta_n \bar{I}^N = -\beta \bar{I}_n^N \Delta t^N + \sqrt{\Delta t^N} g\left(\frac{\Delta_n M^N + \Delta_n Q^N}{\sqrt{\Delta t^N}}\right) \tag{2.8}$$

for an odd function $g \in C^2$ concave over $[0, \infty)$.

In Section 5.1 "Impact of individual transactions" (p. 33), Bouchaud, Farmer, and Lillo (2009)[32] review the empirical results around non-linear price impact at the fill level. The papers establish a locally concave propagator model with $g(x) \propto x^p$ with $p \in [0.2, 0.5]$. Bouchaud et al. (2004)[34] introduce the propagator model with a local concavity $g(x) \propto \log(x)$.

Remark 2.6.8 ($\sqrt{\Delta t^N}$ scaling). *Two different perspectives motivate the scaling in equation 2.8 of Definition 2.6.7:*

1. *The observed market flow has nontrivial quadratic variation. Therefore, any limiting model must be consistent with an Itô term for M. Then, the functional central limit theorems of Jacod and Protter (2012)[122] lead to a consistent limit for the scaling in equation 2.8.*

   Alternatively, one could propose separate models for rough *trades (say, the overall market) and* smooth *trades (say, the agent's flow). However, this approach contradicts the* anonymity assumption:

   > *Price impact has the same functional form, regardless of a fill's origin.*

   Tóth, Eisler, and Bouchaud (2017)[225] empirically verify the anonymity assumption.

2. *One obtains local concavity from a limit order book model, like how Obhizhaeva and Wang (2013)[180] derive linear price impact from a* block-shaped *order book proportional to the Lebesgue measure. Carmona and Webster (2019)[61] derive local concavity from a general order book shape, mathematically formalizing the numerical experiments of Bouchaud, Farmer, and Lillo (2009)[32] and Mastromatteo, Tóth, and Bouchaud (2014)[164]. Such studies also lead to the scaling in equation 2.8 to maintain a bid-ask spread proportional to price volatility.*

Proposition 2.6.9 (Continuous-time limit). *Assume*

- *$M \in \mathcal{S}$ is an Itô process*

  $$dM_t = m_t dt + v_t dW_t$$

  where W is an $\mathcal{F}$-adapted Wiener process, $m \in \mathcal{S}$, and $v \in \mathcal{S}$ is an Itô process

  $$dv_t = a_t dt + b_t dW'_t$$

  for an $\mathcal{F}$-adapted Wiener process W' that is independent of W and with a, b bounded.

- *Q has finite variation.*

- *$\bar{I}^N$ follows a locally concave impact model, as per Definition 2.6.7.*

Then $\bar{I}^N$ *converges to the* linear *impact model*

$$d\bar{I}_t = -\beta\bar{I}_t dt + \lambda_t\left(m_t dt + dQ_t\right) + \zeta_t dW''_t$$

with

$$\lambda_t = \int_{-\infty}^{\infty} g'(x)\phi_{v_t}(x)dx$$

and

$$\zeta_t = \int_{-\infty}^{\infty} g^2(x)\phi_{v_t}(x)dx$$

where ϕ_σ *is the density function of a Gaussian with variance* σ^2 *and* W'' *is an* $\mathcal{F}$*-adapted Wiener process. The convergence is stable in law.*

Proof. The stable convergence in law is an application of Theorem 10.3.2 from Jacod and Protter (2012) ([122], p. 285). See Appendix B for more details. □

Remark 2.6.10 (Time change convergence). *Recall the convergence from Proposition 2.5.3:*

$$\sqrt{\Delta t^N}\sum_{k=0}^{\lceil Nt\rceil}|\Delta_k M^N| \to \sqrt{\frac{2}{\pi}}\int_0^t v_s ds.$$

This result justifies the language market activity *for the term* v*, as it is proportional to the cumulative unsigned trading volumes on the public tape.*

Proposition 2.6.9 outlines a powerful method to derive a stochastic liquidity model λ_t from a tangible market variable: volume predictions. Indeed, Proposition 2.6.9 mechanically links the market activity v with the liquidity parameter λ. Economists call the price impact parameter λ Kyle's lambda referencing the seminal work of Kyle (1985)[143]. λ measures price illiquidity by quantifying how severely prices react to fills and drive up trading costs. The representation

$$\lambda_t = \int_{-\infty}^{\infty} g'(x)\phi_{v_t}(x)dx$$

establishes a one-to-one relationship between the liquidity model λ and the market activity v. For example, the square-root model

$$g(x) \propto \sqrt{x}$$

bears the stochastic liquidity model

$$\lambda_t \propto \frac{1}{\sqrt{v_t}}.$$

Chapter 3 applies the link to leverage volume predictions in optimal execution. Finally, Chapter 7 uses the connection to propose parametric models for stochastic λ. These parametric models are competitive with significantly less parsimonious, non-parametric models of λ.

Practitioners routinely predict v in so-called *volume curves.* Such volume predictions are core signals in trading algorithms, such as VWAP, and are readily available to brokers and their clients. Therefore, the reader computes the straightforward implications of these volume models on transaction costs.

Example 2.6.11 (Linking volume predictions to trading costs)**.** *For instance, in the square-root model, doubling the market activity reduces impact costs by about 30%. In contrast, in the original OW model, an increase in the market's activity does not affect the trader's price impact. Finally, Section 2.7 establishes bounds on volume predictions ruling out price manipulation.*

2.6.4 Global concavity

Additional empirical studies suggest further concavity when trading sizable orders:

> Local, microstructure-level dynamics do not capture substantial orders' price impact.

Bouchaud, Farmer, and Lillo (2009)[32] distinguish between concavity at the fill and order level. Section 5.2 "Impact of aggregate transactions" (p. 34) follows Section 5.1 "Impact of individual transactions" (p. 33) and establishes a square-root function for global concavity.

A globally concave model is the simplest reduced-form model that reflects empirical studies on sizable orders. Alfonsi, Fruth, and Schied (2010)[7] introduced globally concave price impact models to the author's best knowledge.[18] The Handbook refers to the model by Alfonsi, Fruth, and Schied as the AFS or the globally concave model.

Definition 2.6.12 (Globally concave AFS model)**.** *Assume $M, Q \in \mathcal{S}$. Define the AFS model as the reduced-form price impact model*

$$\bar{I}_t = h(\bar{J}_t)$$

[18]Gatheral, Schied, and Slynko (2011)[106] study the model further:

> "In an obvious generalization [...] we describe the evolution of the volume impact process by
>
> $$E_t = \int_{[0,t)} \psi(t-s)dX_s$$
>
> [...] The price process is then given by
>
> $$S_t = S_t^0 + D_t = S_t^0 + F^{-1}(E_t),$$
>
> where S^0 is the unaffected price process" (p. 2, [106]).

The notation from Gatheral, Schied, and Slynko (2011)[106] maps to this section: the "volume impact" (p. 2) process E corresponds to J and the price impact D to I. The function ψ is exponential, and F^{-1} is h.

for an odd function $h \in C^2$ concave over $[0, \infty)$. Let $\bar{J}$ satisfy

$$d\bar{J}_t = -\beta \bar{J}_t dt + \lambda (dM_t + dQ_t)$$

for $\beta, \lambda > 0$ and $\bar{J}_0 = 0$.

Furthermore, denote the anti-derivative of h by H

$$H(x) = \int_0^x h(y)dy.$$

The derivation of the self-financing condition for the AFS model requires a novel microstructure description and is out of scope for this chapter. Like in the locally concave model, one must establish who trades first when M, Q simultaneously jump. For simplicity, Definition 2.6.13 states the self-financing equation in reduced form when M, Q are *continuous* bounded semimartingales.

Definition 2.6.13 (Self-financing equation with global concavity). *Let $M, Q \in \mathcal{S}$ be continuous and $\bar{I}$ be an AFS model as per Definition 2.6.12. Define the fundamental P&L Y as*

$$dY_t = Q_t dS_t - h(\bar{J}_t)dQ_t - \frac{\lambda}{2} h'(\bar{J}_t)d[Q,Q]_t - \lambda h'(\bar{J}_t)d[M,Q]_t.$$

Proposition 2.6.14 (Mapping to impact space with global concavity). *Consider the fundamental P&L Y from Definition 2.6.13. Furthermore, define $J, \bar{\alpha}$ such that $J_0 = 0$, $\bar{\alpha}_0 = 0$,*

$$dJ_t = -\beta J_t dt + \lambda dQ_t$$

and

$$d\bar{\alpha}_t = -\beta \bar{\alpha}_t dt - \lambda dM_t.$$

Let $\alpha_t = \mathbb{E}\left[S_T - S_t \mid \mathcal{F}_t \right]$.

Then

$$\mathbb{E}[Y_T] = \frac{1}{\lambda}\mathbb{E}\left[\int_0^T \beta\left(\alpha_t - h(\bar{J}_t)\right) J_t dt - \int_0^T J_t d\alpha_t - \int_0^T h(\bar{J}_t) d\bar{\alpha}_t\right] + \frac{1}{\lambda}\mathbb{E}\left[\frac{1}{2}\int_0^T h'(\bar{J}_t)d[\bar{\alpha},\bar{\alpha}]_t - H(\bar{J}_T)\right].$$

Proof. Lemma 2.3.3 still applies. Expressing $\mathbb{E}[Y_T]$ in terms of J instead of Q yields

$$\mathbb{E}[Y_T] = \frac{1}{\lambda}\mathbb{E}\left[\int_0^T \left(\alpha_t - h(\bar{J}_t)\right)\left(\beta J_t dt + dJ_t\right) - \frac{1}{2}\int_0^T h'(\bar{J}_t)d[J,J]_t\right] + \frac{1}{\lambda}\mathbb{E}\left[\int_0^T h'(\bar{J}_t)d[\bar{\alpha},J]_t + [\alpha,J]_T\right].$$

This expectation divides into three terms:

(a) The term

$$\frac{1}{\lambda}\mathbb{E}\left[\int_0^T \left(\alpha_t - h(\bar{J}_t)\right)\beta J_t dt\right]$$

is myopic.

(b) The term

$$-\frac{1}{\lambda}\mathbb{E}\left[\int_0^T h(\bar{J}_t)dJ_t + \frac{1}{2}\int_0^T h'(\bar{J}_t)d[J,J]_t - \int_0^T h'(\bar{J}_t)d[\bar{\alpha},J]_t\right]$$

evokes the Itô formula for $H(\bar{J}_T)$. Indeed, by the linearity of the SDEs defining $\bar{J}, J, \bar{\alpha}$, one has

$$\bar{J} = J - \bar{\alpha}.$$

Applying Itô's formula proves

$$H(\bar{J}_T) = \int_0^T h(\bar{J}_t)(dJ_t - d\bar{\alpha}_t) + \frac{1}{2}\int_0^T h'(\bar{J}_t)d[J,J]_t - \int_0^T h'(\bar{J}_t)d[\bar{\alpha},J]_t + \frac{1}{2}\int_0^T h'(\bar{J}_t)d[\bar{\alpha},\bar{\alpha}]_t.$$

Reshuffling terms between the two sides of the equation proves

$$-\int_0^T h(\bar{J}_t)dJ_t - \frac{1}{2}\int_0^T h'(\bar{J}_t)d[J,J]_t + \int_0^T h'(\bar{J}_t)d[\bar{\alpha},J]_t$$

equals

$$-\int_0^T h(\bar{J}_t)d\bar{\alpha}_t + \frac{1}{2}\int_0^T h'(\bar{J}_t)d[\bar{\alpha},\bar{\alpha}]_t - H(\bar{J}_T).$$

(c) By Itô, the remaining term satisfies

$$\int_0^T \alpha_t dJ_t + [\alpha, J]_T = -\int_0^T J_t d\alpha_t.$$

The three terms combine to conclude with the formula

$$\mathbb{E}[Y_T] = \frac{1}{\lambda}\mathbb{E}\left[\int_0^T \beta\left(\alpha_t - h(\bar{J}_t)\right)J_t dt - \int_0^T J_t d\alpha_t - \int_0^T h(\bar{J}_t)d\bar{\alpha}_t\right] + \frac{1}{\lambda}\mathbb{E}\left[\frac{1}{2}\int_0^T h'(\bar{J}_t)d[\bar{\alpha},\bar{\alpha}]_t - H(\bar{J}_T)\right].$$

□

Remark 2.6.15 (Myopic in impact space). *When considering continuous trading strategies Q, M,* global concavity still leads to a myopic control problem in *J. Ackermann, Kruse, and Urusov (2022)[2] propose a method to extend trading problems to larger classes of trading strategies, which one can apply in this setting to include jumps in Q, M. The argument follows four steps:*

(a) Define the objective function for smooth strategies.

(b) Establish the myopic objective in impact space for smooth strategies.

(c) Continuously extend the objective function in impact space *to a larger class of controls.*

(d) Use the inverse map to recover the optimal strategy in trade space.

Using this method, one adds jumps to the self-financing equation from Definition 2.6.13 so that (i) the objective function matches for continuous trading strategies (ii) the trading problem remains well-defined due to being myopic in impact space.

For example, exercise 5 leverages the myopic objective in impact space and solves an AFS model example for power-law global concavity.

Finally, local and global concavity are mathematically unrelated.

(a) Local concavity models how price impact changes with immediate liquidity. Such dynamics play out intraday, most notably with seasonal effects mentioned by Fruth, Schöneborn, and Urusov (2019)[98] and Cont, Kukanov, and Stoikov (2014)[74].

(b) Global concavity models sizable orders. As studied by Bouchaud, Farmer, and Lillo (2009)[32], such orders execute over multiple days, dulling the effect of immediate liquidity.

2.7 Price Manipulation Paradoxes

> "Time-dependent liquidity can potentially lead to price manipulation. In periods of low liquidity, a trader could buy the asset and push market prices up significantly; in a subsequent period of higher liquidity, he might be able to unwind this long position without depressing market prices to their original level, leaving the trader with a profit after such a round trip trade" (p. 1, Fruth, Schöneborn, and Urusov (2013)[97]).

This section highlights three crucial results on price manipulation paradoxes that go beyond the seminal work by Gatheral (2010)[104].

(a) The first part delves into the result of Fruth, Schöneborn, and Urusov (2013, 2019)[97, 98] to describe the potential price manipulation strategy and constraints it implies on price impact models.

(b) The second part extends these constraints to locally concave price impact models.

(c) The third part *translates* these results into bounds on volume predictions. Consequently, Proposition 2.6.9 linking the liquidity factor λ_t to the market activity v_t plays an essential role: it makes the no-price manipulation condition tangible using publicly observable trading volumes.

2.7.1 Constraints on price impact models

The condition

$$2\beta + \gamma' > 0$$

from Corollary 2.5.8 is not a mere technicality: without it, the generalized OW model allows price manipulation strategies. For instance, see Example 2.7.1 and exercise 4.

Example 2.7.1 (A price manipulation example). *Suppose the trader buys Δ_t shares in a jump trade at time t and sells them in another jump trade at time $t' > t$. Then, the corresponding trading profits are*

$$\begin{aligned}
&-\left(S_t + \frac{1}{2}\lambda_t\Delta_t\right)\Delta_t + \left(S_{t'} + e^{-\int_t^{t'}\beta_s ds}\lambda_t\Delta_t - \frac{1}{2}\lambda_{t'}\Delta_t\right)\Delta_t \\
&= (S_{t'} - S_t)\Delta_t + \frac{1}{2}\lambda_t\left(2e^{-\int_t^{t'}\beta_s ds} - 1 - e^{\int_t^{t'}\gamma'_s ds}\right)\Delta_t^2 \\
&= (S_{t'} - S_t)\Delta_t - \frac{1}{2}\lambda_t\left(2\beta_t + \gamma'_t\right)(t' - t)\Delta_t^2 + o(t' - t)
\end{aligned}$$

For sufficiently large trades, the quadratic impact term dominates the linear gains from expected fundamental price changes. Furthermore, for small $t' - t$, the quadratic term is positive if $2\beta_t + \gamma'_t < 0$, as per its first-order expansion in $t' - t$. Consequently, if λ_t decreases too promptly relative to impact decay, then sufficiently large round trips executed sufficiently quickly allow the trader to obtain arbitrary expected profits. These profits are independent of fundamental price changes: they rely solely on price manipulation through the trader's own impact.

Fruth, Schöneborn, and Urusov (2013)[97] first derived the no price manipulation condition for the generalized OW model in "Proposition 8.3 (Price manipulation in the zero spread model)" (p. 27). Figure 2.5 and exercise 4 illustrate this price manipulation paradox.

Unfortunately, the condition $2\beta + \gamma' > 0$ only rules out price manipulation paradoxes, which imply arbitrary profits from price impact. Indeed, the condition does not rule out more general *transaction-triggered price manipulations,*

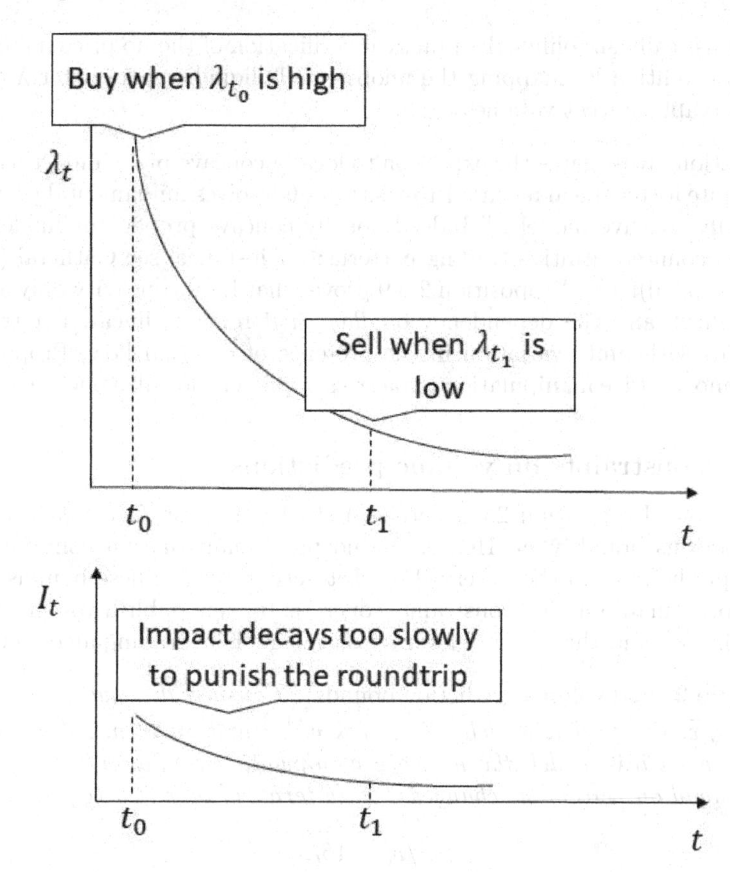

FIGURE 2.5
Potential price manipulation strategy from Example 2.7.1 if λ_t decreases faster than I_t, leading to the condition $2\beta + \gamma' > 0$.

e.g., initial selling to devalue the price for future buying. See "Corollary 8.5 (Transaction-triggered price manipulation in the zero spread model)" (p. 29) by Fruth, Schöneborn, and Urusov (2013)[97] for more details.

2.7.2 Extension to locally concave models

Proposition 2.6.9 augments the analysis of Fruth, Schöneborn, and Urusov (2013, 2019)[97, 98] in two ways.

(a) By proving a correspondence between locally concave price impact models and linear models with stochastic liquidity, it extends the no price manipulation condition to non-linear models.

(b) It drastically simplifies the practical verification of the no price manipulation condition by mapping the unobservable liquidity parameter λ to the observable market volumes v.

This section investigates the extension to locally concave price impact models.

Despite its technical nature, Proposition 2.6.9 solves fundamental problems for locally concave models.[19] Indeed, locally concave propagator models are prone to counter-intuitive trading patterns: for instance, see Gatheral (2010, Section 4 p. 10)[104]. Proposition 2.6.9 proves that local concavity only affects *volatile* flow, and the dependency on *flow drift* remains linear. For trading strategies with finite variation in the presence of external flow, Proposition 2.6.9 removes price manipulation paradoxes implied by locally concave models.

2.7.3 Constraints on volume predictions

Furthermore, Proposition 2.6.9 derives a stochastic push factor λ_t from the instantaneous liquidity v_t. Hence, the no price manipulation conditions on β, λ imply bounds on the observed market activity v_t.[20] These bounds make the no price manipulation constraints concrete: brokers publish volume curves predicting v_t, and their clients leverage such signals in trading algorithms.

Example 2.7.2 (Volume prediction bounds). *Consider the square-root model $g(x) \propto \sqrt{x}$. Under that model, $e^{\gamma_t} = \lambda_t \propto v_t^{-1/2}$ and, in turn, $2\gamma_t' = -v_t'/v_t$. Therefore, in this model, the no price manipulation condition $2\beta + \gamma'$ is an upper bound on percentage changes v'/v in terms of impact decay β:*

$$v_t'/v_t < 4\beta_t.$$

> *Hence, all else being equal, trading activity cannot double in less than a quarter of the impact model's half-life under a square-root price impact model.*

For instance, for a square-root price impact model with a one-hour half-life, the no price manipulation condition states that trading activity cannot double in less than fifteen minutes without creating a price manipulation opportunity.

Price manipulation is asymmetric in liquidity. Negative shocks to market activity, where trading and, in turn, liquidity quickly dries up, do not lead to price manipulation. In contrast, positive liquidity shocks, where trading activity quickly floods the market, potentially invite price manipulation. Therefore, practitioners should take exceptional care when predicting increasing trading

[19] Carmona and Webster (2019)[61] first proved the link between local concavity and linear price impact models. Carmona and Webster motivate local concavity by modeling a general order book shape like Bouchaud, Farmer, and Lillo (2009)[32] and Mastromatteo, Tóth, and Bouchaud (2014)[164].

[20] Muhle-Karbe, Wang, and Webster (2022)[170] first linked Fruth, Schöneborn, and Urusov's (2019) no price manipulation condition to observed market volumes.

volumes: all else being equal, they imply price manipulation strategies. Indeed, such volume curves may trigger counter-intuitive behaviors in trading algorithms that practitioners would want to rule out.

2.8 Summary of Results

This section summarizes reduced-form price impact models and their mathematical properties. Assume that the trader is taking liquidity and that the *no adverse selection* assumption holds.

2.8.1 Generalized OW impact model

Definition 2.8.1 (Generalized OW model). *Let $Q \in \mathcal{S}$ be a trading process. Assume*

- *a stochastic decay $\beta \in \mathcal{S}$ bounded away from zero.*
- *a stochastic push $\lambda = e^{\gamma}$ with $\gamma \in \mathcal{S}$.*
- *that γ is differentiable, with derivative γ'.*
- *the* no price manipulation *condition holds: $2\beta + \gamma'$ is bounded away from zero.*
- *a process $S \in \mathcal{S}$ modeling the unperturbed price.*

Then define the generalized OW model by providing the following two formulas.

The dynamics of the price impact I caused by Q are

$$dI_t = -\beta_t I_t dt + \lambda_t dQ_t.$$

The fundamental P&L Y satisfies the self-financing wealth equation

$$dY_t = Q_t dS_t - I_t dQ_t - \frac{1}{2} d[I, Q]_t$$

in the absence of bid-ask spread.

Remark 2.8.2 (Effect of a time change). *Modeling price impact in a different clock is mathematically equivalent to a time change and acts as a multiplier to the decay parameter β. For example, a constant β in the volume clock equals*

$$\beta_t = \beta v_t$$

in the regular clock, where v_t is the realized market activity, as measured by the volatility of the market trading process M.

Proposition 2.8.3 (Mapping to impact space). *Assume given a generalized OW model. Define the* alpha signal α_t *as*

$$\alpha_t = \mathbb{E}\left[S_T - S_t \middle| \mathcal{F}_t \right].$$

Assume $\alpha \in \mathcal{S}$. Then the following identity holds:

$$\mathbb{E}\left[Y_T\right] = \mathbb{E}\left[\int_0^T \frac{\beta_t + \gamma_t'}{\lambda_t} \alpha_t I_t dt - \int_0^T \frac{2\beta_t + \gamma_t'}{2\lambda_t} I_t^2 dt - \int_0^T \frac{1}{\lambda_t} I_t d\alpha_t - \frac{1}{2\lambda_T} I_T^2 \right].$$

2.8.2 Generalized OW impact model with external impact

Definition 2.8.4 (Generalized OW model with external impact). *Let $Q \in \mathcal{S}$ be the trader's, $M \in \mathcal{S}$ the market's trading process, and $S \in \mathcal{S}$ the unperturbed price.*

Define the generalized OW model with external impact *as the reduced-form model with*

- *the* trader's price impact I *satisfying*

$$dI_t = -\beta_t I_t dt + \lambda_t dQ_t$$

  with $I_0 = 0$.

- *the alpha signal $\bar{\alpha}$ implied from the* external impact

$$d\bar{\alpha}_t = -\beta_t \bar{\alpha}_t dt - \lambda_t dM_t.$$

- *the self-financing equation*

$$dY_t = Q_t dS_t + \left(\bar{\alpha}_t - I_t\right) dQ_t - \frac{1}{2} d[I, Q]_t + d[\bar{\alpha}, Q]_t.$$

Remark 2.8.5 (Locally concave propagator model). *Local concavity is equivalent to a stochastic push factor λ_t that scales with immediate market activity, as measured by the volatility v of M.*

For example,

(a) for the classic square-root propagator model, one has

$$\lambda_t \propto \frac{1}{\sqrt{v_t}}.$$

(b) for the original log propagator model, introduced by Bouchaud et al. (2004)[34], one has

$$\lambda_t \propto \frac{1}{v_t}.$$

The log model matches a linear price impact model on the trading process's instantaneous participation rate $\frac{dQ_t}{v_t}$.

Definition 2.8.6 (Globally concave AFS model). *Assume the following reduced-form price impact model*

$$\bar{I}_t = h(\bar{J}_t)$$

for an odd function $h \in C^2$ concave over $[0, \infty)$. Let $\bar{J}$ satisfy the dynamics:

$$d\bar{J}_t = -\beta \bar{J}_t dt + \lambda (dM_t + dQ_t)$$

for $\beta, \lambda > 0$.

2.8.3 Control problems

Corollary 2.8.7 (Control problem). *Assume*

- *a generalized OW model.*
- *a* deterministic alpha signal $\alpha \in C^2$ *with derivative* α'.

Then the solution to the control problem

$$\sup_{Q \in \mathcal{S}} \mathbb{E}[Y_T]$$

is myopic in I.

The optimal impact state I^* satisfies

$$\forall t \in (0,T), \quad I_t^* = \frac{\beta_t + \gamma_t'}{2\beta_t + \gamma_t'} \alpha_t - \frac{1}{2\beta_t + \gamma_t'} \alpha_t'; \quad I_T^* = 0.$$

One recovers the optimal Q^* via the one-to-one map

$$dQ_t^* = \frac{1}{\lambda_t} (\beta_t I_t^* dt + dI_t^*).$$

Corollary 2.8.8 (Practical formulas). *Consider the control problem and optimal impact state I^* from Corollary 2.8.7.*

Let $Q \in \mathcal{S}$ be a candidate trading process and I its impact. Then the following formulas hold

$$Q_t = \frac{1}{\lambda_t} I_t + \int_0^t \frac{\beta_s + \gamma_s'}{\lambda_s} I_s ds,$$

$$\mathbb{E}[Y_T(Q^*)] = \mathbb{E}\left[\int_0^T \frac{2\beta_t + \gamma_t'}{2\lambda_t} (I_t^*)^2 dt\right],$$

and

$$\mathbb{E}[Y_T(Q)] = \mathbb{E}[Y_T(Q^*)] - \mathbb{E}\left[\int_0^T \frac{2\beta_t + \gamma_t'}{2\lambda_t} (I_t - I_t^*)^2 dt + \frac{1}{2\lambda_T} I_T^2\right].$$

Remark 2.8.9 (Control problem in the presence of external impact). *The problem remains myopic in I, and the impact from M behaves as a signal $-\bar{\alpha}$.*

Remark 2.8.10 (Control problem with global concavity). *The problem remains myopic in impact space.*

2.8.4 Price manipulation bounds

Corollary 2.8.11 (No price manipulation constraint). *For the optimal control problem from Definition 2.8.7 to be well-posed, the no price manipulation condition*

$$2\beta + \gamma' > 0$$

must hold.

Remark 2.8.12 (Volume prediction bounds). *The no price manipulation condition translates into a volume prediction bound for a locally concave price impact model. For example, for the square-root propagator model,*

$$v'/v < 4\beta.$$

> *Under a square-root model, all else being equal, volume predictions cannot double faster than a quarter of the price impact model's half-life.*

2.9 Exercises

Exercise 1 Example impact computations

This exercise provides impact curve examples for two impact models. Assume that $(0, 1)$ represents a full trading day.

Closed-form examples I

Consider the original OW model

$$dI_t = -\beta I_t dt + \lambda dQ_t$$

with $\beta, \lambda > 0$.

1. Assume the trader sends in a Time Weighted Average Price (TWAP) order for the day. Hence, one has $Q_t = Qt$ for a constant Q for all $t \in (0, 1)$. Furthermore, assume that $I_0 = 0$. Compute and plot I_t.
2. Assume S is a martingale. Compute and plot the expected accounting and fundamental P&L for $t \in (0, 1)$.
3. Assume the impact state halves by the start of the next day

$$I_{1+} = \frac{1}{2} I_{1-}.$$

   The trader does not trade on the second day. Compute and plot I, X, and Y for $t \in (1, 2)$.

Numerical examples II

Consider the impact model

$$dI_t = -\beta I_t dt + \frac{\lambda}{v_t} dQ_t$$

with v_t the intraday market activity captured by the volatility of market trades.

Assume the intraday volume profile v follows the deterministic curve

$$v_t = e^{4\cdot(t-0.7)^2}.$$

1. Plot the function v_t to visualize the intraday volume profile.
2. Numerically solve questions 1-3 for this model.

Exercise 2 Closed-form formulas for a constant drift alpha

This exercise solves the optimal control problem for a constant drift alpha. Consider a generalized OW model with parameters $\beta > 0$, $\lambda = e^{\gamma}$ satisfying the no price-manipulation condition $2\beta + \gamma' > 0$. For a given trading process Q, the price impact I is the solution to the SDE

$$dI_t = -\beta_t I_t dt + \lambda_t dQ_t.$$

Furthermore, assume a model for the unperturbed price S and zero initial states $I_0 = 0$ and $Q_0 = 0$.

Assume that the self-financing equation for the trading strategy's fundamental P&L is

$$dY_t = Q_t dS_t - I_t dQ_t - \frac{1}{2} d[I, Q]_t.$$

Assume the unperturbed price is an Itô process

$$dS_t = \mu dt + \sigma dW_t$$

for two constants μ, σ.

1. Compute $\alpha_t = \mathbb{E}\left[S_T - S_t \middle| \mathcal{F}_t\right]$.
2. Write the objective function for a risk-neutral trader in impact space.
3. What is the optimal impact state that maximizes the expected fundamental P&L of the risk-neutral trader? Comment on the behavior for $T - t$ significant and insignificant compared to the timescale $\frac{1}{\beta_t + \gamma'_t}$.

Exercise 3 Trading two consecutive orders

This exercise reconciles two points of view on consecutive trades:

(a) One can treat the second order as the first order's continuation.

(b) One can treat the second order as a separate trade from an external source.

One shows that the two points of view lead to the same optimal trading strategy. Consider a generalized OW model with parameters $\beta > 0$, $\lambda = e^{\gamma}$ satisfying the no price-manipulation condition $2\beta + \gamma' > 0$. For a given trading process Q, the price impact is the solution to the SDE

$$dI_t = -\beta_t I_t dt + \lambda_t dQ_t.$$

Consider a deterministic alpha signal α_t.

1. In the continuation approach, $Q_0 = \tilde{Q}$ and $I_0 = \tilde{I}$ are non-zero and capture the first order's effect. Derive the optimal impact state I_t^0.
2. In the separate approach, $Q_0 = 0$ and $I_0 = 0$. The external impact $-\bar{\alpha}_0 = \tilde{I}$ captures the first order's effect. The first order's external impact satisfies
$$\forall t > 0; \quad d\bar{\alpha}_t = -\beta_t \bar{\alpha}_t dt.$$
Derive the optimal impact state I_t^1.
3. Reconcile the two points of view.

Exercise 4 Absence of price manipulation strategies

This exercise finds a lower bound on the decay parameter β_t of a generalized OW model based on a liquidity profile v_t to rule out price manipulation strategies. Assume the intraday volume profile v follows the deterministic curve

$$v_t = e^{4\cdot(t-0.7)^2}.$$

Local square root model under the calendar clock I

The impact model considered is

$$dI_t = -\beta I_t dt + \frac{\lambda}{\sqrt{v_t}} dQ_t.$$

It is the reduced-form version of the locally concave model with $g(x) \propto \sqrt{x}$.

1. Establish the lower bound on β that guarantees no price manipulation.

2. For $\beta = 0.01$, find a pair of trades that lead to a price manipulation paradox.

Local log model under the volume clock II

The impact model considered is

$$dI_t = -\beta v_t I_t dt + \frac{\lambda}{v_t} dQ_t.$$

It is the reduced-form version of the locally concave model with $g(x) \propto \log(x)$.

1. Establish a lower bound on β that guarantees no price manipulation.
2. For $\beta = 0.01$, find a pair of trades that lead to a price manipulation paradox.

Exercise 5 Globally concave AFS model

This exercise establishes an order-based impact formula from a globally concave AFS model. Consider a globally concave impact model

$$I_t = \text{sign}(J_t)|J_t|^c$$

with local dynamics

$$dJ_t = -\beta J_t dt + \lambda dQ_t$$

for $\beta, \lambda > 0$ and $c \in (0, 1]$.

Consider an order of size $Q > 0$ traded over $[0, T]$. Let S be a martingale, $J_0 = 0$, and $Q_0 = 0$.

1. Map the objective function in impact space.
2. Derive the optimal execution strategy. What is I_T as a function of Q?
3. Consider a TWAP execution. What is I_T as a function of Q?

3

Applications of Price Impact Models

"During the trading process, it is useful to have real time clues about what the performance would be under some changes of execution conditions. What-if *I finish the order just now by aggressively capturing all the needed liquidity in the visible orderbooks?* What-if *I convert my implementation shortfall benchmarked algo to a percentage of volume one? etc."*

– "*What-if* scenarios", Lehalle and Laruelle (2018)[147]

The first two applications of price impact models focus on order scheduling. Traders need to answer questions such as

How much more expensive would a larger order have been?

What would have been the order's arrival slippage if it had traded faster?

How should the trading strategy change with a second order's arrival?

Practitioners call such questions *what-if scenarios* or counterfactuals.

The executions of two orders with different trading parameters are related but not directly comparable. Readers familiar with options or fixed income face similar challenges: instruments with the same underlying but different maturities, strikes, or other parameters are also related but not directly comparable. To solve this comparison problem, practitioners normalize options using a standard unit, *implied volatility.* Implied volatilities arise from a pricing model, such as the seminal Black and Scholes model, and practitioners compare implied to realized volatilities. Price impact models serve the same purpose for algorithmic trading: traders normalize orders using a standard unit, *price impact,* and compare impact to realized alpha. Normalization allows practitioners to make quantifiable, reproducible decisions across what-if scenarios.

This chapter builds upon the mathematical results of Chapter 2, which Section 2.8 summarizes. The first application, *optimal execution*, is the most direct use of price impact models for trading. Section 3.2 covers optimal execution in detail. Chapter 1 introduces the second application and provides a non-technical overview of Transaction Cost Analysis (TCA). Section 3.3 tackles TCA as an experimental framework. Furthermore, a statistical approach

DOI: 10.1201/9781003316923-3

to TCA leverages the mathematical tools of causal inference covered in Chapter 5. Finally, Section 3.4 summarizes this chapter's essential formulas for reference.

3.1 A Pedagogical Example

For traders, this chapter's essential takeaway is that order execution studies the relationship between alpha and price impact:

> Traders answer execution questions by comparing alpha and price impact.

For teaching purposes, this section answers trader questions in the simplest setting: with constant liquidity and alpha signals. Consider the price impact model

$$I'_t = -\beta I_t + \lambda Q'_t$$

with $I_0 = 0$, where Q' is the trader's execution speed. Recall from the pedagogical example of Section 2.1 that the optimal trading strategy for a given alpha is myopic in impact:

$$I^*_t = \frac{1}{2}\left(\alpha_t - \beta^{-1}\alpha'_t\right) \tag{3.1}$$

for $t \in (0, T)$. The corresponding optimal execution speed is

$$Q^{*'}_t = \frac{\beta}{2\lambda}\left(\alpha_t - \beta^{-2}\alpha''_t\right)$$

In this example's constant alpha case, the alpha does not decay over $[0, T]$: it predicts a price move beyond time T. For instance, if T is the end of the trading day, α could anticipate an overnight return. Formula 3.1 simplifies to

$$I^*_t = \frac{1}{2}\alpha.$$

Furthermore, Q_T scales linearly with α, and the trading speed over $(0, T)$ is

$$Q'_t = \frac{\beta}{2\lambda}\alpha$$

This chapter introduces three example execution questions, which the myopic trading formula 3.1 answers.

(a) How much more expensive would a larger order have been?

If the trader doubles the order size, they are implying an alpha twice the original alpha size. For a linear price impact model, the optimal impact state doubles accordingly. Therefore, the total costs quadruple.

(b) What would have been the order's arrival slippage if it had traded faster?

If the trader keeps the order size constant and accelerates the order over $(0, T/2)$, then they decelerate over $(T/2, T)$. For this strategy to be optimal, this implies $\alpha_t' < 0$ over $(0, T/2)$. In trading terms, the trader is front-loading the order because they wish to capture the alpha before it decays. If alpha decay does not drive acceleration, then the impact state is suboptimal: too elevated over $(0, T/2)$ and too low over $(T/2, T)$.

(c) How should the trading strategy change with a second order's arrival?

The answer depends on the second order's intent, measured by its alpha. If an updated alpha does not drive the second order, then the optimal impact state remains unchanged. Therefore, extending the trading interval executes the order while maintaining an optimal impact state throughout.

> If the second order does not increase the alpha, it trades *after* the first order.

Conversely, if the second order increases the alpha prediction, the optimal impact state rises to reflect the alpha signal's update.

> If the second order increases the alpha, it trades *alongside* the first order.

A simple target relationship between alpha and price impact answers three execution questions for constant trading signals. This chapter extends target relationships between alpha and impact to include time-varying and stochastic trading signals, including liquidity and alpha signals.

3.2 Optimal Execution

Section 2.3.1 from Chapter 2 defines optimal execution under simplified assumptions. This section leverages the broader mathematical toolbox of Chapter 2 to solve four practical challenges when scheduling orders.

(a) *How to communicate transaction costs to the portfolio team?*

Trading teams execute orders on behalf of clients, for example, portfolio teams. The first step in optimal execution is for the portfolio team to determine order parameters based on a trade-off between a portfolio's transaction costs and alpha prediction. A pre-trade cost model quantifies this trade-off.

(b) *How to include alpha and liquidity signals in the execution strategy?*

Order instructions only provide basic parameters, for example, the order size and horizon. In addition, trading teams consider alpha signals and liquidity models when scheduling orders.

(c) *How to allow for tactical deviations at the microstructure level?*

The centralized execution schedule cannot react in time to microstructure opportunities due to technological reasons. However, co-located tactical algorithms can benefit the order but require a self-correction method: the execution cannot deviate too much from the centralized strategic schedule.

(d) *How to change the execution strategy when new orders arrive?*

Orders rarely trade in isolation, and the arrival of an overlapping order, especially with wildly different trading parameters, changes the execution strategy.

This section addresses these four challenges in order.

3.2.1 Pre-trade cost model

> "A good understanding of likely costs for a proposed trade can be an important ingredient in making and adjusting trade decisions. Questions such as 'how large a trade can we do', or 'what time horizon should we set', or even 'does this trade have a strong enough alpha to overcome the transaction costs' can all be systematically evaluated using a good pre-trade cost model" (Quantitative Brokers (2019)[39]).

Optimal execution begins with communication between a portfolio and a trading team. The portfolio team proposes a trade based on their alpha, and the trading team executes the order on the market, for instance, buying at a constant speed throughout the day. As outlined by Quantitative Brokers, a pre-trade cost model facilitates this communication: first, the execution team distills the execution problem into a pre-trade model. The portfolio team then proposes their trade, considering the model and their portfolio objectives. Quantitatively minded portfolio teams may automate this process using an Application Programming Interface (API) for the pre-trade cost model rather than rely on verbal communication with the trading team.[1]

The most frequent trade a portfolio team submits is a day order:

---

[1] In an interview with *the wall street lab* (2021)[221], Robert Almgren, co-founder and chief scientist of Quantitative Brokers, highlights the vital role pre-trade cost models play in portfolio managers' decisions:

> "clients do want advice on what will be the expected or the forecast cost of a trade. So, that goes into their decision on whether to trade or how much to trade. If you're a portfolio manager, you have to size your position, maybe you have a strategy that makes money on a small portfolio, but you scale it up by a factor of 10. Suddenly, it's not making money. The reason is the slippage costs are killing you. So, it's very

(a) The team enters the trade at the start of the day $t = 0$ and expects to finish trading by the end of the day $t = T$.

(b) The portfolio team has *no intraday alpha.* If $\tilde{\mathcal{F}}_t$ is the team's information, then $\mathbb{E}\left[S_T - S_t \middle| \tilde{\mathcal{F}}_t\right] = 0$ for all $t \in [0, T]$.

(c) The portfolio team has an *overnight alpha.* Let $\tilde{\alpha}^r$ be the realized overnight return. Assume $\tilde{\alpha} = \mathbb{E}\left[\tilde{\alpha}^r \middle| \tilde{\mathcal{F}}_t\right]$ is constant for all $t \in [0, T]$: the alpha prediction is static throughout the trading day.

3.2.1.1 Idealized optimal execution problem

The trading team simplifies the execution problem into a pre-trade cost model using Definition 3.2.1 and Proposition 3.2.2. This section names the simplified execution problem the *idealized* problem, as the resulting pre-trade cost formulas summarize the execution strategy's complexity. Therefore, the portfolio team's effective daily pre-trade cost model equals a simple one-period quadratic trading penalty.

Definition 3.2.1 (Idealized optimal execution problem). *Assume given a random variable modeling an overnight return $\tilde{\alpha}^r \in E$, an impact model $I \in \mathcal{S}_{1,1}$, and an unperturbed price process $S \in \mathcal{S}$ over an interval $[0, T]$.*

Define the idealized optimal execution problem *as*

$$\sup_{Q \in \mathcal{S}} \mathbb{E}\left[\tilde{\alpha}^r Q_T + \int_0^T Q_t dS_t - \int_0^T I_t dQ_t - \frac{1}{2}[I, Q]_T\right]$$

with the initial condition $Q_0 = 0$.

Proposition 3.2.2 (Solution to the idealized optimal execution problem). *Assume given an idealized optimal execution problem, as per Definition 3.2.1, and that*

- *I follows the generalized OW model with decay and push parameters $\beta, \lambda = e^{\gamma}$ with $\beta, \gamma' \in \mathcal{S}$ satisfying the no price manipulation condition from Corollary 2.5.8.*
- *the initial impact state is neutral, $I_0 = 0$.*
- *S is a martingale over $[0, T]$.*

Then the optimal execution strategy has the following high-level properties:

(a) The optimal order size satisfies

$$Q_T = \frac{\tilde{\alpha}}{\Lambda_T}$$

important to have some sort of model for what would be the expected cost of a particular transaction."

where Λ_T is

$$\frac{1}{\Lambda_T} = \frac{1}{\lambda_T} + \int_0^T \frac{(\beta_t + \gamma_t')^2}{\lambda_t(2\beta_t + \gamma_t')} dt.$$

(b) The order's expected fundamental P&L satisfies

$$\begin{aligned}\mathbb{E}[Y_T] &= \frac{\tilde{\alpha}^2}{2}\mathbb{E}\left[\frac{1}{\Lambda_T}\right] \\ &= \frac{\tilde{\alpha}}{2}\mathbb{E}[Q_T] \\ &= \mathbb{E}\left[\frac{\Lambda_T}{2}Q_T^2\right].\end{aligned}$$

(c) The order's expected arrival slippage satisfies

$$-\mathbb{E}\left[\int_0^T I_t dQ_t + \frac{1}{2}[I,Q]_T\right] = -\mathbb{E}[Y_T].$$

Proof. One modifies Corollaries 2.5.8 and 2.5.9 from Chapter 2 to include the overnight alpha $\tilde{\alpha}$. The optimal target impact state satisfies

$$\forall t \in (0,T), \quad I_t^* = \frac{\beta_t + \gamma_t'}{2\beta_t + \gamma_t'}\tilde{\alpha}; \quad I_T^* = \tilde{\alpha}.$$

The impact state leads to the terminal position at optimum

$$\begin{aligned}Q_T &= \frac{1}{\lambda_T}I_T^* + \int_0^T \frac{\beta_t + \gamma_t'}{\lambda_t} I_t^* dt \\ &= \tilde{\alpha}\left(\frac{1}{\lambda_T} + \int_0^T \frac{(\beta_t + \gamma_t')^2}{\lambda_t(2\beta_t + \gamma_t')} dt\right) \\ &= \frac{\tilde{\alpha}}{\Lambda_T}.\end{aligned}$$

The optimal trade's expected fundamental P&L in impact space satisfies

$$\begin{aligned}\mathbb{E}[Y_T] &= \mathbb{E}\left[\int_0^T \frac{2\beta_t + \gamma_t'}{2\lambda_t}(I_t^*)^2 dt + \frac{1}{2\lambda_T}(I_T^*)^2\right] \\ &= \mathbb{E}\left[\int_0^T \frac{(\beta_t + \gamma_t')^2}{2\lambda_t(2\beta_t + \gamma_t')}\tilde{\alpha}^2 dt + \frac{1}{2\lambda_T}\tilde{\alpha}^2\right] \\ &= \frac{\tilde{\alpha}^2}{2}\mathbb{E}\left[\frac{1}{\lambda_T} + \int_0^T \frac{(\beta_t + \gamma_t')^2}{\lambda_t(2\beta_t + \gamma_t')} dt\right] \\ &= \frac{\tilde{\alpha}^2}{2}\mathbb{E}\left[\frac{1}{\Lambda_T}\right].\end{aligned}$$

Identity (c) derives from the observation

$$\mathbb{E}[Y_T] = \mathbb{E}\left[\tilde{\alpha}Q_T - \int_0^T I_t dQ_t - \frac{1}{2}[I,Q]_T\right].$$

□

3.2.1.2 Communication with the portfolio team

> "As a common feature of linear-quadratic functions, the optimum is reached at the point where the quadratic penalties amount to one-half the linear term. **This means that it is optimal to give away half of the forecast-driven gross pnl to impact cost**" (p. 193, Isichenko (2021)[121]).

Proposition 3.2.2 provides two takeaways that the trading team can impress on the portfolio team.

(a) The portfolio team pays half their alpha in impact costs for an optimally sized order. Section 3.3 proposes testable predictions for TCA based on this relation between alpha and impact.

(b) The trading team can provide a daily pre-trade cost model without sharing their intraday liquidity model. This pre-trade cost model only requires estimates of the daily liquidity parameter Λ_T^{-1} instead of the underlying models β, λ.

Point (a) is essential for portfolio managers: managers who only track the bid-ask spread will be astonished that price impact shrinks their paper P&L by half. Paper P&L refers to expected P&L in the absence of transaction costs. In an interview with Institutional Investor (2022)[119], Yves Lemperiere, head of alpha research at CFM, emphasizes price impact for sizable orders.

> "For sizeable trades, not only do you pay your fraction of the spread, but you move the price up when you buy. [...] There's a whole range of complexity in modeling transaction cost – in short, just saying that it's a quarter of the spread or half the spread is naive."

Even portfolio managers aware of price impact may be surprised that *the ideal order size* pays half its paper profits in price impact.

Point (b) is vital for brokers to protect their intellectual property: the trading team can provide their clients with a pre-trade cost model that reflects the team's view on liquidity without sharing the underlying models.[2]

[2] An execution team's ability to confidently share its liquidity view is a competitive advantage. For example, in the wall street lab interview of Almgren (2021)[221], Almgren highlights a service his company provides to clients:

> "the other thing we can do is we can identify market regimes. So, intraday, the market goes through various states of liquidity and volatility and we can provide that information."

Indeed, Lehalle and Laruelle (2018)[147] stress that communication around order execution is two-sided:

> "But the communication between the portfolio managers and "their" dealing desk has to go both ways: Managers sending metaorders on the one hand and traders sending back information about the state of the liquidity" (p. 245).

3.2.1.3 Implied alpha

A trader inverts the optimal trade size equation to define an order's *implied alpha*

$$\tilde{\alpha} = \Lambda_T Q_T.$$

Of course, other considerations besides alpha could affect the order size Q. For example, a portfolio manager may downsize an order based on risk considerations, despite having a robust overnight alpha signal. Therefore, the implied alpha only estimates the signal behind the portfolio team's trade to the first order. Finally, traders can complement the implied alpha with statistical models based on historical orders.

This chapter's implied alpha measure, specific to the generalized OW model, is a case of *alpha profiling.* Wikipedia [237] defines alpha profiling.

> "Alpha profiling models learn statistically-significant patterns in the execution of orders from a particular trading strategy or portfolio manager and leverages these patterns to associate an optimal execution schedule to new orders. In this sense, it is an application of statistical arbitrage to best execution."

Criscuolo and Waelbroeck (2012)[79] link alpha profiling to price impact.

> "Alpha profiling can be used to classify trade urgency. This enables one to select an execution schedule that minimizes total cost, counting both impact and alpha decay" (p. 56).

3.2.1.4 The square-root law

The reader extends Proposition 3.2.2 for sizable orders by replacing the OW model with the globally concave AFS model introduced in Section 2.6.4 of Chapter 2. The derivations are involved and covered by exercises 6 and 7. For the standard square-root model

$$h(x) \propto \operatorname{sign}(x)\sqrt{|x|},$$

the idealized execution problem changes.

(a) The portfolio team pays two-thirds of their paper P&L in impact costs.

(b) The total transaction costs are proportional to $|Q|^{3/2}$.

(c) The order's implied alpha is

$$\tilde{\alpha} \propto \text{sign}(Q)\sqrt{|Q|}.$$

3.2.2 Including alpha signals in the execution strategy

> "[T]raders and trading algorithms also strive for using short term price predictors in their dynamic order execution schedules. [...] Consequently, one of the main challenges in the area of optimal trading with price impact deals with the question of how to incorporate short term predictive signals into a stochastic control framework" (Neuman and Voß (2022, p. 2)[177]).

In practice, trading teams add *alpha signals* to their algorithms to refine the order's execution. For example, Almgren's presentation *Real Time Trading Signals* (2018)[9] lists three types of alpha models.[3]

- "Prices overshoot and relax
- Related assets tend to move together
- Presence of imbalance in the market" (p. 11)

3.2.2.1 Alpha latency

Practitioners incorporate alpha signals into live trading algorithms depending on their latency requirements.

(a) When possible, traders include signals directly in the core execution problem. Proposition 3.2.3 solves the execution schedule with an alpha signal. It is a particular case of the general result by Neuman and Voß (2022)[177], which includes instantaneous quadratic costs. Bacidore (2020)[12] informally characterizes signals to include in the core execution schedule:

> "where the alpha is relatively small over the life of each child order, but meaningful over the life of the parent, the alpha could be incorporated in the core algorithm" (p. 144).

Practitioners refer to short-term alpha signals, modeled by α and distinct from $\tilde{\alpha}$, as *trading alphas*.

[3] Sciulli (2021)[208] mentions the same list in his interview with Man Group CIA Sandy Ratray:

> "there are for sure three types of models most short term traders will try to capture [...] One is definitely reversion [...] The other one, it revolves around information around the order book [...] And then finally, the other one is also cross asset information."

(b) High-frequency signals require a decentralized approach to take advantage of low-latency infrastructure, such as co-located servers. However, strategic execution schedules are, by nature, centralized algorithms. Therefore, traders implement low-latency signals separately and incorporate them differently into the core execution problem. Section 3.2.3 provides details and examples regarding signals with latency requirements. Isichenko (2021)[121] describes such high-frequency signals as "micro-alphas" (p. 221).

> "Given the extreme time sensitivity of micro alpha [...], a competitive advantage is gained by using low-latency, or high-frequency, strategies and co-located access infrastructure" (p. 222).

Furthermore, Section 3.2.3 measures the value of tactical deviations from the strategic schedule, such as those initiated by decentralized micro-alphas.

3.2.2.2 Reactive execution schedule

Proposition 3.2.3 extends the optimal execution problem to react to dynamic liquidity and alpha signals. Given any β, λ, and α, traders can implement the optimal trading strategy without using a numerical optimizer, leading to a fast, transparent, and reproducible implementation. Section 3.2.5 implements an example.

Proposition 3.2.3 (Solution with trading alpha). *In Proposition 3.2.2, replace the assumption that S is a martingale with the deterministic trading alpha assumption from Corollary 2.4.1 in Chapter 2. Then the optimal execution strategy satisfies the following optimality equations.*

(a) The target impact state for the optimal execution problem is

$$\forall t \in (0,T), \quad I_t^* = \frac{\beta_t + \gamma_t'}{2\beta_t + \gamma_t'}(\tilde{\alpha} + \alpha_t) - \frac{1}{2\beta_t + \gamma_t'}\alpha_t'; \quad I_T^* = \tilde{\alpha}.$$

(b) The corresponding order size is

$$Q_T = \frac{\tilde{\alpha}}{\Lambda_T} + Q_T(\alpha)$$

where

$$Q_T(\alpha) = \int_0^T \frac{\beta_t + \gamma_t'}{\lambda_t(2\beta_t + \gamma_t')}\left((\beta_t + \gamma_t')\,\alpha_t - \alpha_t'\right) dt$$

is the contribution of the trading alpha α_t to the order size.

Proof. One modifies Corollary 2.5.9 from Chapter 2 to include the overnight alpha $\tilde{\alpha}$, leading to the target impact state I^*. The equation for the terminal position is

$$Q_T = \frac{1}{\lambda_T} I_T^* + \int_0^T \frac{\beta_t + \gamma_t'}{\lambda_t} I_t^* \, dt.$$

It is linear, and the trading alpha's contribution to the order size is

$$\begin{aligned} Q_T(\alpha) &= \int_0^T \frac{\beta_t + \gamma_t'}{\lambda_t} \left(\frac{\beta_t + \gamma_t'}{2\beta_t + \gamma_t'} \alpha_t - \frac{1}{2\beta_t + \gamma_t'} \alpha_t' \right) dt \\ &= \int_0^T \frac{\beta_t + \gamma_t'}{\lambda_t (2\beta_t + \gamma_t')} \left((\beta_t + \gamma_t') \, \alpha_t - \alpha_t' \right) dt. \end{aligned}$$

□

As per Remark 2.4.3 of Chapter 2, the optimal trading strategy is myopic in the impact state and incorporates price predictions through the intraday alpha's *level* α_t and *decay* α_t'. Traders also refer to an alpha's decay as its trajectory. For instance, Bacidore (2020)[12] describes how execution algorithms react to intraday alpha signals.

> "How aggressively should the trader trade? This depends on both the *level* and the *trajectory* of the alpha. [...] [If] the alpha is not expected to occur until later in the day, the trader does not need to rush execution to capture the alpha" (p. 45).

3.2.2.3 How trading alphas affect order size

The execution problem solved in Propositions 3.2.2 and 3.2.3 presents one practical hurdle: the terminal order size is a random variable. The uncertain order size affects real-life applications:

(a) Proposition 3.2.2 establishes a daily pre-trade cost model for the portfolio team. The trading team can offer a model that leads to deterministic order sizes in two ways:

- by using deterministic, time-varying liquidity models β, λ.
- by providing the average daily liquidity parameter $\mathbb{E}\left[\Lambda_T^{-1}\right]$. The portfolio team estimates their order size and expected transaction costs using the formulas

$$\mathbb{E}\left[Q_T\right] = \tilde{\alpha} \mathbb{E}\left[\Lambda_T^{-1}\right]$$

and

$$\begin{aligned} \mathbb{E}\left[\int_0^T I_t dQ_t + \frac{1}{2}[I, Q]_T\right] &= \frac{1}{2} \mathbb{E}\left[\Lambda_T^{-1}\right] \tilde{\alpha}^2 \\ &= \frac{1}{2\mathbb{E}\left[\Lambda_T^{-1}\right]} \mathbb{E}\left[Q_T\right]^2 \end{aligned}$$

from Proposition 3.2.2.

(b) The case of Proposition 3.2.3 depends heavily on the trading team's discretion in over- or under-filling the order. Bacidore (2020)[12] describes the discretion problem in Section 8.3 "Practical considerations" (p. 111).

> "Note that trades can be submitted as 'must complete' orders, meaning the algorithm must aim to complete the entire basket [...]. Others are submitted on a 'best efforts' basis, which means that, ideally, the trader would like the algorithm to complete the trade. But leaving residual quantities is permitted" (p. 111).

Remark 3.2.4 covers trading discretion in detail.

Remark 3.2.4 (Two ways to calibrate implied alpha). *The trading team calibrates an order's implied alpha in two ways, depending on whether the trading team executes the order on a best-effort or must-complete basis.*

(a) On a best-effort basis, the trader uses the implied alpha $\tilde{\alpha} = \Lambda_T Q$ in the control problem from Proposition 3.2.3. $Q_T(\alpha)$ quantifies the over- or under-filling the trading alpha α causes.

(b) On a must-complete basis, a trader implies a new *alpha to guarantee that $Q_T = Q$. The equation*

$$Q_T = \frac{\tilde{\alpha}}{\Lambda_T} + Q_T(\alpha)$$

yields the modified implied alpha

$$\tilde{\alpha} = \Lambda_T \left(Q - Q_T(\alpha)\right)$$

for deterministic liquidity parameters. For stochastic liquidity parameters, one goes back to first principles and derives the Lagrange multiplier for the terminal constraint.

3.2.3 Allowing for tactical deviations at the microstructure level

> "Opportunistic algorithms are designed to take advantage of favorable market conditions. Liquidity-driven algorithms designed to search for liquidity across several market venues play a key role in this process" (Cesari, Marzo, and Zagaglia (2012, p. 11)[67]).

Traders use venue-specific data, such as dark pool fills or limit order book events, to predict price changes on microseconds and shorter timescales. For example, Kolm, Turiel, and Westray (2021)[131] "achieve state-of-the-art predictive accuracy by training simpler 'off-the-shelf' artificial neural networks on stationary inputs derived from the order book" (p. 1).[4]

---

[4]Two papers provide further instances of microstructure signals. First, Zovko (2017)[243] study trading in dark pools.

Traders implement brief, venue-specific alpha signals, called microstructure alphas, outside the strategic execution schedule. The signal's short life forces traders to act immediately on the venue, and latency rules out direct coordination with the central execution. Practitioners refer to microstructure alphas implemented outside the centralized execution as *opportunistic trading algorithms*. An opportunistic algorithm outputs a *tactical deviation* from the strategic schedule. Section 1.2.4 of Chapter 1 introduces the distinction between intended and realized trades. In the context of optimal execution, the strategic schedule reflects the *intended* and the tactical deviations the *realized* trades of the team.[5]

Practitioners monitor tactical deviations, measure their contribution to a trading strategy's P&L, and propose self-correction algorithms to maintain a well-defined optimal execution problem.

> "When new information invalidates the hypotheses behind the original plan, it is essential to monitor the market response and re-estimate the optimal strategy to maintain optimality in the course of execution. The entire framework is an application of adaptive prediction and predictive control to the trade-execution problem" (Criscuolo and Waelbroeck (2012, p. 48)[79]).

> "**Systematic Deviation:** [...] If a prediction of short term price move is possible, it can leverage that information by making a systematic decision to get ahead or stay behind the schedule and take advantage of that price information. This has to be done carefully to avoid incurring any additional market impact or risk of catching up that will cancel out any possible benefit of the move. The schedule also needs to have a component that will move the strategy back on schedule" (Velu, Hardy, and Nehren (2020, p. 351)[227]).

> "The method we describe in this paper colours the dark liquidity by dynamically detecting evidence of causality between dark and lit fills and provides reactive information for a trader to adjust their trading" (p. 7).

Second, Baldacci and Manziuk (2022)[16] use a Bayesian approach to dynamically react to changing conditions on trading venues and decide which venue to execute a fill on.

> "If one assumes constant parameters over the trading period, he believes in the quality of his parameters' estimation. In this case, his strategy is not robust to changes in price dynamics or the platforms' behavior. For example, the trading period may occur when another market participant is executing a buy (or sell) metaorder on one or several venues" (p. 2).

[5] In the *wall street lab interview of Almgren* (2021)[221], Almgren highlights the need to capture tactical liquidity outside the execution problem.

> "You basically need to go in, so you can write your differential equations, that's fine. But when you actually go to trade, you need to be much more opportunistic. So, the way this came up is, we looked at a lot of executions, and let's say, you draw your schedule of how much you'd like to execute each hour through the day. Okay, that's your mathematical optimum, and then in this second hour, the market offers you a chance to do a large amount of the trade. But let's say there's suddenly a lot of liquidity. So, I would say you should take it. Why would you not take it?"

Devising systems that track trading schedules and self-correct when realized trades diverge is a notable sub-field of algorithmic trading.[6] For example, Bulthuis et al. (2017)[45] consider a "model to incorporate a benchmark trading schedule, and show that our model is capable of generating optimal strategies that very closely follow any given schedule while minimizing trading cost. This allows us to understand a tradeoff between following schedule and seeking profits, and thus, evaluate how profitable it may be to deviate from schedule" (p. 3).

3.2.3.1 Quantifying deviation's impact

Under the generalized OW model, price impact measures a tactical deviation's effect. Definition 3.2.5 quantifies the difference in prices the trading deviation causes, and Proposition 3.2.6 quantifies its P&L.[7]

Definition 3.2.5 (Trade and impact deviation)**.** *If $Q \in \mathcal{S}$ is the intended trading schedule and $Q^r \in \mathcal{S}$ the realized trading process, then define the* trade deviation *as*

$$\delta Q = Q^r - Q.$$

Define the price impact deviation *as*

$$\delta I = I\left(Q^r\right) - I\left(Q\right).$$

Proposition 3.2.6 (Impact deviation in the generalized OW model)**.** *Assume given*

1. *a generalized OW model with liquidity parameters $\beta, \lambda = e^{\gamma}$ with $\beta, \gamma' \in \mathcal{S}$ satisfying the no price-manipulation condition from Corollary 2.5.8.*

[6]The most frequent tracking logic involves *trading bands*, also called *no-trade zones.* Guilbaud, Mnif, and Pham (2013)[112] identify the "shape of the policy" (p. 17):

> "We can see three zones: a buy zone (denoted BUY on the graph), a sell zone (denoted SELL on the graph) and a no trade zone (denoted NT on the graph)" (p. 17).

Lataillade and Chaouki (2020)[145] extend the result to include an alpha signal, which they call a "predictor" (p. 1).

> "we show why the shape of its optimal solution is necessarily a band [...] [and] derive path-integral equations for the upper and lower edges of the band. We then restrict ourselves to the case of a predictor following an Ornstein-Uhlenbeck dynamics and obtain explicit solutions in this case, for which we can derive the asymptotic behavior as a function of the predictor's value" (p. 2).

[7]Waelbroeck et al. (2012)[231] introduce the related concept of "impact anomaly" in the context of "real-time order flow analysis" (p. 4).

> "Furthermore, the system constantly looks at differences between the results predicted by the strategy and the actual results. For example, impact anomaly is the difference between actual return and expected return given a pre-trade model" (p. 4).

2. the solution Q^* to the optimal execution problem from Proposition 3.2.3.

3. a realized trade $Q^r \in \mathcal{S}$, or equivalently, a trade deviation $\delta Q = Q^r - Q^*$.

Then the price impact deviation satisfies the SDE

$$d\delta I_t = -\beta_t \delta I_t dt + \lambda_t d\delta Q_t$$

with $\delta I_0 = 0$.

Furthermore, the expected fundamental P&L of the realized trades satisfies

$$\mathbb{E}\left[Y_T(Q^r)\right] - \mathbb{E}\left[Y_T(Q^*)\right] = -\mathbb{E}\left[\int_0^T \frac{2\beta_t + \gamma_t'}{2\lambda_t}\left(\delta I_t\right)^2 dt + \frac{1}{2\lambda_T}\left(\delta I_T\right)^2\right].$$

Proof. The impact deviation SDE stems from the price impact SDE's linearity. Corollary 2.5.9 from Chapter 2 derives the expected fundamental P&L of the deviation. □

Traders compare the expected opportunity costs derived in Proposition 3.2.6 to tactical deviations' expected profits to determine their merit. Exercise 8 presents an example. Under the generalized OW model, the self-correction behavior is myopic in impact space: the centralized algorithm reacts to an impact deviation by trading back toward the target impact state.

3.2.3.2 Block trades and auctions

A particular deviation stems from the optimal execution's border conditions: the generalized OW model, due to its myopia, leads to jumps at times $t = 0, T$. In general, algorithms implement jump trades outside a limit order book, for example, using an auction or a dark pool. Therefore, traders should view the optimal strategy as an idealized target and implement deviations close to the border conditions. A trading algorithm achieves the target for t far from the border conditions but requires a pragmatic solution around $t = 0, T$. Two practical approaches follow.

(a) Block trades or trading auctions implement the jumps. Trading auctions are appealing for day orders: $t = 0$ corresponds to the open auction and $t = T$ to the closing auction of the trading day.

(b) Instantaneous quadratic costs smooth out the trading trajectory. Gârleanu and Pedersen (2016)[102] study optimal trading under the OW model with instantaneous quadratic costs. Chen, Horst, and Tran (2019)[70] explicitly solve the generalized OW model with instantaneous quadratic costs, fit the model and implement a simulation of the optimal trading strategy on real-life data.

3.2.4 Changing the execution strategy when new orders arrive

> "[O]ptimizing the separate costs can lead to a significant understatement of the real cost of trading whilst also adversely impacting order scheduling" (Bordigoni et al. (2021, p. 1)[29]).

When the portfolio team submits a second order, the trading team answers two questions:

(a) How does the first order's trading path influence the second order's execution?

Practitioners must answer this question *even if the first order finished trading*, as the first order impacted the prices on which the second order trades.

> "[C]ommon approaches suffer from a type of myopia: impact is only measured for the current transaction. In many cases, orders are correlated and the impact of the first order will affect the execution of future orders" (Harvey et al. (2022, p. 1)[116]).

(b) How to optimally trade the combined order?

Combining two orders may seem straightforward when they share trading parameters but becomes challenging when they have competing parameters. For example, how should one execute a small, fast order when the team is also working on a large, slow order?

When comparing and combining two orders' execution, the general answer expresses all trading into impact space. Such a normalization allows traders to compare combined alphas and impact states. This section illustrates the approach with three concrete instances.

Example 3.2.7 (Trading a follow-up order). *The case of an order submitted after a previous order completes is straightforward in impact space: the optimal trading strategy is myopic under the generalized OW model. Bordigoni et al. (2021)[29] solve the two-order case for a more general class of propagator models using methods from* optimal transport, *a rich mathematical field.*

Consider a trading process Q modeling the execution of an order over $[T_1, T_2]$ with implied alpha $\tilde{\alpha}$. Assume that a previous order with implied alpha $\tilde{\alpha}^0$ finished trading at $T_0 < T_1$ and denote by Q^0 its trading process. Under the generalized OW model, $I_{T_1} \neq 0$ captures the previous order's price impact. Mathematically, one establishes the impact caused by the earlier trading process Q^0 and the new process Q:

$$\forall t \in (T_1, T_2), \quad I_t^* \left(Q^0 + Q\right) = \frac{\beta_t + \gamma_t'}{2\beta_t + \gamma_t'} \left(\tilde{\alpha} + \alpha_t\right) - \frac{1}{2\beta_t + \gamma_t'} \alpha_t'.$$

The optimality equation for the new order's target impact satisfies

$$\begin{aligned} I_t^*(Q) &= \frac{\beta_t + \gamma_t'}{2\beta_t + \gamma_t'}(\tilde{\alpha} + \alpha_t) - \frac{1}{2\beta_t + \gamma_t'}\alpha_t' - I_t(Q^0) \\ &= \frac{\beta_t + \gamma_t'}{2\beta_t + \gamma_t'}(\tilde{\alpha} + \alpha_t) - \frac{1}{2\beta_t + \gamma_t'}\alpha_t' - e^{-\int_{T_0}^t \beta_s ds}\tilde{\alpha}^0. \end{aligned}$$

Example 3.2.8 (Combining simple overlapping orders). *If the same team sends two overlapping orders, the second order's implied alpha $\tilde{\alpha}$ acts as an independent update to the first implied alpha $\tilde{\alpha}^0$. Therefore, the optimality condition is*

$$I_t^* = \frac{\beta_t + \gamma_t'}{2\beta_t + \gamma_t'}(\tilde{\alpha} + \tilde{\alpha}^0 + \alpha_t) - \frac{1}{2\beta_t + \gamma_t'}\alpha_t'.$$

A trader adds up the two implied alphas. However, one should not *double-count the trading alpha α_t. Indeed, two orders' execution speeds are not additive in the presence of trading alpha: both orders cannot simultaneously leverage the same signal for execution.*

Remark 3.2.9 (Combining correlated overlapping orders). *The situation where two different teams submit overlapping orders is intricate. Even if the trader correctly implies the alpha from each source, the implied alphas are not additive: the signals are unlikely to be independent.*

Waelbroeck et al. (2012)[231] outline two reasons portfolio teams submit correlated orders: their orders could be rooted in a behavioral trend or based on shared information.

> *"[D]ifferent portfolio managers are reasonably likely to think coherently, either because a particular investment style is currently in fashion, or because their reasoning is influenced by analysts that publish reports" (p. 57).*

If both teams use identical signals to submit trades, the optimal strategy is to trade only one order to avoid double counting the shared alpha. In the general case, using statistical learning, Criscuolo and Waelbroeck's (2012)[79] alpha profiling combines the two implied alphas' predictive power.

The two Examples 3.2.7 and 3.2.8 and Remark 3.2.9 use price impact and alpha to summarize two orders' execution. The optimal strategy targets a tractable, myopic relationship between impact and alpha: once the trader measures the two orders' impact and alpha, they disregard any further interactions.

3.2.5 A simulation example

This section empirically simulates the Handbook's optimal execution framework, tying together results from other chapters. Section 4.2.3 of Chapter 4 describes the simulated trading environment proposed by Waelbroeck et al.

(2012)[231]. Section 7.4 of Chapter 7 empirically fits and compares price impact models from the literature on the S&P 500 stocks.

The price impact model used in this section is a reduced-form limit of a locally concave OW model

$$dI_t = -\beta I_t dt + \frac{\lambda\sigma}{\sqrt{\mathrm{ADV}\cdot v_t}} dQ_t$$

where σ and ADV are the stock's volatility and average daily volume. v_t is an estimator of the stock's instantaneous trading activity.

Remark 3.2.10 (A useful heuristic). *An essential empirical observation for the chosen model is that $\gamma_t' \ll \beta$. Refer to this approximation as* the heuristic. *Under the heuristic, the following formulas simplify:*

$$\frac{1}{\Lambda_T} \approx \frac{1}{\lambda_T} + \int_0^T \frac{\beta_t}{2\lambda_t} dt$$

and

$$\forall t \in (0,T), \quad I_t^* \approx \frac{1}{2}\left(\tilde{\alpha} + \alpha_t - \beta_t^{-1}\alpha_t'\right); \quad I_T^* = \tilde{\alpha}.$$

This section implements trading strategies for exact formulas and approximations under the heuristic. The heuristic $\gamma_t' \ll \beta$ does not negate the importance of a stochastic liquidity model λ_t: the heuristic allows for a simplified optimality condition for I^* but still requires λ_t to recover the trading process Q^*.

The simulation emulates a trader executing daily orders. In three steps, the simulation

(a) generates a daily order for each stock. The trader quotes the order size as an *implied alpha* $\tilde{\alpha}$ of 10bps, 50bps, and 100bps, with a default of 50bps.

(b) generates a synthetic trading alpha for use in the execution. Exercise 12 in Chapter 4 describes the generation of synthetic alphas with a given correlation to realized returns. The correlation ρ of the intraday alpha with future returns is 1%, 5%, and 10%, with a default of 5%.

(c) compares four execution strategies:

- a TWAP algorithm, trading at a constant speed throughout the day. TWAP is the optimal risk-neutral strategy under the Almgren and Chriss (2001)[10] model.
- the optimal execution under the *generalized* OW model *without intraday alpha.*
- the optimal execution under the *generalized* OW model *with intraday alpha.*
- the *heuristic* of the optimal execution strategy *with intraday alpha.*

TABLE 3.1
Average performance metrics across implied alphas $\tilde{\alpha}$.

$\tilde{\alpha}$	strategy	size (% of ADV)	impact (bps)	slippage (bps)
50bps	TWAP	14	37	37
50bps	no alpha	15	27	26
50bps	with alpha; $\rho = 1\%$	15	27	25
50bps	with alpha; $\rho = 5\%$	15	27	24
50bps	heuristic; $\rho = 5\%$	15	27	24
50bps	with alpha; $\rho = 10\%$	15	27	23
10bps	TWAP	3	7.5	7.0
10bps	no alpha	3	5.3	4.4
10bps	with alpha; $\rho = 5\%$	3	5.6	-0.1
10bps	heuristic; $\rho = 5\%$	3	5.6	1.4
100bps	TWAP	27	75	74
100bps	no alpha	30	53	52
100bps	with alpha; $\rho = 5\%$	30	52	52
100bps	heuristic; $\rho = 5\%$	30	52	52

Table 3.1 summarizes simulation metrics, repeated over a year on the S&P 500 universe to provide 125000 simulated executions. The historical S&P 500 data provides realized price and market volume paths. One also calibrates the impact model using historical data and simulates perturbed prices. Figure 3.1 plots the distribution of the slippage due to price impact across various strategies for the default case where the order's implied alpha $\tilde{\alpha}$ is 50bps. Five broad observations follow.

(a) The TWAP execution pays three-quarters of the implied alpha in impact costs.

(b) The optimized executions pay half of the implied alpha in impact costs.

(c) The contribution of the intraday alpha is notable for tiny implied alphas. For example, for an implied alpha of 10bps, roughly twice the bid-ask spread, the strategy without alpha pays 4bps of price impact while the intraday alpha contributes 4bps of savings. An implied alpha of 10bps corresponds to a 3% ADV order.

(d) The trader may suggest executing the 30% ADV order over multiple days. In this scenario, a globally concave price impact model, such as the AFS model, is suited, and intraday signals matter less to the execution.

(e) Figure 3.1 shows, for moderate implied alphas, the effect of liquidity and intraday alphas on the distribution of target impact states of the optimal execution strategy. Price impact measures trading speed in return units. Dynamic signals lead to fluctuations in speeds equivalent to 5bps

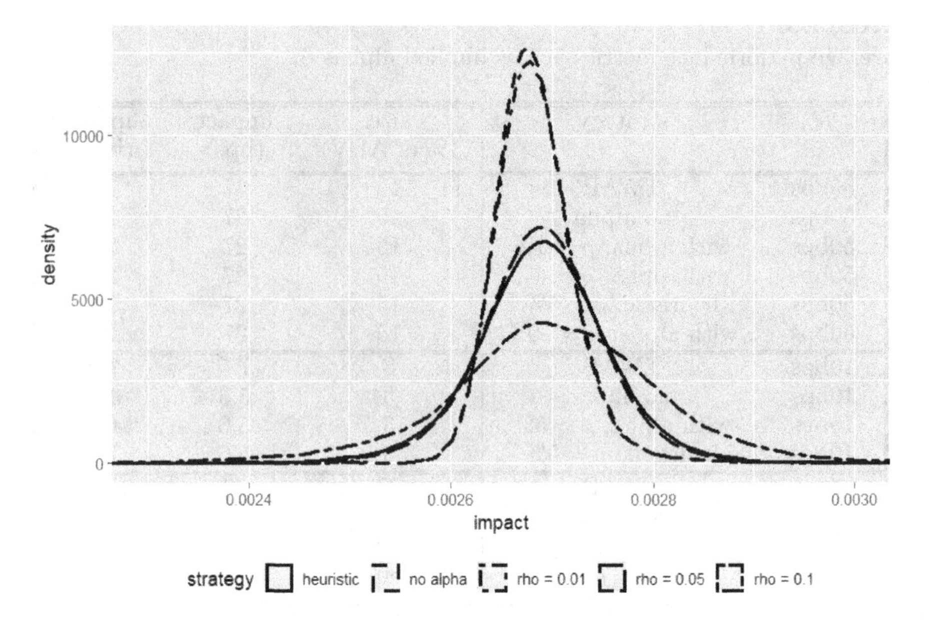

FIGURE 3.1
Distribution of slippage due to price impact for an order with an implied alpha $\tilde{\alpha}$ of 50bps.

of impact. These strategic fluctuations in trading rates reduce the order's total slippage by 4bps.

3.2.6 Summary

Optimal execution has four steps and is rooted in the language of price impact and alpha. Each phase relates impact and alpha through closed-form formulas and allows for non-parametric methods, such as machine learning models, for all execution inputs. These inputs include intraday alpha signals α_t and volume curves v_t. In addition, the formulas simplify further under the heuristic $\gamma' \ll \beta$. Optimal execution's four steps are:

(a) *Communicate a pre-trade cost model to the portfolio team.*

Under the generalized OW model, transaction costs equal half an order's expected alpha. The implied alpha relationship

$$\mathbb{E}\left[Q_T\right] = \mathbb{E}\left[\frac{\tilde{\alpha}}{\Lambda_T}\right]$$

provides the optimal order size given an overnight alpha $\tilde{\alpha}$ and a daily liquidity parameter

$$\frac{1}{\Lambda_T} = \frac{1}{\lambda_T} + \int_0^T \frac{(\beta_t + \gamma_t')^2}{\lambda_t(2\beta_t + \gamma_t')} dt.$$

The formula's heuristic version is

$$\frac{1}{\Lambda_T} \approx \frac{1}{\lambda_T} + \int_0^T \frac{\beta_t}{2\lambda_t} dt.$$

(b) *Establish a strategic schedule based on alpha signals and liquidity models.*

The most straightforward strategic schedules are TWAP, VWAP, or Participation of Volume (PoV) algorithms: the former track a price, the latter a volume schedule. Exercises 6 and 7 outline examples.

Under the generalized OW model, the strategic schedule tracks a target impact state I^* that considers non-parametric alpha signals and liquidity models via the optimality condition

$$\forall t \in (0,T), \quad I_t^* = \frac{\beta_t + \gamma_t'}{2\beta_t + \gamma_t'} (\tilde{\alpha} + \alpha_t) - \frac{1}{2\beta_t + \gamma_t'} \alpha_t'; \quad I_T^* = \tilde{\alpha}.$$

The formula's heuristic version is

$$\forall t \in (0,T), \quad I_t^* \approx \frac{1}{2} \left(\tilde{\alpha} + \alpha_t - \beta_t^{-1} \alpha_t'\right); \quad I_T^* = \tilde{\alpha}.$$

(c) *Implement microstructure tactics as deviations from the strategic schedule.*

Microstructure opportunities, such as venue-specific signals, require tactical deviations from the strategic schedule. Therefore, traders quantify the opportunity cost of deviating from the schedule and compare those costs to the deviations' benefits.

Under the generalized OW model, the expected cost of a deviation δI from the target impact state I^* is

$$E\left[\int_0^T \frac{2\beta_t + \gamma_t'}{2\lambda_t} (\delta I_t)^2 \, dt + \frac{1}{2\lambda_T} (\delta I_T)^2\right].$$

The formula's heuristic version is

$$E\left[\int_0^T \frac{\beta_t}{\lambda_t} (\delta I_t)^2 \, dt + \frac{1}{2\lambda_T} (\delta I_T)^2\right].$$

(d) *Adapt the schedule when a new order arrives.*

Orders rarely trade in isolation, and execution strategies should consider the price impact of previous or simultaneous trading.

Under the generalized OW model, the target relationship between alpha and impact remains unchanged upon the new order's arrival:

$$\forall t \in (0,T), \quad I_t^* = \frac{\beta_t + \gamma_t'}{2\beta_t + \gamma_t'}(\tilde{\alpha} + \alpha_t) - \frac{1}{2\beta_t + \gamma_t'}\alpha_t'; \quad I_T^* = \tilde{\alpha}.$$

A shift in I and α measures the new order's effect. The same statement applies to the heuristic.

3.3 Transaction Cost Analysis (TCA)

"The examples presented above are intended to help illustrate some ways that Deutsche Bank looks to extend post-trade analysis and the evaluation of strategy customizations. They are part of a larger collection of methods used to analyze specific client flows and execution objectives. It is important to acknowledge that when we run these analyses, it's done only for a subset of orders that are affected by a controlled A/B experiment with routine changes, so we can really quantify their impact." (*Extended Transaction Cost Analysis (TCA)* by Sotiropoulos and Battle (2017, p. 5)[215])

Section 1.3.1 in Chapter 1 introduces TCA's role in modern sell- and buy-side trading teams. A mock TCA report provides examples of metrics and actions to expect from the analysis. The section also relates TCA to MiFID II and the regulatory push for best execution and increased transparency of trading decisions. This section complements best execution regulation with a trading research perspective.

Figure 3.2 illustrates TCA's crucial role in trading algorithms' research and development cycle. Components of trading algorithms, such as alpha and liquidity signals, go through frequent improvement cycles. Upgrading a model involves research and back tests. Once the trading team is confident in the new model's value, the team deploys the signal in production, and a controlled live trading experiment measures its performance. Finally, a TCA report provides a comprehensive assessment of the model upgrades. TCA is vital to the research and development cycle, as the report *reconciles* the new model's predicted and live behavior. Based on the TCA report, quantitative researchers revisit the model, and the improvement cycle continues.

Practitioners apply TCA to evaluate all components of modern trading algorithms, including brokers, scheduling strategies, opportunistic algorithms,

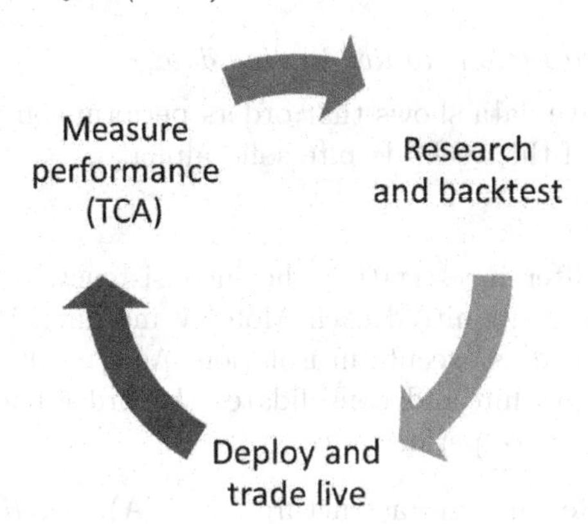

FIGURE 3.2
High-level schema for trading algorithms' research and development cycle.

and trading latency.[8] For example, this section evaluates scheduling strategies and diagnoses problems and biases within the trading infrastructure. Trading biases play a glaring role in TCA experiments. Chapter 6 describes trading biases statistically and leverages the mathematics of causal inference introduced in Chapter 5.

3.3.1 TCA best practices

Traders use TCA to *organize, analyze, and present trading experiments.* A practical mindset for a TCA practitioner emulates that of an experimental researcher: probe the trading system, question inconsistencies, and propose actions to fix the underlying causes. This analogy with the scientific method suggests four TCA steps:

(a) *Propose assumptions.*

For example, a trader assumes that the system submits orders one at a time, that executions are isolated, and that higher alpha drives performance.

(b) *Submit experiments and predictions on their outcome.*

For example, a simulation experiment predicts that orders perform equally well throughout the week.

[8] For instance, Markov (2019)[162] applies TCA to evaluate brokers using a Bayesian framework.

> "We present a formulation of transaction cost analysis (TCA) in the Bayesian framework for the primary purpose of comparing broker algorithms using standardized benchmarks" (p. 1).

(c) *Confront the predictions to live trading data.*

For example, live data shows that orders perform conspicuously worse at the beginning of the week, despite solid alpha.

(d) *Action.*

For example, after investigating the inconsistency, a trader discovers a backlog of orders submitted each Monday morning, invalidating the assumption that orders execute in isolation. As a result, the team modifies the trading algorithm and consolidates the order backlog into a single execution to enhance performance.

Remark 3.3.1 (Research Infrastructure for TCA). *Trading experiments, as outlined above, are highly iterative and call for reproducibility across iterations to maintain trust in the conclusions. Therefore, trading teams require a robust research infrastructure to generate TCA reports promptly and on demand.*

Section A.4 in Appendix A provides an implementation of TCA in a kdb+ database. Appendix A also contains references with further details on setting up a research and trading infrastructure for TCA.

3.3.1.1 Control for basic trading parameters

Two best practices emphasize the importance of the basic trading parameters listed in Section 1.2.5 of Chapter 1 in TCA. Both consider trading parameters when analyzing TCA results and ideally use a causal graph as per Chapter 5 to avoid under- or over-controlling for variables.

(a) Statistical models should control for relevant trading parameters and stock characteristics. Markov (2019)[162] provides examples of control variables for TCA on stocks.

> "Transaction cost analysis (TCA) studies the relation between execution costs (price impact model) and a set of stock and order characteristics such as average daily trading volume, order size, market cap, volatility, spread, participation rate, momentum, trading strategy and presence of limit prices" (p. 1).

(b) Live trading experiments should control for relevant trading parameters and stock characteristics. Albuquerque, Song, and Yao (2020)[5] provide an in-depth analysis of the SEC's tick size experiment, one of the most extensive A-B experiments to date, and highlight this methodology.

> "[T]he SEC's stratified random sampling procedure creates a laboratory-like experiment in an actual financial market that eliminates selection issues. The procedure provides a control group of stocks built as part of the random assignment of securities to the pilot program, thus removing any discretion from the econometrician" (p. 701).

"An important feature of the pilot program is the stratified random sampling procedure used to determine the stocks to be allocated to each group. The stratification is over three variables-share price, market capitalization and trading volume-and yields 27 possible categories (e.g., low price, medium market capitalization, and high volume). The pilot securities were randomly selected from the 27 categories to form the three test groups and the control group" (p. 703).

3.3.1.2 TCA predictions

Three mathematical results from Chapter 2 yield testable predictions under the generalized OW model.

(a) *Return decomposition.*

The observed returns decompose into alpha, the trader's impact, and the external impact:

$$p_T - p_t = \underbrace{(S_T - S_t)}_{alpha} + \underbrace{(I_T - I_t)}_{impact} \underbrace{- (\bar{\alpha}_T - \bar{\alpha}_t)}_{external\ impact} . \tag{3.2}$$

Proposition 2.6.4 in Chapter 2 shows traders can view external impact as an alpha signal with a sign flip. For simplicity, this section does not explicitly model external impact.

(b) *Transaction costs under the optimal strategy.*

Without intraday alpha, the optimal trading strategy pays half the trade's expected profits in impact costs

$$\mathbb{E}\left[\int_0^T I_t dQ_t + \frac{1}{2}[I,Q]_T\right] = \frac{1}{2}\mathbb{E}\left[\tilde{\alpha}Q_T\right]. \tag{3.3}$$

(c) *Optimal impact heuristic*

When $\gamma_t' \ll \beta_t$, the optimal target impact state satisfies

$$I_t^* \approx \frac{1}{2}\left(\tilde{\alpha} + \alpha_t - \beta_t^{-1}\alpha_t'\right). \tag{3.4}$$

The following sections provide examples and templates for a systematic approach to TCA. Each instance links to one of the "Possible Measurement Biases" (p. 238) listed by Bouchaud et al. (2018)[37] and analyzes a scenario under the generalized OW model.

3.3.2 An experiment to size orders correctly

Portfolio teams rely on a pre-trade cost model to size orders. Section 3.2.1 applies the generalized OW model to pre-trade cost models and derives the implied alpha relationship between an order's size and its overnight alpha. In addition, this section describes an experiment to uncover statistical inconsistencies in orders' implied alphas.

Bouchaud et al. (2018)[37] identify a bias when estimating an order's price impact based on its size and name it *prediction bias.*

> "Traders with strong short-term price-prediction signals may choose to execute their metaorders particularly quickly, to make the most of their signal before it becomes stale or more widely-known. Therefore, the strength of a prediction signal may itself influence the subsequent impact path" (p. 238).

Assumption 3.3.2 describes an experiment where long-term alpha drives trading.

Assumption 3.3.2 (No trading alpha assumptions). *Assume orders for which the following three properties hold.*

(a) The order's alpha realizes after the order stops trading,

$$\alpha_t = 0.$$

(b) The generalized OW model for price impact holds.

(c) The order follows the optimal execution strategy.

In their experiments on consecutive orders, Harvey et al. (2022)[116] use assumption (a) to generate predictions:

> "we make the simplifying assumption that the trades contain no alpha, which we believe is a reasonable approximation in practice when the holding period is long with respect to the execution horizon. That is, if the trade is designed to capture alpha that will be realized over one month and the execution happens in one day, it is safe to assume that the alpha will not be realized in the execution day" (p. 15).

Prediction 3.3.3 (Predictions in the absence of intraday alpha). *Under the* no trading alpha assumptions, *the following predictions hold:*

- The arrival slippage equals half the order's long-term alpha.
- The target impact state has low volatility when $\gamma_t' \ll \beta_t$.

  Indeed, when $\gamma_t' \ll \beta_t$, assumptions (a), (b), and (c) imply that

$$I_t^* \approx \frac{1}{2}\tilde{\alpha}.$$

The two predictions are complementary: they constrain price impact's average and volatility over the order's execution.

Action 3.3.4 (Actions when alpha and impact do not match). *Practitioners act on a breakdown between alpha and impact. For example, in the mock report in Figure 3.3:*

(a) The portfolio team can increase order size in the high volatility group: their realized alpha is higher than their implied alpha. Therefore, a larger order would grow profits despite the pronounced price impact.

(b) The trading team can investigate the execution in the medium volatility group. Indeed, excessive price impact may indicate a faulty algorithm.

Order set	Grouping	Order slippage	Realized overnight alpha	Implied alpha	Price impact (avg)	Price impact (vol)
No trading alpha	Low volatility	25bps	50bps	50bps	25bps	5bps
	Medium volatility	30bps	60bps	60bps	40bps	15bps
	High Volatility	35bps	80bps	70bps	35b	

Impact is too high compared to alpha

Increase order size / implied alpha to match realized alpha

FIGURE 3.3
Mock TCA report for trades without trading alpha.

3.3.3 Clean-up costs for partial executions

"[C]onsider the situation of a buy order sent to a broker where the algorithm only trades when the price is below the arrival price. Of course, under such a strategy orders will not be fully executed. But those that are will see a negative slippage. Consequently, without applying the appropriate clean-up cost one will incorrectly infer that the trading algorithm significantly outperforms versus average costs" (Kolm and Westray (2021, p. 3)[132]).

Partial execution describes orders where the intended and realized sizes do not match, for example, when the order ends partially unfilled. This scenario arises when the trading team has discretion over the size, as described in Remark 3.2.4 in Section 3.2.2. Partial execution biases TCA if traders do not correctly account for the remainder. The bias stems from the uncertainty on the size, or equivalently the implied alpha, to compare the order with: should the trader compare intended or realized order sizes? Both lead to biases.

- Comparing the partial execution to completed orders of the same *intended* size *advantages* the partially executed order, as it mechanically has less price impact.
- Comparing the partial execution to completed orders of the same *realized* size *disadvantages* the partially executed order, as it exhibits more alpha slippage.
- Furthermore, using the realized size introduces a potential look-ahead bias.

Bouchaud et al. (2018)[37] mention "implementation bias" in their list of "Possible Measurement Biases" (p. 238) and provide the same example as Kolm and Westray (2021)[132]:

> "we have assumed that both the volume Q and execution horizon T are fixed before a metaorder's execution begins. In reality, however, some traders may adjust these values during execution, by conditioning on the price path. In these cases, understanding metaorder impact is much more difficult. For example, when examining a buy metaorder that is only executed if the price goes down, and abandoned if the price goes up, impact will be negative. This implementation bias is expected to be stronger for large volumes" (p. 238, [37]).

Practitioners compare orders based on their intended size and apply a clean-up cost model to account for partial execution. This section re-implements the clean-up cost method of the paper *What happened to the rest? A principled approach to clean-up costs in algorithmic trading* by Kolm and Westray (2021)[132] for the generalized OW model.[9]

Definition 3.3.5 (Clean-up cost of an order). *Consider an order traded over an interval* $[0, T]$ *with intended size* $Q^i_T \in \mathbb{R}$ *and realized size* $Q^r_T \in E_T$*. Denote by* $Q^r \in \mathcal{S}$ *the realized trading process over time* $[0, T]$*.*

Define a clean-up strategy Q *of the realized trading process* Q^r *as a hypothetical extension over time* $(T, \tau]$ *for a deterministic* $\tau > T$ *that finishes the intended order*

$$\forall t \in [0, T], \quad Q_t = Q^r_t; \quad Q_\tau = Q^i_T.$$

Assume given a price impact model I*. Define the clean-up strategy's associated clean-up cost as the expected costs over* $(T, \tau]$*:*

$$\mathbb{E}\left[\int_T^\tau I_t dQ_t + \frac{1}{2}\int_T^\tau d[I, Q]_t \middle| \mathcal{F}_T\right].$$

[9]Practitioners also refer to clean-up costs as *opportunity costs*. For example, Harvey et al. (2022)[116] define opportunity costs in the following way:

> "Execution costs relate to already executed trades while opportunity costs relate to unexecuted transactions, i.e., past trading opportunities that were unrealized. Faster execution generally increases execution costs and lowers opportunity costs" (p. 2).

The clean-up strategy is an optimal execution problem on the remainder, regardless of the original order's execution strategy. The clean-up horizon τ matters: practitioners may choose τ as a fraction of the original order's trading horizon T.

Proposition 3.3.6 (Implied clean-up cost). *Let $Q^r \in \mathcal{S}$ be a realized trading process with intended size $Q^i_T \neq Q^r_T$ and $\tau > T$ a clean-up horizon. Assume given*

(a) a generalized OW model I with liquidity parameters $\beta, \lambda = e^\gamma$ with $\beta, \gamma' \in \mathcal{S}$ satisfying the no price manipulation condition from Corollary 2.5.8.

(b) an unperturbed martingale price $S_t \in \mathcal{S}$ over $t \in [T, \tau]$.

(c) an initial impact state $I_T \in E_T$ for the clean-up strategy.

The clean-up costs equal

$$\frac{\Lambda}{2}\tilde{Q}^2 + \frac{\Lambda}{\lambda_T} I_T \tilde{Q} + \frac{1}{2\lambda_T}\left(\frac{\Lambda}{\lambda_T} - 1\right) I_T^2$$

where $\tilde{Q} = Q^i_T - Q^r_T$ is the remainder to clean up,

$$\frac{1}{\tilde{\Lambda}} = \frac{1}{\lambda_\tau} + \int_T^\tau \frac{(\beta_t + \gamma'_t)^2}{\lambda_t(2\beta_t + \gamma'_t)} dt$$

and

$$\Lambda = \mathbb{E}\left[\tilde{\Lambda} \middle| \mathcal{F}_T\right]$$

is the clean-up period's expected liquidity parameter.

Proof. Proposition 3.3.6 extends Proposition 3.2.2 to include a non-zero initial impact state for $t = T$. The optimal target impact state satisfies

$$\forall t \in (T, \tau), \quad I_t^* = \frac{\beta_t + \gamma'_t}{2\beta_t + \gamma'_t}\tilde{\alpha}; \quad I_T^* = \tilde{\alpha}$$

where one solves the implied alpha $\tilde{\alpha}$ such that $Q_\tau = Q^i_T$. The latter satisfies

$$\begin{aligned} Q_\tau - Q_T &= \frac{1}{\lambda_\tau} I_\tau^* - \frac{1}{\lambda_T} I_T + \int_T^\tau \frac{\beta_t + \gamma'_t}{\lambda_t} I_t^* dt \\ Q^i_T - Q^r_T &= \frac{1}{\tilde{\Lambda}}\tilde{\alpha} - \frac{1}{\lambda_T} I_T \end{aligned}$$

where

$$\frac{1}{\tilde{\Lambda}} = \frac{1}{\lambda_\tau} + \int_T^\tau \frac{(\beta_t + \gamma'_t)^2}{\lambda_t(2\beta_t + \gamma'_t)} dt.$$

The constraint on Q_τ implies

$$\tilde{\alpha} = \tilde{\Lambda}\left(Q^i_T - Q^r_T + \frac{1}{\lambda_T} I_T\right).$$

The expected cleanup costs C equal

$$
\begin{aligned}
C &= \mathbb{E}\left[\int_T^{\tau} \frac{2\beta_t + \gamma_t'}{2\lambda_t} \left(I_t^*\right)^2 dt + \frac{1}{2\lambda_\tau} \left(I_\tau^*\right)^2 - \frac{1}{2\lambda_T} I_T^2 \,\middle|\, \mathcal{F}_T\right] \\
&= \mathbb{E}\left[\frac{1}{2\tilde{\Lambda}} \tilde{\alpha}^2 - \frac{1}{2\lambda_T} I_T^2 \,\middle|\, \mathcal{F}_T\right] \\
&= \mathbb{E}\left[\frac{1}{2} \tilde{\Lambda} \left(Q_T^i - Q_T^r + \frac{1}{\lambda_T} I_T\right)^2 - \frac{1}{2\lambda_T} I_T^2 \,\middle|\, \mathcal{F}_T\right] \\
&= \mathbb{E}\left[\frac{1}{2} \tilde{\Lambda} \left(Q_T^i - Q_T^r\right)^2 + \frac{\tilde{\Lambda}}{\lambda_T} I_T \left(Q_T^i - Q_T^r\right) + \frac{1}{2\lambda_T} \left(\frac{\tilde{\Lambda}}{\lambda_T} - 1\right) I_T^2 \,\middle|\, \mathcal{F}_T\right].
\end{aligned}
$$

□

The implied cost from Proposition 3.3.6 provides a closed-form formula for a clean-up strategy under straightforward assumptions. The first term,

$$
\frac{\Lambda}{2} \tilde{Q}^2
$$

equals the expected impact costs from a pre-trade cost model applied to the order remainder. The second term,

$$
\frac{\Lambda}{\lambda_T} I_T \tilde{Q} + \frac{1}{2\lambda_T} \left(\frac{\Lambda}{\lambda_T} - 1\right) I_T^2
$$

tracks additional costs due to the remainder trading on a non-zero initial impact state. Moreover, Kolm and Westray (2021)[132] define and implement their methodology for a propagator model with a general decay function and alpha signal.

From a TCA perspective, traders use clean-up costs to penalize remainders and reduce implementation bias. For example, Figure 3.4 provides a mock A-B test example across orders with and without discretion. Indeed, a clean-up cost model suggests that the *best-effort* algorithm underperforms the *must-complete algorithm* in low-volatility periods. Therefore, traders may investigate the best-effort algorithm to tweak its behavior when volatility is gentle and improve its performance.

Order set	Grouping	Order slippage	Clean-up cost	Total cost
Must complete (no discretion)	Low volatility	25bps	0bps	25bps
	High volatility	70bps	0bps	70bps
Best effort (with discretion)	Low volatility	15bps	20bps	35bps
	High volatility	40bps	30bps	70bps

Must complete outperform best effort orders in low volatility

FIGURE 3.4
Mock TCA report for an A-B test of two execution strategies: must-complete orders have no discretion, and best-effort orders have discretion over order size.

3.3.4 An experiment for consecutive orders

> "[In] many cases, orders are correlated and the impact of the first order will affect the execution of future orders" (Harvey et al. (2022, p. 1)[116]).

Section 3.2.4 addresses scheduling consecutive orders. This section tackles TCA for consecutive orders and summarizes the paper *Quantifying long-term impact* by Harvey et al. (2022)[116]. First, Bouchaud et al. (2018)[37] identify the bias inherent to consecutive orders.

> "**Issuer bias:** Another bias may occur if a trader submits several dependent metaorders successively. If such metaorders are positively correlated and occur close in time to one another, the impact of the first metaorder will be different to the impact of subsequent metaorders" (p. 239).

Then, Harvey et al. (2022)[116] compare three approaches to mitigate issuer bias. The authors define and implement all three in a simulation and real-life experiment "to a set of proprietary order flow and execution data from Man Group" (p. 15). For the three approaches, consider Example 3.3.7.

Example 3.3.7 (Two consecutive orders). *Consider two time intervals* $[0, T_0]$ *and* $[T_1, T_2]$ *with* $T_1 > T_0$. *Let* $Q \in \mathcal{S}$ *be a trading process that trades an order of size* Q^0 *over* $[0, T_0]$, *does not trade over* (T_0, T_1), *and trades an order of size* Q^1 *over* $[T_1, T_2]$.

Assume $I \in \mathcal{S}_{1,1}$ *is a price impact model with* $I_0 = 0$ *and define the realized trading costs of* Q

$$C = \int_0^{T_2} (P_t - P_0) dQ_t. \tag{3.5}$$

It equals the arrival slippage of a hypothetical stitched order $Q = Q^0 + Q^1$.

(a) *Stitching orders together.*

For Example 3.3.7, this approach explicitly implements formula 3.5, leading to a model-free estimate of the strategy's realized trading costs. In practice, trading systems do not provide clean data for consecutive orders: one relies on a heuristic to determine orders to *stitch* together into a consistent trading strategy.[10] Harvey et al. (2022)[116] provide two examples where standard stitching heuristics fail for specific strategies. Harvey et al. refer to arrival slippage with the language "implementation shortfall" and "IS".

[10] Bershova and Rahklin (2013)[25] document a stitching heuristic applied to the TCA of proprietary AllianceBernstein LP orders. MDV stands for Median Daily Volume.

> "If the aggregated order imbalance for that symbol exceeded 5% MDV over multiple consecutive days, the aggregated daily buys orders and sell orders were stitched together to form multi-day orders" (p. 45).

> "Unfortunately, this approach is problematic in several situations. Firstly, suppose that trades are correlated with price moves such as in a mean reversion strategy. In this case, prices may be drifting down over multiple metaorders leading to a longer sequence of buys, followed by a sequence of sells when the price is moving up [...]. This would likely lead to low, or even negative, implementation shortfall when benchmarked to the starting metaorder price. Importantly, this shortfall may have nothing to do with market impact" (p. 10).

> "Finally, consider the case when a perfectly correlated sequence (such as in a very large composite trade) is interspersed with small metaorders of opposite side (buy/sell), [...] a situation that often arises due to volatility adjustment or short term alpha. This type of pattern would drastically reduce the IS effect as measured by stitching together" (p. 10).

(b) *Measuring the impact state before the second order.*

Remark 3.2.7 describes how scheduling strategies change behavior by monitoring the *impact state at the order's start.* In Example 3.3.7, this approach singles out the starting impact state I_{T_1} of the second execution:

$$\begin{aligned} C &= \int_0^{T_0} (P_t - P_0) dQ_t + \int_{T_1}^{T_2} (P_t - P_0) dQ_t \\ &= \underbrace{\int_0^{T_0} (P_t - P_0) dQ_t}_{C_0} + \underbrace{\int_{T_1}^{T_2} (P_t - P_{T_1}) dQ_t + I_{T_1} Q^1}_{\tilde{C}_1} + \underbrace{(S_{T_1} - S_0)\, Q^1}_{Z}. \end{aligned}$$

The first term defines the first order's realized trading costs as its arrival slippage

$$C_0 = \int_0^{T_0} (P_t - P_0) dQ_t.$$

The following term modifies the second order's arrival slippage to consider the starting impact state of its execution

$$\tilde{C}_1 = \int_{T_1}^{T_2} (P_t - P_{T_1}) dQ_t + I_{T_1} Q^1.$$

Traders frequently assume $Z = (S_{T_1} - S_0)\, Q^1$ to be mean zero. Harvey et al. (2022)[116] describe this method as *cleaning up* arrival prices.

> "Using the propagator model, the market impact of each realized trade can be subtracted from the observed prices to obtain *cleaned prices*, i.e. the prices that would have been observed without the impact of trading. These hypothetical prices can be used to calculate the cleaned P&L of the strategy" (p. 11).

This methodology relies on a price impact model instead of a stitching heuristic to estimate the consecutive orders' trading costs. In summary, the correction term

$$I_0 Q$$

adjusts an order's arrival slippage to consider past trades' impact. Using an impact model carries model risk but, unlike the stitching method, does not present implementation challenges when applied to specific strategies, such as mean-reversion strategies.

(c) *Measuring the impact caused by the first order on expected future orders.* The last approach outlined in Harvey et al. (2022)[116] flips the measurement problem to provide a bias-free solution with reduced model risk: the Expected Future Flow (EFF) model. In Example 3.3.7, instead of estimating how the second order is impacted by previous trades, the EFF model predicts how the first order *impacts future orders.*

$$\begin{aligned}
\mathbb{E}[C] &= \mathbb{E}\left[\int_0^{T_2} (P_t - P_0) dQ_t\right] \\
&= \mathbb{E}\left[\int_0^{T_2} (Q_{T_2} - Q_t)\, dP_t\right] \\
&= \mathbb{E}\left[\int_0^{T_0} (Q_{T_0} - Q_t)\, dP_t + \int_0^{T_0} \mathbb{E}\left[Q_{T_2} - Q_{T_0} \mid \mathcal{F}_t\right] dP_t\right] \\
&\quad + \mathbb{E}\left[\int_{T_0}^{T_2} (Q_{T_2} - Q_t)\, dP_t\right] \\
&= \underbrace{\mathbb{E}\left[\int_0^{T_0} (P_t - P_0)\, dQ_t + \int_0^{T_0} \mathbb{E}\left[Q_{T_2} - Q_{T_0} \mid \mathcal{F}_t\right] dP_t\right]}_{\tilde{C}_0} \\
&\quad + \underbrace{\mathbb{E}\left[\int_{T_0}^{T_2} (P_t - P_{T_0})\, dQ_t\right]}_{C_1}.
\end{aligned}$$

Define the expected trading costs attributed to the first order:

$$\tilde{C}_0 = \int_0^{T_0} (P_t - P_0) dQ_t + \int_0^{T_0} \underbrace{\mathbb{E}\left[Q_{T_2} - Q_{T_0} \mid \mathcal{F}_t\right]}_{\text{EFF}} dP_t.$$

The EFF statistically distinguishes itself from method (b): one can implement EFF without a price impact model. Instead, the EFF prediction $\mathbb{E}\left[Q_{T_2} - Q_{T_0} \mid \mathcal{F}_t\right]$ carries the model risk. Harvey et al. (2022)[116] describe two methods to build an EFF model.

> "The bigger challenge is to estimate the EFF [...]. The EFF might be deduced from the specific trading strategy or it might need to be estimated from historical data" (p. 12).

The first order's realized slippage applies to the *predicted future order quantity* and penalizes the order for its *impact* on subsequent trades. As with method (b), model risk drives EFF's validity.

Harvey et al. (2022) compare all three approaches and the basic method of ignoring the bias. They test the methods using simulation and actual execution data. The basic approach is biased on all data and strategies, and stitching is unbiased for "Neutral" strategies but presents problems for "Reversion" and "Momentum" strategies. On the other hand, the two model-based approaches (b) and (c) are unbiased across all data and strategies. The section's remainder formalizes the experiment.

Assumption 3.3.8 (No trading alpha). *Assume given consecutive orders for which the no trading alpha assumption holds.*

Prediction 3.3.9 (No trading alpha predictions). *Under the no trading alpha assumption, all three methods estimate identical total costs for consecutive orders.*

While the expected total slippage for the consecutive orders match, the three methods attribute the costs between two orders differently.

(a) The stitching method does not provide a breakdown between the first and second order.

(b) The method based on price impact attributes the additional costs to the second order.

(c) The method based on EFF attributes the additional costs to the first order.

Action 3.3.10 (Actions based on the experiment). *Methods (b) and (c) imply different actions, as Figure 3.5 illustrates.*

(a) The actions of method (b) enhance the second order's execution. Attributing the additional cost to the second order incentivizes it to consider the non-zero initial impact state.

(b) The actions of method (c) slow down the first order. Attributing the additional cost to the first order incentivizes it to consider the expected future order flow.

Order set	Grouping	Slippage first order	Slippage second order	Initial impact cost	EFF slippage	True slippage
Consecutive orders	Neutral strategy	25bps	25bps	0bps	0bps	25bps
	Momentum strategy	25bps	25bps	12bps	12bps	37bps
	Reversal strategy	25bps	25bps	-12bps	-12bps	13bps

Bias in individual order slippage

Either impact or EFF adjustment fixes the bias

FIGURE 3.5
Mock TCA report for sequential orders.

3.3.5 A simulation to improve high-touch trading

Before algorithmic trading, practitioners assumed that orders from the same organization traded in isolation from each other. This isolation assumption separated traders' contributions.

> "If institutional trading desks only had to worry about individual non-overlapping orders that are never modified, a detailed attribution [...] would suffice to enhance optimization of portfolio performance. But, in practice, trading desks have to deal with complex workflows that challenge standard post-trade methodologies" (Waelbroeck and Gomes (2017, p. 6)[230]).

Institutional trading desks, named high-touch desks, trade extensively based on human decisions. By its nature, high-touch trading is qualitative, opaque, and non-reproducible: explaining and evaluating high-touch trades is challenging in the context of best execution requirements such as MiFID II. Moreover, the frequency and interconnectedness of trading make isolated orders the exception rather than the norm. Hence, the complexity of modern trading systems compounds high-touch trading's opaqueness. Waelbroeck and Gomes (2017)[230] are critical of "behavioral trading practices that in some cases hurt rather than help portfolio performance" (p. 1) and propose a simulation-based solution:

> "To align the incentives of the trading desk with optimal portfolio performance, we propose a new execution quality framework" (p. 1).

> "We show how to compute a baseline price by simulating the actions of a trading machine [...] following strictly the benchmark plan. Replacing the execution results [...] provides a measure of portfolio performance that is isolated from trader decisions" (p. 1).

Waelbroeck and Gomes's simulation experiment compares the realized high-touch and the simulated trades of a reproducible, automated strategy. The simulation includes a price impact model to provide simulated arrival slippages and applies the simulation framework of Waelbroeck et al. (2012)[231]. Section 4.2.3 of Chapter 4 covers the simulation framework of Waelbroeck et al. (2012)[231]. This section outlines how a trading team can leverage a simulation experiment to benefit their high-touch trades.

As Waelbroeck and Gomes argue, isolated orders are scarce in modern trading. Assumption 3.3.11 weakens isolated trading into an achievable assumption.

Assumption 3.3.11 (Neutral border conditions assumption). *Assume orders such that*

$$I_0 = 0$$

and the expected future order flow at submission time, as predicted by an EFF model, is zero.

The weaker assumption 3.3.11 requires the proper calibration of impact and EFF models, as defined in Section 3.3.4.

Action 3.3.12 (Quantifying qualitative trading). *TCA decomposes order slippage into alpha, impact, and external impact and provides a simulated benchmark. The* neutral border condition *assumptions allow traders to compare themselves to an automated process and interpret the behavioral differences. Under the* neutral border conditions assumption *3.3.11, current trading is unaffected by previous decisions and does not need to consider future orders. Therefore, the trader controls their slippage and can focus on their own decisions. Without assumption 3.3.11, high-touch traders must consider past and future choices outside their control.*

(a) If the trader outperforms the simulated benchmark strategy, the team can execute more high-touch flow. For example, a trader anticipating external impact could surpass the automated system.

(b) If the trader performs poorly, the simulated strategy indicates which metrics they may misjudge. For example, a trader could overestimate alpha and underestimate price impact.

3.4 Summary of Results

This section summarizes formulas for order execution and TCA. Assume a generalized OW model

$$dI_t = -\beta_t I_t dt + \lambda_t dQ_t$$

with parameters $\beta, \lambda = e^{\gamma}$ such that $\gamma' \ll \beta$. Let $[0, T]$ represent a trading day and define

$$\frac{1}{\Lambda_T} = \frac{1}{\lambda_T} + \int_0^T \frac{\beta_t}{2\lambda_t} dt$$

as the day's liquidity parameter.

3.4.1 Optimal execution without intraday alpha

Consider a day order with overnight alpha $\tilde{\alpha}$. Assume no intraday alpha is present during the execution.

3.4.1.1 Pre-trade cost model

The optimal execution strategy has the following properties:

(a) The daily liquidity parameter estimates the optimal order size

$$Q_T = \frac{\tilde{\alpha}}{\Lambda_T}.$$

(b) The optimal impact state is half the alpha

$$I_t^* = \frac{1}{2}\tilde{\alpha}.$$

(c) The order captures half the overnight alpha in its expected fundamental P&L

$$\mathbb{E}\left[Y_T\right] = \frac{\tilde{\alpha}}{2}\mathbb{E}\left[Q_T\right].$$

(d) The order pays half the overnight alpha in trading costs.

3.4.1.2 Implied alpha

For a single order, assuming no intraday alpha and a risk-neutral trader, one implies its overnight alpha

$$\tilde{\alpha} = \Lambda_T Q_T.$$

Alpha profiling statistically combines an order's implied alpha with the trader's past execution data to predict returns.

3.4.1.3 The case of sizable orders

For sizable orders, the AFS model with square-root concavity applies. Hence, $I_t = \text{sign}(J_t)\sqrt{|J_t|}$ with

$$dJ_t = -\beta J_t dt + \lambda dQ_t.$$

Under the AFS model,

(a) The implied alpha relationship is concave

$$\tilde{\alpha} \propto \text{sign}(Q_T)\sqrt{|Q_T|}.$$

(b) The optimal impact state is two-thirds of the alpha

$$I_t^* = \frac{2}{3}\tilde{\alpha}.$$

(c) The order captures a third of the overnight alpha in its expected fundamental P&L.

(d) The order pays two-thirds of the overnight alpha in trading costs.

3.4.2 Implied alpha's TCA implication

If a trader assumes no intraday alpha and isolated, fully executed orders, the following three TCA metrics align:

(a) The order's implied alpha

(b) The order's realized alpha

(c) The order's optimal impact state

If the expected realized and implied alpha markedly differ, the portfolio team sized the order inappropriately. Likewise, if the impact state and implied alpha disagree, the trading team executed the order sub-optimally.

3.4.3 Clean-up costs

When evaluating orders that did not fully execute, penalize them using a *clean-up cost* model. A clean-up cost model computes the expected arrival slippage of the order remainder $\tilde{Q}$ based on the final impact state I_T at the end of the partial execution.

For the generalized OW model, the clean-up cost formula is

$$C = \frac{1}{2}\Lambda\tilde{Q}^2 + \frac{\Lambda}{\lambda_T}\tilde{Q}I_T + \frac{1}{2\lambda_T}\left(\frac{\Lambda}{\lambda_T} - 1\right)I_T^2.$$

The first term equals a pre-trade cost model where the initial impact state is zero. The next terms correct the clean-up costs for the partial execution's price impact. Finally, one extends the formula to the concave AFS model for sizable remainders.

3.4.4 Intraday and low-latency alphas

Consider the same day order in the presence of an intraday alpha α_t with derivative (alpha decay) α_t'.

3.4.4.1 Intraday alpha

The optimal execution strategy reacts to the intraday alpha.

(a) The optimal order size shifts by

$$Q_T(\alpha) = \int_0^T \frac{1}{2\lambda_t} (\beta_t \alpha_t - \alpha_t')\, dt.$$

(b) The optimal impact state is

$$I_t^* = \frac{1}{2} \left(\tilde{\alpha} + \alpha_t - \beta_t^{-1} \alpha_t'\right).$$

3.4.4.2 The cost of tactical algorithms

Assume given an optimal execution schedule Q^* or, equivalently, an optimal impact state I^*. If a low-latency algorithm tactically deviates from the schedule by δQ, then it deviates the impact by

$$d\delta I_t = -\beta_t \delta I_t dt + \lambda_t d\delta Q_t.$$

This deviation adds

$$-\mathbb{E}\left[\delta Y\right] = \mathbb{E}\left[\int_0^T \frac{\beta_t}{\lambda_t} \left(\delta I_t\right)^2 dt + \frac{1}{2\lambda_T} \left(\delta I_T\right)^2\right]$$

in expected opportunity costs. Therefore, $-\mathbb{E}\left[\delta Y\right]$ are the minimum gains the tactical algorithm must add to be profitable to the strategic schedule.

3.4.5 Optimal execution for multiple orders

3.4.5.1 Combining order executions

When considering multiple orders, the optimal impact formula remains the same:

$$I_t^* = \frac{1}{2} \left(\tilde{\alpha} + \alpha_t - \beta_t^{-1} \alpha_t'\right).$$

Two observations follow.

(a) One should not double-count the trading alpha: the team must split α_t between the two orders.

(b) Combining the two orders' execution is equivalent to combining the two orders' implied alphas. Therefore, the optimal impact formula turns execution into a purely statistical *alpha profiling* problem.

3.4.5.2 TCA for consecutive orders

When considering two successive orders, penalize the second order based on the price impact caused by the first order. If C, Q are the second order's arrival slippage and size, the modified slippage is

$$\tilde{C} = C + I_0 Q$$

where I_0 is the impact at the second order's start. This modification increases the measured slippage for same-sided orders and decreases the measured slippage for offsetting orders.

3.5 Exercises

Exercise 6 Optimal execution for large orders

This exercise solves an optimal execution problem under a globally concave AFS price impact model. As argued in Section 2.6.4 of Chapter 2, the AFS model is particularly relevant when submitting sizable orders. In that regime, the instantaneous liquidity conditions are of second order, and price impact's concavity drives trading costs. The interval $[0, T]$ represents a single trading day. Consider

$$I_t = \operatorname{sign}(J_t)|J_t|^c$$

with local dynamics

$$dJ_t = -\beta J_t dt + \lambda dQ_t$$

for $c \in (0, 1]$.

Assume that $I_0 = 0$.

1. Under this model, solve the idealized optimal execution problem for the target impact state I^*. Assume given a deterministic alpha α_t and an overnight alpha $\tilde{\alpha}$.
2. Solve the same optimal execution problem without trading alpha. What is the relationship between the order size Q_T and the implied overnight alpha $\tilde{\alpha}$?
3. Suggest a testable TCA prediction under this model.
4. Assume that $I_0 > 0$ from a previous order. Solve the idealized optimal execution problem for the target impact state I^* without trading alpha. How does the order size Q_T change given I_0?

Exercise 7 Participation of volume (PoV) algorithms

Practitioners define a PoV algorithm as an execution strategy satisfying

$$dQ_t = r v_t dt$$

for a constant r called the order's *participation rate*. v_t is the instantaneous market activity. Assume $\alpha_t = 0$. The intuition is that the strategic schedule tracks a constant *fraction* of the visible market trades to reduce its footprint on the public tape.

1. Consider an original OW model under the volume clock. Show that, for $t \in (0, T)$, the optimal execution strategy follows a PoV algorithm.

2. Consider a globally concave AFS model under the volume clock, with

$$h(x) = \text{sign}(x)|x|^c$$

and $c \in (0, 1]$.
Show that, for $t \in (0, T)$, the optimal execution strategy follows a PoV algorithm. How does the optimal *participation rate* r depend on the concavity parameter c for a given implied alpha?

Exercise 8 Profitability of tactical deviations

Consider trading under an idealized optimal execution problem and the original OW model

$$dI_t = -\beta I_t dt + \lambda dQ_t.$$

Assume given $\beta, \lambda, \alpha, \tilde{\alpha}$.

Define δQ as the tactical deviation from the corresponding optimal trading strategy Q^*. The tactical deviations mean revert, satisfying the SDE

$$d\delta Q_t = -\gamma \delta Q_t dt + \eta dW_t$$

with $\delta Q_0 = 0$. Define δI as the deviations in impact space.

1. Compute $\mathbb{E}\left[(\delta Q_t)^2\right]$ and $\mathbb{E}\left[\delta Q_t \delta I\right]$ in the steady state where $t \gg \max\left(\frac{1}{\beta}, \frac{1}{\gamma}\right)$.
2. In steady state, quantify the expected difference in impact state $\mathbb{E}\left[(\delta I_t)^2\right]$ generated from the deviations.
3. In steady state, quantify the deviation's cost. Then, comment on the minimum microstructure-level profits to employ this tactical deviation.

Exercise 9 Post-trade reversion TCA

This exercise extends the experiment from Section 3.3.2 and makes predictions about the expected returns *after* the order ends. Assume the following:

(a) A trader submits a day order with implied alpha $\tilde{\alpha}$.

(b) The no intraday alpha assumptions 3.3.2 hold.

(c) No order trades the next day.

1. Make a prediction for the next day's return.
2. Assume that a correlated alpha follows the next day. How does the relationship change?
3. Lift the assumption that no order trades the second day. Modify the analysis.
4. Suggest actions should the original prediction from question 1 not hold.

Exercise 10 Clean-up costs for large orders

This exercise studies clean-up costs under the globally concave AFS price impact model. As argued in Section 2.6.4 of Chapter 2, this model is particularly relevant when submitting sizeable orders. Consider

$$I_t = \operatorname{sign}(J_t)|J_t|^c$$

with local dynamics

$$dJ_t = -\beta J_t dt + \lambda dQ_t$$

for $c \in (0, 1]$.

1. Extend the implied clean-up costs from Proposition 3.3.6 to this model.

4

Further Applications of Price Impact Models

"Eventually, however, [price impact] costs always catch up and overtake expected gains. This effect of diminishing returns with ever larger orders is what limits the 'capacity' of investment vehicles, and it is the principal reason why successful funds are often closed to further investment."

– "Price impact and the capacity of a trading strategy", Capital Fund Management (2019)[161]

Price impact applications extend beyond scheduling orders and evaluating trading strategies. This chapter revisits the last three applications introduced in Chapter 1: statistical arbitrage, liquidity risk management, and portfolio consolidation. These applications follow a common theme:

The primary obstacle to scaling up portfolios is price impact.

Practitioners consider price impact when scaling issues arise in portfolio management. For example, statistical arbitrage portfolios are, to first order, constrained by their ability to match their alpha signals' turnover. Price impact determines a portfolio's ability to turn over and directly affects alpha signals' profits in a statistical arbitrage portfolio. Another example arises when portfolio managers leverage or deleverage their portfolio: liquidity risk, as measured by price impact, determines the cost of scaling up or down the portfolio.

This chapter builds upon the mathematical results of Chapter 2, which Section 2.8 summarizes. The first application, *statistical arbitrage*, sub-divides into three self-contained problems: building an alpha signal based on external price impact, establishing a statistical loss function that considers liquidity, and back testing strategies using price impact. The second application, *portfolio and risk management*, tackles price impact's effect on a portfolio's mark-to-market P&L focusing on times of stress. The last application, *portfolio consolidation*, takes a game-theoretic approach to trading and price impact.

DOI: 10.1201/9781003316923-4

4.1 A Pedagogical Example

For traders, this chapter's essential takeaway is that price impact goes beyond order execution: it affects portfolio scale. For teaching purposes, this section describes one example of price impact's effect on portfolio P&L in the simplest setting: with constant liquidity parameters β, λ and no intraday alpha.

Consider two orders of size $\Delta Q = Q_T - Q_0$ executed over $[0, T]$:

(a) the position-entering trade satisfies $Q_0 = 0$ and $Q_T = \Delta Q$. The order's implied alpha is

$$\tilde{\alpha} = \frac{2\lambda}{2 + \beta T} \Delta Q.$$

(b) the position-exiting trade satisfies $Q_0 = -\Delta Q$ and $Q_T = 0$. The order's implied alpha is

$$\tilde{\alpha} = \frac{2\lambda}{2 + \beta T} \Delta Q.$$

The optimal trading strategies for both orders are equal. For $t \in (0, T)$,

$$I_t^* = \frac{1}{2}\tilde{\alpha}; \quad Q_t' = \frac{\beta}{2\lambda}\tilde{\alpha}.$$

From an execution perspective, the two orders are indistinguishable, and the trading costs are symmetric: their expected arrival slippage is half their implied alpha for both orders.

However, the portfolio's mark-to-market P&L for position-entering and -exiting trades are *asymmetric*. Indeed, the price impact of a *position-entering* trade mechanically inflates the position's value by I_T, leading to an additional accounting profit of $I_T Q_T > 0$. However, this mechanical P&L move behaves differently for *position-exiting* trades: $I_t Q_t < 0$ for $t < T$ and $I_T Q_T = 0$.

> Trading costs are *symmetric* between position-entering and -exiting trades. Mark-to-market P&L, on the other hand, is *asymmetric*.

Figure 4.1 simulates the asymmetry between position-entering and -exiting trades.

Price impact causes this asymmetry and leads to phantom accounting P&L: mark-to-market, a position-entering trade looks profitable. However, these mechanical profits dissipate when closing the position. Section 4.3 extends results on accounting P&L with price impact and applies them to liquidity risk management.

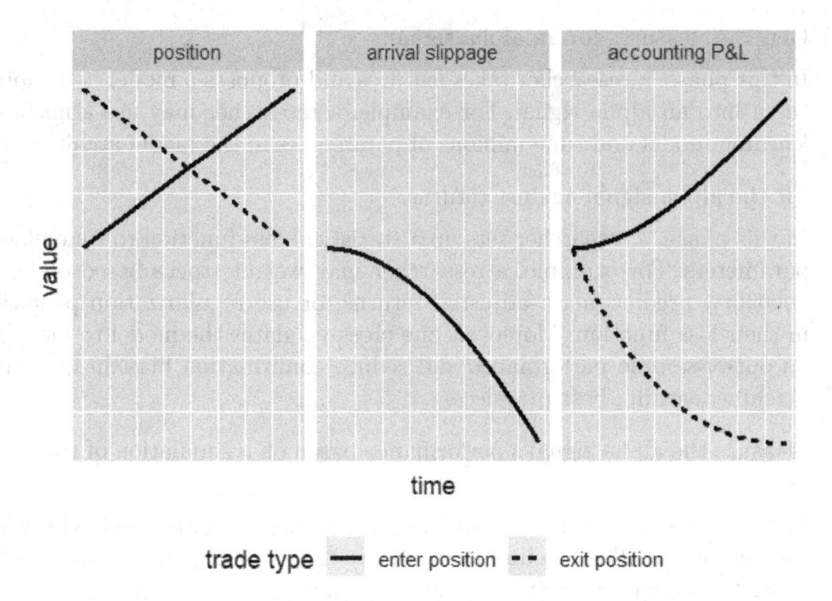

FIGURE 4.1
Expected P&L simulation for two identical trades, one entering and another exiting a position. The arrival slippages are equal. However, due to price impact's effect on positions, the position-entering and -exiting trades have radically different mark-to-market P&L.

4.2 Statistical Arbitrage

In statistical arbitrage, three phases structure the research process, and price impact models contribute to all three phases.[1,2]

---

[1] Tulchinsky et al. (2015)[226] list extended steps under Chapter 5 "How to Develop an Alpha" (p. 27).

> "Alpha logic → Information in the Form of Data → Idea → Mathematical Expression → Apply Operations → Final Robust Alpha → Translate into Positions in Financial Instrument → check Historical PnL, Other Performance Measurements" (p. 27).

[2] Price impact plays a similar role in statistical arbitrage as in optimal execution. Three papers outline this parallel and evaluate the corresponding strategies using simulations. To the author's best knowledge, Gârleanu and Pedersen (2016)[102] first applied the OW model to statistical arbitrage and mapped alpha signals into trading strategies. Casgrain and Jaimungal (2019)[65] build trading strategies with alpha signals and price impact models for both use cases.

> "We presented two examples one where the trader wishes to completely liquidate a large position (the optimal execution problem) and the other where the trader uses the latent states to generate a statistical arbitrage trading strategy" (p. 26).

(a) Generate features for an alpha signal.

In this phase, a researcher takes features and proposes a model to combine them into an alpha signal. For example, a researcher may use a machine learning model and tune millions of parameters in the model architecture.

(b) Fit the alpha signal against returns.

In this phase, a researcher sets up a statistical loss function to tune model parameters. For example, a researcher may weight stocks based on their volatility, predict factor-adjusted returns, or use regularization penalties in their loss function. Moreover, one cross-validates the model to measure its out-of-sample performance and reduce confirmation bias, in line with machine learning best practices.

(c) Measure the alpha signal's performance based on a simulation of the trading strategy.

In this phase, a researcher simulates the trading strategy's behavior with the new signal. The simulation translates the alpha signal into trades and bears two concrete benefits to the researcher.

- The simulation quantifies the alpha signal's performance into P&L. Researchers communicate these simulated P&L numbers to stakeholders but rely on robust statistical measures for daily work.
- Practitioners compare simulation to production trades. For example, a researcher contrasts the new trading strategy's simulation with the strategy it replaces. In another instance, a trader deploys the strategy in production and reconciles its live and simulated behavior.

As outlined in this section, price impact models can provide *alpha signals based on external impact, a liquidity-aware statistical loss function, and simulations for new trading strategies.*

4.2.1 Using external impact as an alpha signal

> "The idea that it is order-flow that must be predicted, even if uninformed, resonates well with the intuition of finance professionals and allows one to understand why statistical regularities might exist and be exploited by quant firms. Indeed, flow data is quite popular among statistical arbitrage funds" (Bouchaud (2022, p. 10)[31]).

Bergault, Drissi, and Guéant (2021)[23] also propose a mathematical framework based on a tradeoff between alpha and price impact and apply their framework to both use cases.

> "Using examples based on data from the foreign exchange and stock markets, we eventually illustrate our results and discuss their implications for both optimal execution and statistical arbitrage" (p. 1).

Section 2.6 from Chapter 2 introduces external impact and proves its correspondence with alpha in Proposition 2.6.4. The link between external impact and alpha is not novel, and practitioners have used external flow alphas since algorithmic trading's inception. However, viewing external flow under the lens of price impact presents multiple benefits. Bouchaud (2022)[31] names such an approach "the order-driven paradigm" (p. 10) and illustrates its virtues.

> "The order-driven paradigm also allows one to resolve some paradoxes, like for example that it is surprisingly easier to find predictive signals for large cap. stocks than for small cap. stocks, probably because the former are more actively traded and that the order flow reveals more statistical regularities" (p. 10).

4.2.1.1 Cont, Cucuringu, and Zhang's alpha signal

Cont, Cucuringu, and Zhang (2021)[73] provide an alpha signal example leveraging a price impact model in their paper *Price Impact of Order Flow Imbalance: Multi-level, Cross-sectional and Forecasting.* Statistically, the contribution of Cont, Cucuringu, and Zhang is to *distinguish* between predicting *contemporaneous* and *future* returns using a price impact model.

> "**Forecasting future returns.** Finally, we investigate the performance of forward-looking price-impact and cross-impact models, i.e. using OFIs to forecast future returns, which has received a lot less attention in the literature, as it is an inherently much more challenging problem than explaining contemporaneous returns" (p. 3).

Cont, Cucuringu, and Zhang leverage two advanced features in building their price impact model. The literature documents both parts.

(a) The term "multi-level OFIs" (p. 1) refers to a price impact model leveraging *further limit order book events*: the method includes events beyond the best bid and ask. Cont, Kukanov, and Stoikov (2013)[74], Eisler, Bouchaud, and Kockelkoren (2018)[87], and Kolm, Turiel, and Westray (2021)[131] also study price predictors based on limit order book events.

(b) "Cross-impact" (p. 1) refers to a price impact model predicting the effect of trading one stock on the price of related stock. Section 7.5 of Chapter 7 covers this class of models.

Kolm, Turiel, and Westray (2021)[131] study a broad class of standard architectures for the same problem as Cont, Cucuringu, and Zhang (2021)[73]. Kolm, Turiel, and Westray also outline comprehensive applications for alpha models of the limit order book.

> "There are several practical applications for the finding in this article. First, the short-term forecasts can be incorporated into execution algorithms, either as alphas or indirectly in the order submission logic

that decides whether to place limit or market orders. Second, the alpha forecasts will be useful in market making systems. In addition, the alphas can be used to design high-frequency trading strategies. Of course, for the latter use case, proper modeling of transaction costs and application of an appropriate execution strategy is critical to their profitability" (p. 31).

4.2.1.2 Model architecture

This section outlines the model architecture proposed by Cont, Cucuringu, and Zhang (2021)[73]. Cont, Cucuringu, and Zhang leverage four computational layers to generate their predictions, which Figure 4.2 illustrates. The starting features contain all the events across the first ten order book levels, aggregated over ten-second bins.

(a) The OFI layer normalizes the events of a given order book level, making the features comparable across event types and levels. See Sections 7.2 and 7.3.1 in Chapter 7 for a definition and implementation of OFI.

(b) The integrated OFI layer aggregates OFI features across multiple order book levels into one feature using Principal Component Analysis (PCA).

(c) The cross-impact layer interacts integrated OFI of every stock with every other stock, leading to a fully connected layer. A Lasso regularization penalizes the number of interactions to maintain a sparse model.

(d) A time-kernel layer lags the features with a lag of one or two in the predictive use case.

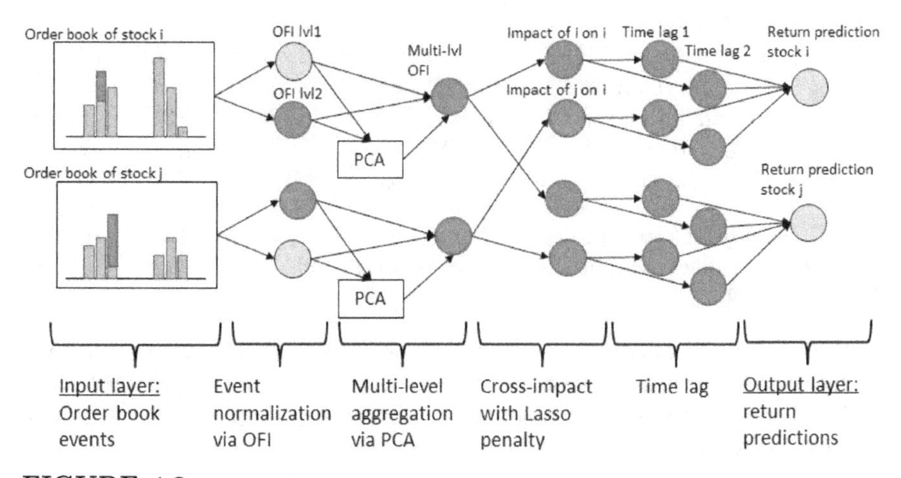

FIGURE 4.2
Model architecture in Cont, Cucuringu, and Zhang (2021)[73] containing an input layer, four computational layers, and an output layer.

The same model predicts contemporaneous returns when removing the time-kernel layer. Cont, Cucuringu, and Zhang (2021)[73] find that such a model consistently provides an out-of-sample R^2 of 80%. This high R^2 for contemporaneous price impact models is in line with other price impact studies, for example, Su et al. (2021)[218]. The empirical analysis in Section 7.4.2 of Chapter 7 estimates an out-of-sample R^2 closer to 40%-50% for more straightforward models. Cont, Cucuringu, and Zhang (2021)[73] find that the cross-impact layer drives the value when *forecasting future returns.*

> "The results reveal that involving cross-sectional OFIs can increase both in-sample and out-of-sample R^2. We subsequently demonstrate that this increase in out-of-sample R^2 leads to additional profits, when incorporated in common trading strategies" (p. 33).

4.2.1.3 Extensions

Cont, Cucuringu, and Zhang (2021)[73] propose "Future research directions" (p. 33). One direction grows the layers' sizes inside the model. A second direction focuses the model on the closing auction and the trading patterns leading into it.

> "Another interesting direction pertains to performing a similar analysis as in the present paper, but for the last 15-30 minutes of the trading day, where a significant fraction of the total daily trading volume occurs" (p. 33).

Remark 4.2.1 (The closing auction). *Cont, Cucuringu, and Zhang (2021)[73] cite Banerji (2020)[17] to motivate their focus on the end of the trading day. One motivation is the high proportion of trades on or near the closing auction, as outlined by Banerji (2020)[17].*

> *"Closing auctions have grown in volume over the past decade, in part because of the rising popularity of index funds, whose managers passively track indexes like the S&P 500 rather than actively seeking to pick stocks. These types of investments often use closing prices as a benchmark, leading their managers to execute trades at the end of the session.*
>
> *As index funds have fueled a frenzy of trading at the close, other big investors have shifted much of their trading to the end of the day, taking advantage of the growing presence of big market participants."*

Banerji (2020)[17] highlights adaptations made by models tracking trade imbalances, in line with the proposal of Cont, Cucuringu, and Zhang (2021)[73].

> *"Some traders use these 'imbalance' messages as cues to start placing bets ahead of the closing bell in anticipation of what they think the final prices will be. For example, if there appears to be heavy demand for a*

> *stock at the end of the day, a trader might opt to buy minutes before the bell, expecting gains.*
>
> *'People can bring those into their models and be predictive about where they think the market is going to close,' said Rob Bernstone, a managing director at Credit Suisse Group AG"*

Exercise 21 in Chapter 6 outlines a causal model for trading on the closing auction.

4.2.2 Adjusting regression techniques for liquidity

Researchers use signals in control problems, for example, mean-variance optimization, to generate P&L for a portfolio. Alpha signals predict future unperturbed returns to raise P&L. Mathematically, an alpha prediction satisfies the equation

$$\alpha_t = \mathbb{E}\left[S_T - S_t \middle| \mathcal{F}_t \right]. \tag{4.1}$$

Practitioners face two challenges when implementing an alpha model to satisfy equation 4.1.

(a) Alpha signals predict *unperturbed returns.*

Realized prices are unsuited for alpha research: price impact of past trading strategies perturb history. Waelbroeck et al. (2012)[231] describe methods to estimate alpha signals that "analyze many (typically hundreds of) drivers coming from both fundamental and technical information. These drivers include but are not limited to how the market reacts to news, momentum since the open, overnight gaps, and how the stock is trading relative to the sector" (p. 4). Waelbroeck et al. use the unperturbed price within their alpha fitting framework:

> "the system may estimate corrected market prices that would have been observed had price impact not existed" (p. 14).

(b) Researchers fit alpha signals using least-squares regressions. However, the standard least-squares loss function does not map to portfolio P&L.

This section aligns the least-squares loss function to the alpha signal's expected P&L using the regret anticipated by a price impact model.

Proposition 4.2.2 computes the trading strategy's expected P&L against the realized alpha. This expected P&L is the regret anticipated by the price impact model for the strategy.

Proposition 4.2.2 (Objective function in impact space). *Assume given*

- *an unperturbed price process $S \in \mathcal{S}$,*
- *a trading process $Q \in \mathcal{S}$, and*

- *a generalized OW model I with liquidity parameters $\beta, \lambda = e^{\gamma}$ where $\beta, \gamma' \in \mathcal{S}$ satisfy the no price manipulation condition from Corollary 2.5.8.*

Then the expected P&L in impact space satisfies

$$\mathbb{E}\left[Y_T\right] = \mathbb{E}\left[\int_0^T \frac{1}{\lambda_t}\left((\beta_t + \gamma_t')\alpha_t^r I_t - \frac{2\beta_t + \gamma_t'}{2} I_t^2\right) dt + \int_0^T \frac{1}{\lambda_t} I_t d\mu_t - \frac{1}{2\lambda_T} I_T^2\right]$$

where $\alpha_t^r = S_T - S_t$ is the realized alpha *over the horizon $(t, T]$ and μ is the drift of S.*

Proof. Proposition 4.2.2 is a direct consequence of Proposition 2.8.3 from Chapter 2. □

The next step simplifies the anticipated regret and expresses it in terms of a candidate signal, setting up the researcher's statistical loss function. For a candidate alpha signal α, the corresponding trading strategy is

$$\forall t \in (0, T), \quad I_t(\alpha, \alpha') = \frac{\beta_t + \gamma_t'}{2\beta_t + \gamma_t'}\alpha_t - \frac{1}{2\beta_t + \gamma_t'}\alpha_t'$$

and $I_T = 0$ by Corollary 2.8.7 from Chapter 2. One extends the definition of $I_t(\alpha, \alpha')$ to allow for stochastic signals. In particular, applying the optimal impact state to the realized alpha α^r and alpha drift μ^r defines the optimal look-ahead impact state $I_t(\alpha^r, \mu^r)$.

Corollary 4.2.3 (Anticipated P&L regret). *Consider the objective function from Proposition 4.2.2. Let*

(a) $I_t(\alpha^r, \mu^r)$ and $Y_T(\alpha^r, \mu^r)$ be the corresponding optimal look-ahead impact state and P&L.

(b) α be a candidate alpha signal such that $\alpha_T = 0$ and let $I_t(\alpha, \alpha')$ and $Y_T(\alpha, \alpha')$ be its target impact state and P&L.

Then one has

$$\mathbb{E}\left[Y_T(\alpha, \alpha')\right] = \mathbb{E}\left[Y_T(\alpha^r, \mu^r)\right] - \mathbb{E}\left[\int_0^T \frac{2\beta_t + \gamma_t'}{2\lambda_t}\left(I_t(\alpha, \alpha') - I_t(\alpha^r, \mu^r)\right)^2 dt\right].$$

Proof. Corollary 4.2.3 is a direct consequence of Corollary 2.5.9 from Chapter 2. □

By Corollary 4.2.3, a researcher who minimizes the least-square loss *in impact space*

$$\min_{\alpha} \mathbb{E}\left[\int_0^T \frac{2\beta_t + \gamma_t'}{2\lambda_t}\left(I_t(\alpha, \alpha') - I_t(\alpha^r, \mu^r)\right)^2 dt\right]$$

minimizes their anticipated P&L regret. This result only holds for the generalized OW model due to its myopic relationship between price impact and alpha. Two observations follow.

(a) The loss function's weighting scheme is

$$\frac{2\beta_t + \gamma_t'}{2\lambda_t}.$$

(b) The least-squares penalty in impact space leads to a balance between predicting the *alpha level* α and *decay* α'.

Remark 4.2.4 (Machine learning implementation)**.** *The anticipated P&L regret fits naturally within the layers of a machine learning model, as illustrated in Figure 4.3. Three steps implement the method.*

(a) A pre-computation node establishes the optimal trading strategy $I(\alpha^r, \mu^r)$ with perfect look-ahead bias.

(b) Upstream layers, for example, a neural network, translate the statistical model's features into two nodes: the alpha level α and the decay α'.

(c) The final layer translates the alpha level and decay into impact space, for which weighted least-squares is the optimal P&L maximizing loss function. An alpha researcher can add regularization penalties to the least-squares loss function.

This layered approach allows the efficient use of backpropagation.

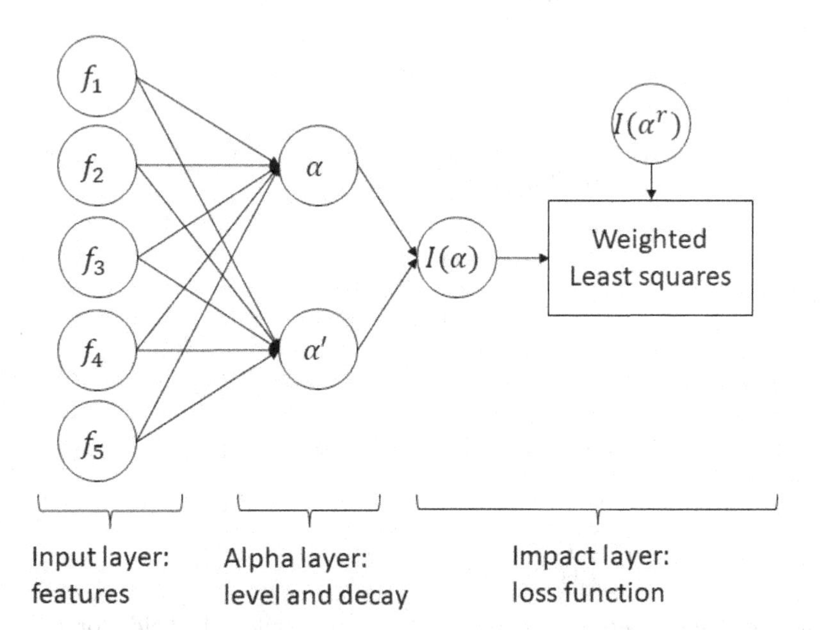

FIGURE 4.3
Implementation of the anticipated P&L regret function in a layered machine learning architecture allowing for backpropagation. See Remark 4.2.4.

An alpha researcher can also recalibrate their alpha model based on different parameters for the generalized OW model. Such a sensitivity analysis assesses how changing the price impact model changes an alpha signal's expected P&L.

4.2.3 Using price impact for simulation

> "The powerful aspect of having a market simulator is that it effectively enables the re-running of different realities where our interventions with the market either do, or do not, occur" (Powrie, Zhang, and Zohren (2021, p. 6)[189]).

The causality challenge, defined in Section 1.1.3 of Chapter 1, drives the need to simulate new prices when back testing a trading strategy. For example, Kolm and Westray (2021)[132] back test a clean-up strategy using a price impact model to simulate new prices. Kolm and Westray also mention the broader applications of such a market simulator.

> "[Price impact] is particularly important when backtesting trading strategies by simulation" (p. 8).

> "While we describe the methodology in the context of cleaning up orders in algorithmic trading, it is straightforward to modify it for other settings where market data needs to be properly adjusted for realized impact. For example, when simulating trading strategies on historical data, the same methodology can be used to adjust returns based on one's own realized trading" (p. 16).

4.2.3.1 Waelbroeck's simulation environment

This section re-implements the simulation environment of Waelbroeck et al. (2012)[231]. First, Waelbroeck et al. provide an example of their general simulation environment on a single order, described in Example 4.2.5.

Example 4.2.5 (Example in Waelbroeck et al. (2012)[231]). *This example establishes, via simulation, whether trading a specific order at a slower speed than the historical speed would have achieved a higher P&L.*

> *"The customer chose a 20% participation rate, and one observes the P/(L) of [an execution with] 20% [participation rate]. The customer has a loss of 15bps.*
>
> *Would the customer have had a lower loss if he had picked a 10% participation rate? To answer that question entails simulating the P/(L) the customer would have gotten if the customer had picked the 10% participation rate.*

And for that one may:

i) *take the observed prices [...]*
ii) *subtract from observed prices the impact of the execution at 20% to see what the impact-free price is [...]*
iii) *calculate the average impact-free price for the execution at 10% [...]*
iv) *to get the P/(L) at 10% one then needs to add to the impact-free price the impact of the execution at 10%*

If one did not take impact into account, the only thing that one would notice is that 10% takes more time, and if the stock moves away the customer will incur more losses. If the impact is not taken into account, one will not see how much is saved in impact by lowering the speed" (p. 3).

Waelbroeck et al. (2012)[231] generalize their methodology in one sentence.

"To make a good assessment of alternative strategies, one may wish to first subtract out the impact of those strategies to then be able to simulate accurately alternative strategies" (p. 3).

Definition 4.2.6 formalizes the simulation framework of Waelbroeck et al., and Figure 4.4 illustrates an example.

Definition 4.2.6 (Realized data, simulated trades). *Assume given the following inputs:*

(a) Realized data

Let $Q^r, P^r \in \mathcal{S}$ be the realized trading process and prices.

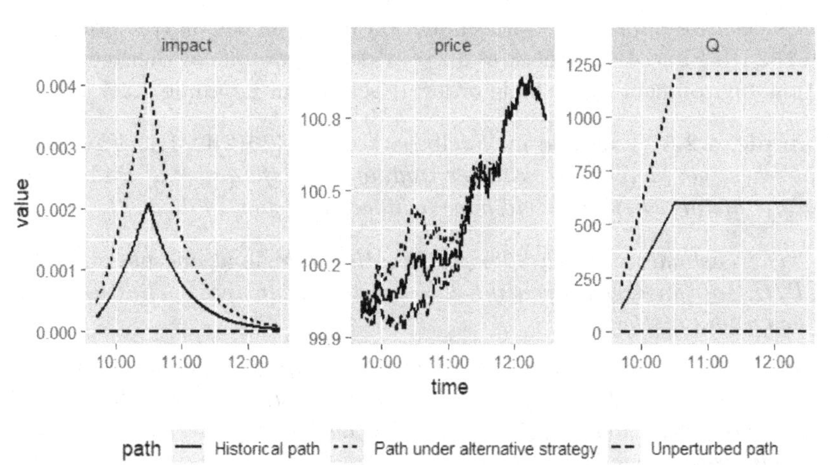

FIGURE 4.4
Simulated example of realized, unperturbed, and simulated prices.

(b) *Simulated trades*

Let $Q \in \mathcal{S}$ be a trading process representing novel trades. The trading process Q simulates an alternative scenario to the realized trading process Q^r.

(c) *A price impact model*

Let $I \in \mathcal{S}_{1,1}$ be an impact model.

Define the following outputs:

$$S_t = P_t^r - I_t\left(Q^r\right)$$

is the unperturbed price, consistent with Definition 2.2.6 from Chapter 2.

$$P_t(Q) = P_t^r - I_t\left(Q^r\right) + I_t\left(Q\right)$$

is the simulated price under the new trading process Q.

4.2.3.2 Business applications of a market simulator

A trading team implements Definition 4.2.6 to simulate the market response to a hypothetical process Q. Almgren highlights four uses for such *a market simulator* in his presentation *Using a Simulator to Develop Execution Algorithms* (2016)[8].

- "Historical
  - rerun scenarios for algo improvement
  - backtest for potential clients
- Real-time
  - clients can connect to "test-drive" algos
- Algorithm development
  - test new signals on historical orders
  - multi-market legging trades
- Real-time splitting for testing
  - compare simulator executions with real" (p. 41)

There are four related applications of a market simulator within a trading firm:

(a) The firm's *accounting system and trading User Interface (UI)*

For example, Brookfield (2015)[40] describes an overlay for traders to visualize and compare realized and simulated market data.

> "The overlay may cover the underlying market data feed snapshot or may be displayed in conjunction with the original recording of market data such that the trader can appreciate a change caused by the trade activity" (p. 9).

The ability to compute unperturbed and simulated prices upstream in a firm's infrastructure opens the use of the firm's accounting systems and UI.

(b) Stress test scenarios

Section 4.3 describes how one simulates market reactions during a stress test. These simulations inform risk and portfolio managers about the liquidity risk they carry and a hypothetical hedge's trading costs.

(c) The team's *back test engine*

Researchers use price impact during two stages of a back test.

- They can simulate an optimal trading strategy Q based on alpha signals, as covered in Section 3.2 of Chapter 3.
- They can simulate market prices and P&L for an arbitrary process Q.

  This second use of price impact for back testing is *independent* of the first one: regardless of the origin of the hypothetical Q, impact models simulate resulting prices and P&L. For example, Brookfield's (2015)[40] market data simulators "provide widespread, systematic back/replay testing of tradable objects and trading strategies, **including custom algorithms.**" (p. 1) Emphasis was added.

(d) In *production*

> "In addition to historical simulation, it is useful to continuously simulate real-time production trading. This process is called *paper trading*" (p. 231, Isichenko (2021)[121]).

While the causality principle states that only one strategy can trade live in the market, a trading team can implement a real-time simulation alongside the realized process. Indeed, comparing a real-time simulation to realized trading helps monitor the live algorithm.

> "The benefits of paper trading include [...] [m]onitoring for a deviation of production data from paper trading. A significant difference would indicate a bug or a need for recalibration of the execution model or trading costs" (p. 231, [121]).

For example, Corollary 2.5.9 from Chapter 2 quantifies the expected opportunity cost when the realized trades Q^r deviate from the schedule Q^*:

$$\mathbb{E}\left[Y(Q^r)\right] - \mathbb{E}\left[Y(Q^*)\right] = -\mathbb{E}\left[\int_0^T \frac{2\beta_t + \gamma_t'}{2\lambda_t}\left(I_t(Q^r) - I_t(Q^*)\right)^2 dt\right]$$

where, for simplicity, $I_T(Q^r) = I_T(Q^*)$.

4.3 Portfolio and Risk Management

"The practice of valuation by marking-to-market with current trading prices is seriously flawed" (Caccioli, Bouchaud, and Farmer (2012, p. 1)[53]).

Section 1.3.3 in Chapter 1 introduces liquidity stress tests and outlines how regulatory changes after the 2008 sub-prime mortgage crisis led to the modern emphasis on estimating realistic liquidation costs during fire sales. This section's crucial message is that price impact distorts a portfolio's accounting P&L and perceived risk. To the author's best knowledge, Bank and Baum (2004)[18] were the first to use a price impact model to distinguish a portfolio's mark-to-market value and its liquidation value. In addition, Caccioli, Bouchaud, and Farmer (2012)[53] empirically quantify price impact's distorting effect on accounting P&L for applications in risk management.

> "Standard risk-management methods give no warning of this problem, which easily occurs for aggressively leveraged positions in illiquid markets. We propose an alternative accounting procedure based on the estimated market impact of liquidation that removes the illusion of profit" (p. 1).

Cont and Schaanning (2017)[75] extend the results to measure *systemic* liquidity risk across the financial sector. This section re-implements Caccioli, Bouchaud, and Farmer's (2012)[53] liquidity risk framework under the generalized OW model. Mathematically, one proves that price impact distorts accounting P&L and

- artificially inflates a portfolio's accounting P&L at constant turnover and leverage.
- mechanically increases P&L whenever turnover or leverage increases, leading to phantom P&L called *position inflation.*
- mechanically decreases P&L whenever turnover or leverage decreases, putting this P&L at risk during stress periods.

Whether these mark-to-market losses become permanent or recover depends on proper portfolio and risk management. The section concludes by computing a portfolio's expected liquidation costs during a fire sale.

4.3.1 How price impact distorts accounting P&L and perceived risk

Per Section 2.2.2 of Chapter 2, traders measure their P&L in two ways: the *accounting* and the *fundamental* P&L. The two measures coincide when the trade *closes*, as the distinction is what price marks the stock position. The same distinction between accounting and fundamental P&L arises for portfolios; the same closing property holds: a portfolio's accounting and fundamental P&L coincide when the stock position is zero. While a trade or position's *closing* P&L are equal under both methodologies, their paths differ. This section's vital result is that the *accounting P&L*'s path is a biased estimate of the portfolio's closing P&L.

Accounting P&L's bias is methodological rather than statistical: under a known price impact model, marking one's P&L to market prices does not reflect a position's true value. This leads to a tradeoff between using the model-free but biased accounting P&L or a model-dependent but bias-free fundamental P&L. This tradeoff between bias and model-risk is frequent in quantitative finance. For instance, options and fixed income practitioners must balance the use of market- and model-driven prices when evaluating portfolios of thinly traded assets.

Definition 4.3.1 introduces a portfolio's *trading footprint*. Example 4.3.4 outlines how the trading footprint contributes to a bias when predicting a position's closing P&L.

Definition 4.3.1 (Trading footprint). *Let $Q \in \mathcal{S}$ be a portfolio's trading process and $X, Y \in \mathcal{S}$ its accounting and fundamental P&L.*

Define the portfolio's trading footprint *at time t as the gap between the portfolio's accounting and fundamental P&L*

$$X_t - Y_t.$$

The trading footprint splits into *internal* and *external* footprints. A trader controls their internal footprint, making it straightforward to measure and forecast. On the other hand, the external footprint depends on other market participants' trades and is challenging to predict. For example, Section 6.4 of Chapter 6 provides a causal model for crowding.

Definition 4.3.2 (Internal and external trading footprint in the generalized OW model). *Assume given a generalized OW model with external impact as per Definition 2.8.4 from Chapter 2. Then the* trading footprint *of a position Q decomposes into*

$$X_t - Y_t = F_t + \bar{F}_t$$

where

$$F_t = I_t Q_t; \quad \bar{F}_t = -\bar{\alpha}_t Q_t.$$

Define F *as the* internal *and* $\bar{F}$ *as the external trading footprint of the portfolio. The internal footprint reflects the effect of the portfolio's trades. The external trading footprint reflects the market's effect on the portfolio.*

For simplicity, this section focuses on a portfolio's internal footprint and ignores its external footprint. Section 4.3.3 revisits the external footprint to model fire sales.

4.3.1.1 Expected closing P&L

This section replicates an essential argument in Bank and Baum (2004)[18]: the best model for a portfolio's actual value considers its eventual closing P&L. For a given position Q_t, consider the next time T such that $Q_T = 0$. T is the position's *closing time.* At time T, the trading process's P&L is unambiguous and risk-free. Therefore, traders expect their current P&L to estimate their expected closing P&L. Under broad assumptions, this section proves the statement

> A position's expected closing P&L equals its *fundamental* P&L minus its liquidation costs.

An immediate result follows: even considering eventual liquidation costs, accounting P&L is a biased estimator of a position's actual value.

Proposition 4.3.3 (Expected closing P&L of a position)**.** *Consider a trading process* Q *and stopping time* T *such that* $Q_T = 0$*. The stopping time* T *is the position's* closing time. *At time* $t < T$*, the position's expected closing P&L is*

$$\begin{aligned}\mathbb{E}\left[X_T \middle| \mathcal{F}_t\right] = X_t - & \underbrace{F_t}_{trading\ footprint} \\ & + \underbrace{\mathbb{E}\left[\int_t^T Q_s dS_s \middle| \mathcal{F}_t\right]}_{future\ alpha} - \underbrace{\mathbb{E}\left[\int_t^T I_s dQ_s + \frac{1}{2}\int_t^T d[I,Q]_s \middle| \mathcal{F}_t\right]}_{future\ transaction\ costs}.\end{aligned}$$

Proof. At closing time T, one has $X_T = Y_T$. Hence,

$$\begin{aligned}\mathbb{E}\left[X_T \middle| \mathcal{F}_t\right] &= \mathbb{E}\left[Y_T \middle| \mathcal{F}_t\right] \\ &= Y_t + \mathbb{E}\left[\int_t^T Q_s dS_s \middle| \mathcal{F}_t\right] - \mathbb{E}\left[\int_t^T I_s dQ_s + \frac{1}{2}\int_t^T d[I,Q]_s \middle| \mathcal{F}_t\right] \\ &= X_t - F_t \\ &\quad + \mathbb{E}\left[\int_t^T Q_s dS_s \middle| \mathcal{F}_t\right] - \mathbb{E}\left[\int_t^T I_s dQ_s + \frac{1}{2}\int_t^T d[I,Q]_s \middle| \mathcal{F}_t\right].\end{aligned}$$

□

Bank and Baum (2004)[18] derive a position's expected closing P&L in Lemma 2.2 (p. 8). They name the expected closing P&L the "asymptotically *realizable* or *real wealth*" (p. 7). Example 4.3.4 provides a closed-form formula under simple assumptions and justifies the frequent use of terminal quadratic costs on positions in stochastic control problems: they estimate the position's distant future closing P&L, or, in Bank and Baum's language, the *asymptotically realizable wealth.*

Example 4.3.4 (Estimating a position's closing P&L). *Three terms make accounting P&L a biased estimator of a position's closing P&L.*

(a) The portfolio's trading footprint *depends on past trades, is measurable at time t and mechanically goes to zero when the position closes.*

(b) The portfolio's expected future alpha capture *depends on future trading opportunities.*

(c) The portfolio's expected future transaction costs *depend on future trades.*

Practitioners can easily measure a portfolio's trading footprint in real-time. For the last two terms, one makes assumptions about future trading opportunities. For example, the portfolio may liquidate in the far future over a duration τ without encountering another alpha opportunity:

$$\mathbb{E}\left[\int_t^T Q_s dS_s \middle| \mathcal{F}_t\right] = 0$$

and

$$\mathbb{E}\left[\int_t^T I_s dQ_s + \frac{1}{2}\int_t^T d[I,Q]_s \middle| \mathcal{F}_t\right] \to \mathbb{E}\left[\frac{\Lambda_\tau}{2}\right] Q_t^2$$

as $T \to \infty$, where $\mathbb{E}\left[\frac{\Lambda_\tau}{2}\right] Q_t^2$ is the pre-trade cost model defined in Proposition 3.2.2 of Chapter 3.

Under this liquidation without alpha, *the expected closing P&L converges to*

$$\mathbb{E}\left[X_T \middle| \mathcal{F}_t\right] \to Y_t - \mathbb{E}\left[\frac{\Lambda_\tau}{2}\right] Q_t^2$$

as $T \to \infty$. Traders estimate the position's closing P&L by subtracting the anticipated liquidation costs $\mathbb{E}\left[\frac{\Lambda_\tau}{2}\right] Q_t^2$ from the fundamental P&L.

4.3.1.2 P&L bias examples

Lemma 4.3.5 establishes the P&L bias caused by the trading footprint when entering a position. Example 4.3.6 and Corollary 4.3.7 illustrate this bias in concrete trading scenarios.

Lemma 4.3.5 (Internal trading footprint when increasing leverage). *Consider a trading process $Q \in \mathcal{S}$ and a generalized OW model I such that*

- $I_0 = 0$.
- $|Q|$ *monotonically increases over* $[0, T]$.

Then the portfolio's internal trading footprint is positive:

$$\forall t \in (0, T] \quad F_t > 0.$$

Proof. Without loss of generality, one can assume $Q > 0$ and prove that $I_t > 0$ for $t \in (0, T]$. Define Z as the solution to the random ODE

$$dZ_t = \beta_t Z_t dt$$

with $Z_0 = 1$. The solution equals

$$Z_t = e^{\int_0^t \beta_s ds} > 0.$$

Then, one has

$$\begin{aligned} d(I_t Z_t) &= I_t dZ_t + Z_t dI_t \\ &= \beta_t I_t Z_t dt - \beta_t Z_t I_t dt + \lambda_t Z_t dQ_t \\ &= \lambda_t Z_t dQ_t. \end{aligned}$$

Hence, $I_t Z_t$ increases and $I_t Z_t > 0$ for $t \in (0, T]$. It follows that $I_t > 0$ for $t \in (0, T]$. □

Example 4.3.6 (Round-trip trade). *Consider a round-trip trade with three phases:*

(a) a leveraging phase $[0, T_1]$ *where* Q *monotonically increases from* $Q_0 = 0$ *to* $Q_{T_1} = L$ *for a constant* $L > 0$.

By Lemma 4.3.5, if $I_0 = 0$*, the internal footprint is positive for* $t \in (0, T_1]$ *and contributes to the accounting P&L overestimating the position's closing P&L. Call this* position inflation.

Unfortunately, even if the fundamental price is a martingale and captures no alpha, *the trade's* accounting P&L *has positive expectation during the leveraging phase. However, the* fundamental P&L *has negative expectation due to transaction costs. This P&L gap resolves when the round-trip completes.*

(b) a holding phase (T_1, T_2) *where* Q *remains constant.*

I slowly decays over the holding phase: the position's accounting P&L remains inflated *but slowly* deflates.

(c) *a deleveraging phase* $[T_2, T]$ *where* Q *monotonically decreases from* $Q_{T_2} = L$ *to* $Q_T = 0$.

The trading footprint mechanically liquidates to zero by $t = T$. *Therefore, the portfolio loses the inflated P&L upon exiting the position. Furthermore, the deleveraging phase's trading footprint is* not *symmetric with the leveraging phase's: temporarily, the trading footprint may turn negative when price impact over-shoots during aggressive liquidations.*

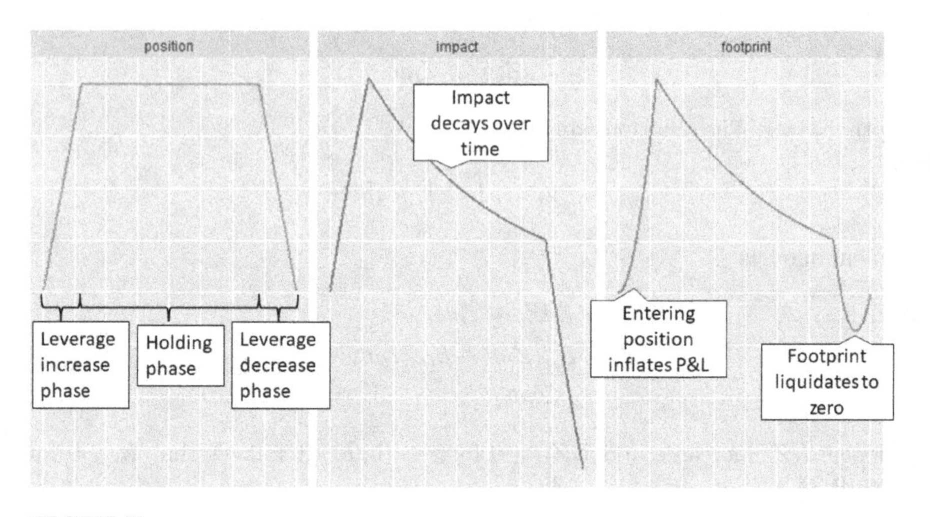

FIGURE 4.5
Example 4.3.6 of a trading footprint on a round-trip trade.

The internal footprint leads to provable, mechanical *inflation* of a portfolio's accounting P&L, as illustrated in Figure 4.5. Caccioli, Bouchaud, and Farmer (2012)[53] remark that traders should anticipate this inflated P&L *before entering the trade* rather than only observing the accounting loss when *the position closes.*

> "Whereas mark-to-market valuation only indicates problems with excessive leverage after they have occurred, our method makes them clear before positions are entered" (p. 3).

Corollary 4.3.7 highlights this P&L asymmetry between position-entering and -exiting trades on the closed-form example of Obizhaeva and Wang.

Corollary 4.3.7 (The Obizhaeva and Wang Model)**.** *Consider the standard Obizhaeva and Wang (2013)[180] execution problem as per Definition 2.3.1 in Chapter 2 with general initial condition* Q_0 *and terminal constraint* $Q_T = Q_0 + Q$.

Then, the expected fundamental and accounting P&Ls for the optimal strategy are

$$\mathbb{E}[Y_T] = -\frac{\lambda}{2+\beta T}Q^2,$$
$$\mathbb{E}[X_T] = \frac{2\lambda}{2+\beta T}Q_0 Q - \mathbb{E}[Y_T].$$

In particular, consider two cases.

- *For a position-entering order,*

$$Q_0 = 0; \quad Q_T = Q.$$

- *For a position-exiting order,*

$$Q_0 = -Q; \quad Q_T = 0.$$

(a) The optimal trades and arrival slippages are identical for both orders, and the expected fundamental P&L equals

$$\mathbb{E}[Y_T] = -\frac{\lambda}{2+\beta T}Q^2.$$

(b) The optimal trading strategy has expected accounting P&L

$$\mathbb{E}[X_T] = -\mathbb{E}[Y_T] > 0$$

for a position-entering order and

$$\mathbb{E}[X_T] = \mathbb{E}[Y_T] < 0$$

for a position-exiting order.

Proof. Corollary 4.3.7 follows from Corollary 2.3.6 in Chapter 2. □

This closed-form example illustrates how a position-entering trade *appears profitable* even without alpha. Mark-to-market P&L leads to "the illusion of profit" (p. 1, Caccioli, Bouchaud, and Farmer (2012)[53]), and the portfolio only realizes its actual P&L, driven by trading costs, over time.

The reader may question if these profits are genuinely an illusion: can a smart trader lock in gains? The answer is no: if such a strategy exists, then the impact model exhibits a price manipulation strategy. Indeed, Gatheral (2010)[104] and Fruth, Schöneborn, and Urusov (2019)[98] rule out price manipulation strategies for comprehensive classes of price impact models.

Traders can use the round-trip Example 4.3.6 and Corollary 4.3.7 to build intuition on position inflation. In practice, the trading footprint is *measurable in real-time*: a trading infrastructure computes F_t based on live positions and recent trade history. Caccioli, Bouchaud, and Farmer (2012)[53] accentuate the ease with which risk management systems could consider price impact. Emphasis was added.

"The procedures that we suggest have the key virtue of being extremely easy to implement. They are based on quantities such as volatility, trading volume, or the spread, that are easy to measure. Risk estimates can be computed for the typical expected behavior or for the probability of a loss of a given magnitude - **anything that can be done with standard risk measures can be easily replicated to take impact into account, with little additional effort**" (p. 15).

4.3.1.3 P&L bias in steady state

Portfolio managers act on liquidity risk when it diverges from a reasonable benchmark. Therefore, a team must propose a standard liquidity profile for their portfolio held at constant size and turnover, also called the portfolio's steady state. For instance, one establishes a bar for a portfolio's trading footprint based on historical data or a portfolio model. Proposition 4.3.8 provides an example portfolio model in steady state for the original OW model.

Proposition 4.3.8 (Trading footprint and transaction costs for stationary portfolios). *Let $Q \in \mathcal{S}$ be a trading process representing a portfolio. Assume the following:*

(a) I follows the original OW model with constant *push and decay parameters $\lambda, \beta > 0$.*

(b) Both I and Q are weakly stationary:

$$\forall t > 0, \quad \mathbb{E}\left[I_t^2\right] = I_0^2 \quad \text{and} \quad \mathbb{E}\left[Q_t^2\right] = Q_0^2.$$

Then the expected trading footprint is

$$\mathbb{E}\left[F_t\right] = e^{-\beta t} F_0 + \frac{1}{\lambda} I_0^2,$$

and the expected running transaction costs per unit of time are

$$\frac{1}{T}\mathbb{E}\left[\int_0^T I_t dQ_t + \frac{1}{2}[I, Q]_T\right] = \frac{\beta}{\lambda} I_0^2.$$

If F_t is weakly stationary, then $\mathbb{E}\left[F_t\right] = \frac{1}{\lambda} I_0^2$.

Proof. By Itô's formula,

$$\begin{aligned}
dF_t &= I_t dQ_t + Q_t dI_t + d[I, Q]_t \\
&= \frac{1}{\lambda} I_t \left(\beta I_t dt + dI_t\right) + Q_t \left(-\beta I_t dt + \lambda dQ_t\right) + d[I, Q]_t \\
&= \frac{\beta}{\lambda} I_t^2 dt + \frac{1}{\lambda} I_t dI_t - \beta F_t dt + \lambda Q_t dQ_t + d[I, Q]_t \\
&= -\beta F_t dt + \frac{\beta}{\lambda} I_t^2 dt + \frac{1}{2\lambda} d\left(I_t^2\right) + \frac{\lambda}{2} d\left(Q_t^2\right).
\end{aligned}$$

The stationarity assumption implies

$$\frac{d\mathbb{E}\left[F\right]_t}{dt} = -\beta\mathbb{E}\left[F_t\right] + \frac{\beta}{\lambda}I_0^2.$$

Solving the ODE proves the result for F. For the transaction costs, recall that, in impact space,

$$\int_0^T I_t dQ_t + \frac{1}{2}[I, Q]_T = \int_0^T \frac{\beta}{\lambda}I_t^2 dt + \frac{1}{2\lambda}\left(I_T^2 - I_0^2\right).$$

□

In steady state, running transaction costs and trading footprint relate:

$$\begin{aligned}\mathbb{E}\left[\int_0^T I_t dQ_t + \frac{1}{2}[I, Q]_T\right] &= \frac{\beta T}{\lambda}I_0^2 \\ &= \beta T \mathbb{E}\left[F_t\right].\end{aligned}$$

Turnover and size drive transaction costs and position inflation for a portfolio in steady state. Trading teams can extend the sensitivity analysis for a broader range of models via simulation, using either historical or synthetic portfolios. Example 4.3.9 simulates a synthetic portfolio.

Example 4.3.9 (Simulation of a stationary portfolio). *Consider a statistical arbitrage portfolio driven by factors or alpha signals. Such a portfolio continuously updates its positions with small trades considering its drivers. The Ornstein-Uhlenbeck model provides a straightforward simulation of such a systematic portfolio.*

Consider the Ornstein-Uhlenbeck model for an n-dimensional portfolio Q

$$dQ_t = -\frac{1}{\tau}Q_t dt + \sigma dW_t,$$

where $\tau > 0$, σ is an $n \times k$-dimensional volatility matrix, and W is a k-dimensional Brownian motion. Intuitively, each Brownian motion component models an independent factor or alpha signal driving the portfolio's construction. One simulates paths of this process for a portfolio of $n = 1000$ stocks and computes the following quantities:

- *the portfolio size,*
- *the daily portfolio turnover,*
- *the daily transaction costs per dollar traded, and*
- *the trading footprint per dollar held in the portfolio.*

In this simulation, $Q \notin S$: the Ornstein-Uhlenbeck model is unbounded. One extends Proposition 4.3.8 to square-integrable semimartingales.

Figure 4.6 illustrates the simulation. A researcher maintains the portfolio's size constant, varies τ to increase the portfolio's turnover, and observes the turnover's effect on the portfolio's running transaction costs and trading footprint. Turnover increases trading, and the average impact state across stocks grows. Furthermore, each stock's price impact correlates with its position. Therefore, significant turnover inflates positions, even maintaining the portfolio size constant.

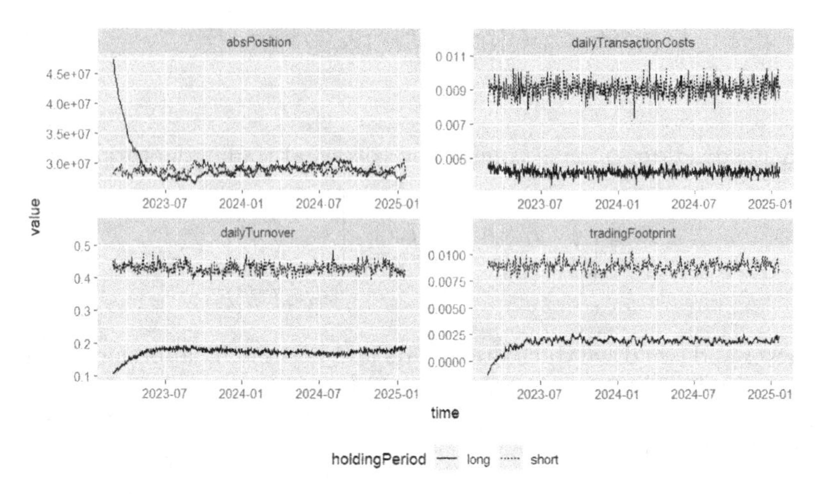

FIGURE 4.6
Simulation of an Ornstein-Uhlenbeck portfolio in steady state from Example 4.3.9.

A portfolio manager can compare their trading footprint and transaction costs with their alpha capture to estimate their fundamental and accounting P&L.

4.3.2 General implications and actions

Section 4.3.1 shows that a portfolio's trades mechanically inflate its *accounting* P&L, and managers can measure these artificial profits in real-time. Caccioli, Bouchaud, and Farmer (2012)[53] make a case for monitoring a portfolio's trading footprint to control its *liquidity risk.* This section's implications and actions build on this observation.

> "From the point of view of a regulator or a risk manager this makes it clear that an alternative to mark-to-market accounting is badly needed. Neglecting impact allows huge positions on illiquid instruments to appear profitable when it is actually not the case" (p. 2).

4.3.2.1 Portfolio management implications

From a portfolio management perspective, managers run their portfolios in steady state. Therefore, they do not perceive position inflation until the regime shifts. As a result, portfolio managers look out for three changes when assessing their liquidity risk.

(a) Deleveraging a portfolio involves significant trading costs **and** deflates remaining positions by shrinking prices. Managers can use a price impact model to explain mechanical accounting losses and focus on their investment thesis's fundamental P&L.

(b) Leveraging a portfolio inflates mark-to-market P&L. These temporary profits are not sustainable once the portfolio settles. Managers can use a price impact model to explain mechanical accounting profits and set expectations for trading costs in the new steady state.

(c) Increasing turnover boosts a portfolio's trading footprint: faster trading increases price impact. Conversely, a portfolio manager who stops trading mechanically loses accounting P&L as impact reverts. Managers can explain P&L moves mechanically due to turnover.

Example 4.3.10 (Portfolio regime shift). *Figure 4.7 provides a mock report a portfolio manager can use to manage expectations around a proposed portfolio change. Three observations follow.*

(a) Increasing the portfolio size leads to 195k temporary accounting profits expected to revert over 20 days. The portfolio increases its transaction costs by 5bps in the new regime.

(b) Reducing the portfolio's turnover leads to 90k of temporary accounting losses expected to revert over 10 days. In addition, the portfolio decreases its transaction costs by 20bps in the new regime.

Leverage scenario	Steady-state portfolio size	Steady-state daily turnover	Steady-state footprint (bps; total)	Steady-state transaction costs	Temporary P&L during transition	Half-life of transition P&L
Current size and turnover	30mil	30%	70bps; 210k	35bps	-	-
Increased size	45mil	30%	90bps; 405k	40bps	+195k	20 days
Lower turnover	30mil	15%	40bps; 120k	15bps	-90k	10 days
Increased size & lower turnover	45mil	15%	50bps; 225k	35bps	+15k	1 day

Proposed leverage scenario maintains similar trading footprint and steady-state transaction costs

FIGURE 4.7
Mock report of a proposed turnover and size change for a portfolio. See Example 4.3.10.

(c) *Simultaneously increasing the portfolio's size and decreasing turnover leads to minimal temporary P&L. The transaction costs are unchanged. On the one hand, a larger portfolio increases its total gains if it maintains its alpha. On the other hand, lower turnover may prevent the portfolio from capturing faster-moving signals.*

4.3.2.2 Liquidity risk implications

From a risk manager's and regulator's perspective, the trading footprint is P&L at risk. Risk management techniques, from hedging to reducing turnover, deflate a portfolio's trading footprint and lead to temporary accounting losses and permanent trading costs. Reduced market liquidity compounds these losses. Caccioli, Bouchaud, and Farmer (2012)[53] point out the importance of not just computing hedging costs but also considering the effect on mark-to-market P&L. The "previous disasters" (p. 15) refer to "LTCM" (p. 15) and "Lehman Brothers" (p. 15).

> "The method of valuation that we propose here could potentially be used both by individual risk managers as well as by regulators. Had such procedures been in place in the past, we believe that many previous disasters could have been avoided" (p. 15).

Example 4.3.11 (Liquidity stress test). *Figure 4.8 provides a mock liquidity report for a stress test. Three observations follow.*

(a) *A risk manager quantifies temporary accounting losses by measuring the trading footprint at the start and end of the stress tests. Even if the portfolio does not trade, the rest of the market sells, leading to temporary losses on the position.*

(b) *The accounting P&L includes permanent trading costs and temporary position deflation. Risk managers compare the two losses across hedging scenarios.*

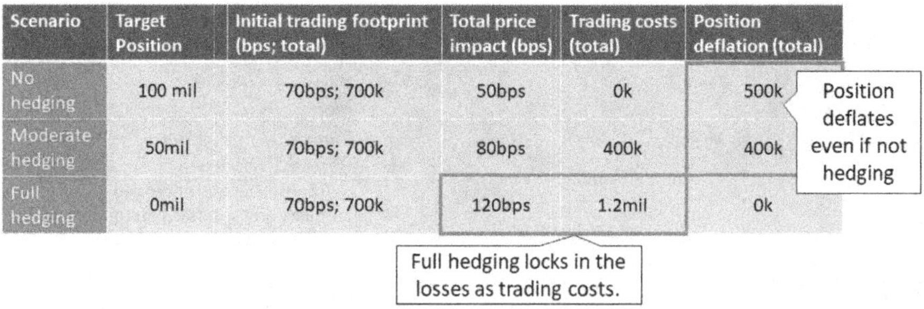

Scenario	Target Position	Initial trading footprint (bps; total)	Total price impact (bps)	Trading costs (total)	Position deflation (total)
No hedging	100 mil	70bps; 700k	50bps	0k	500k
Moderate hedging	50mil	70bps; 700k	80bps	400k	400k
Full hedging	0mil	70bps; 700k	120bps	1.2mil	0k

Position deflates even if not hedging

Full hedging locks in the losses as trading costs.

FIGURE 4.8
Mock liquidity report for a stress test. See Example 4.3.11.

(c) *When the portfolio fully hedges its position, it locks in losses as trading costs.*

4.3.2.3 Senior management implications

From a senior management perspective, position inflation leads to misaligned incentives for traders and portfolio managers. Senior management can consider a portfolio's trading footprint when evaluating traders or portfolio managers. This approach avoids providing incentives to portfolio managers that temporarily inflate mark-to-market P&L. Caccioli, Bouchaud, and Farmer (2012)[53] estimate the trading footprint an order of ten times the stock's ADV generates. The numbers range from two to nine percent for *liquid* stocks, for example, AAPL. Such hypothetical returns are temporary: they revert once the trader or portfolio manager stops inflating their position.

Practitioners can measure a portfolio's trading footprint and simulate hypothetical trading scenarios for risk management applications. Section 4.3.3 provides a closed-form example for fire sales.

4.3.3 Simulating fire sales

> "Financial regulators across the world are monitoring the collapse of the New York-based billionaire Bill Hwang's personal hedge fund.
>
> The sudden liquidation of Hwang's Archegos Capital Management sparked a fire sale of more than $20bn assets that has left some of the world's biggest investment banks nursing billions of dollars of losses" (Neate and Makortoff (2021)[173]).

Fire sales describe a situation where stock prices dislocate from fundamentals and result from a feedback loop in trading. For example, a sharp loss in a stock may force a fund to liquidate a position to satisfy margin calls or investor redemptions. Consequently, the portfolio manager trades based on exogenous commitments rather than a fundamental view of the stock. These trades' impact leads to further price drops unrelated to the company's economics.

Example 4.3.12 (Archegos Capital Management). *Neate and Makortoff (2021)[173] provide a fire sale example on the heavily leveraged hedge fund Archegos.*

> *"The sudden collapse of Archegos was said to have been triggered by a sharp drop in the share price of the US media giant ViacomCBS last week. The fund had a big exposure to Viacom – via loans – and it was forced to unwind its position, which caused the price to drop further. Archegos was also forced to sell stakes in other media companies and a host of Chinese tech companies."*

Federal prosecutors charged both Archegos's founder and former CFO with securities fraud and market manipulation, as reported by Ramey, Pulliam, and Chung in the Wall Street Journal (WSJ) (2022)[193].

> *"We allege that the defendants and their co-conspirators lied to banks to obtain billions of dollars, that they then used to inflate the stock price of a number of publicly traded companies" (Damian Williams, United States Attorney for the Southern District of New York).*

Regardless of the initial trigger, the common element of every fire sale is price impact. Portfolio and risk managers use fire sale models to predict the depth and length of price dislocations and set up hedging strategies to manage their liquidity risk. Definition 4.3.13 formalizes a liquidation model for stress testing.

Definition 4.3.13 (Liquidation model). *Assume the following inputs:*

- *a globally concave AFS price impact model $\bar{I}$ as per Definition 2.6.12 from Chapter 2*
$$\bar{I}_t = h(J_t - \bar{\alpha}_t)$$
where h is a power-law
$$h(x) = sign(x)|x|^c.$$
$J, \bar{\alpha}$ satisfy the SDEs
$$dJ_t = -\beta J_t dt + \lambda dQ_t$$
and
$$d\bar{\alpha}_t = -\beta \bar{\alpha}_t dt - \lambda dM_t.$$
Recall that $M \in \mathcal{S}$ is the market flow, and $-\bar{\alpha}$ is the market's contribution to price impact.
- *a set of initial conditions: $Q_0, J_0, \bar{\alpha}_0$.*
- *a hedging fraction $\eta \in [0, 1]$ and hedging timescale T to reduce the position Q.*

Define a liquidation model for stress testing as

(a) a trading process $Q \in \mathcal{S}$ such that $Q_T = (1 - \eta) \cdot Q_0$,

(b) its expected accounting and fundamental P&L $\mathbb{E}[X_T]$ and $\mathbb{E}[Y_T]$,

(c) and a half-life τ for the trading footprint at time T.

4.3.3.1 Liquidation without fire sale

Proposition 4.3.14 derives liquidation costs under neutral market conditions. Proposition 4.3.16 extends the liquidation cost model under adverse conditions to stress-test a fire sale scenario. To lighten the notation, extend the function $x \mapsto x^c$ to be odd.

Proposition 4.3.14 (Constant speed, no fire sale). *Consider a liquidation model's inputs. Assume the unperturbed price is a martingale and $\bar{\alpha} = 0$.*
Then the liquidation model satisfies

(a) the target impact state

$$\forall t \in (0,T), \quad J_t^c = \Lambda_T^c \left(\lambda^{-1} J_0 - \eta Q\right)^c; \quad J_T^c = (1+c)\Lambda_T^c \left(\lambda^{-1} J_0 - \eta Q\right)^c$$

where $\Lambda_T = \frac{\lambda}{(1+c)^{1/c} + \beta T}$.

(b) the trading process

$$\forall t \in (0,T), \quad dQ_t = \frac{\beta}{\lambda} \Lambda_T^c \left(\lambda^{-1} J_0 - \eta Q\right)^c dt.$$

(c) the expected fundamental P&L

$$\mathbb{E}\left[Y_T\right] = Y_0 + \frac{1}{(1+c)\lambda} |J_0|^{1+c} - \Lambda_T^c \left|\lambda^{-1} J_0 - \eta Q\right|^{1+c}$$

and trading footprint

$$F_T = (1+c)\Lambda_T^c \left(\lambda^{-1} J_0 - \eta Q\right)^c (1-\eta) Q.$$

(d) the half-life τ

$$\tau = \frac{\log 2}{\beta c}.$$

Proof. Proposition 2.6.14 applies with $\bar{\alpha} = 0$. The proposition leads to

$$\begin{aligned} \mathbb{E}\left[Y_T - Y_0\right] &= \frac{1}{\lambda} \mathbb{E}\left[H(J_0) - H(J_T) - \int_0^T \beta h(J_t) J_t dt\right] \\ &= -\frac{1}{\lambda} \mathbb{E}\left[\frac{1}{1+c}\left(|J_T|^{1+c} - |J_0|^{1+c}\right) + \int_0^T \beta |J_t|^{1+c} dt\right]. \end{aligned}$$

Adding a Lagrange multiplier θ yields the objective function

$$\frac{1}{\lambda} \mathbb{E}\left[\frac{1}{1+c}|J_0|^{1+c} + \theta J_T - \frac{1}{1+c}|J_T|^{1+c} + \int_0^T \beta\left(\theta J_t - |J_t|^{1+c}\right) dt\right].$$

Therefore, the optimal impact state J_t^* satisfies

$$\forall t \in (0,T), (J_t^*)^c = \frac{\theta}{1+c}; \quad (J_T^*)^c = \theta.$$

The constraint leads to the equation

$$\begin{aligned}
-\lambda\eta Q &= J_T - J_0 + \beta \int_0^T J_t dt \\
&= \theta^{1/c}\left(1 + \frac{\beta T}{(1+c)^{1/c}}\right) - J_0.
\end{aligned}$$

Therefore, one has

$$\begin{aligned}
\theta &= (1+c)\left(\frac{J_0 - \lambda\eta Q}{(1+c)^{1/c} + \beta T}\right)^c \\
&= (1+c)\Lambda_T^c\left(\lambda^{-1}J_0 - \eta Q\right)^c.
\end{aligned}$$

The expected fundamental P&L is

$$\begin{aligned}
\mathbb{E}\left[Y_T - Y_0\right] &= -\frac{1}{\lambda}\mathbb{E}\left[\frac{1}{1+c}\left(|J_T|^{1+c} - |J_0|^{1+c}\right) + \int_0^T \beta|J_t|^{1+c}dt\right] \\
&= \frac{1}{(1+c)\lambda}|J_0|^{1+c} - \frac{1}{(1+c)\lambda}|\theta|^{1+1/c}\left(1 + \beta T(1+c)^{-1/c}\right) \\
&= \frac{1}{(1+c)\lambda}|J_0|^{1+c} - \frac{1}{1+c}(1+c)^{-1/c}\Lambda_T^{-1}|\theta|^{1+1/c} \\
&= \frac{1}{(1+c)\lambda}|J_0|^{1+c} - (1+c)^{-1-1/c}\Lambda_T^{-1}(1+c)^{1+1/c}\Lambda_T^{1+c} \\
&\quad \left|\lambda^{-1}J_0 - \eta Q\right|^{1+c} \\
&= \frac{1}{(1+c)\lambda}|J_0|^{1+c} - \Lambda_T^c\left|\lambda^{-1}J_0 - \eta Q\right|^{1+c}.
\end{aligned}$$

When the process stops trading, the impact reverts

$$J_{T+\Delta t}^c = e^{-\beta c\Delta t}J_T^c,$$

which leads to the formula for the half-life τ. □

Figure 4.9 simulates the liquidation described in Proposition 4.3.14 across hedging scenarios and initial impact states. Two observations follow.

(a) When the initial trading footprint is significant, position deflation drives the accounting P&L.

The trader profits from the non-zero initial impact state and is active even when $\eta = 0$.

(b) When the initial impact state is minor, transaction costs drive the accounting P&L.

Liquidation trades still affect the remaining position's trading footprint. For $\eta > 0.5$ the accounting P&L reaches a plateau as trading footprint and fundamental P&L balance.

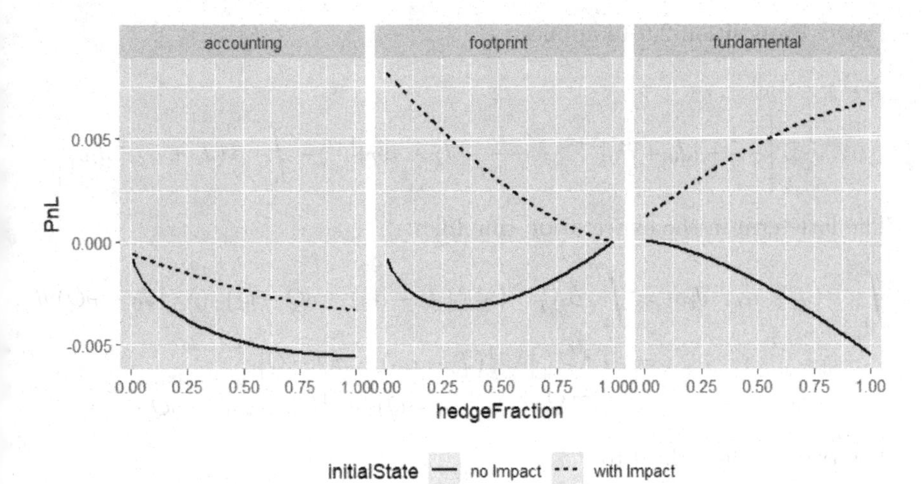

FIGURE 4.9
Simulated liquidation P&L across hedging fractions η and initial impact states. See Proposition 4.3.14.

4.3.3.2 Liquidation with fire sale

> "Liquidation by a large actor, or the concurrent liquidation by a set of smaller investors, can depress valuations, with resulting losses causing further liquidations and eventually cascading losses among all participants" (p. 2, Capital Fund Management (2019)[161]).

A risk manager simulating a fire sale needs a model of external impact. A simple approach assumes that similar hedging trades drive external impact. For the Nash equilibrium case, see Section 4.4.2, Cont and Schaanning (2017)[75], and Fu, Horst, and Xia (2022)[100].

Definition 4.3.15 (A fire sale model)**.** *Assume the rest of the market hedges $-\bar{Q} \in \mathbb{R}$ over horizon T. Define the fire sale model of external price impact $-\bar{\alpha}_t \in \mathcal{S}$ as*

$$\forall t \in (0,T) \quad \bar{\alpha}_t = \Lambda_T \left(\lambda^{-1}\alpha_0 + \bar{Q}\right); \quad \bar{\alpha}_T = (1+c)^{-1/c}\Lambda_T \left(\lambda^{-1}\alpha_0 + \bar{Q}\right).$$

Proposition 4.3.16. *Assume a fire sale model as per Definition 4.3.15. The expected fundamental P&L satisfies*

$$\lambda\mathbb{E}\left[Y_T - Y_0\right] = \frac{1}{1+c}\left|J_0 + \bar{\alpha}_0\right|^{1+c} - \Lambda_T^c \left|\lambda^{-1}(J_0 - \bar{\alpha}_0) - \eta Q - \bar{Q}\right|^{1+c}$$
$$- \beta T \Lambda_T^{1+c}(\lambda^{-1}\bar{\alpha}_0 + \bar{Q})\left(\lambda^{-1}(J_0 - \bar{\alpha}_0) - \eta Q - \bar{Q}\right)^c.$$

and the terminal trading footprint is

$$F_T + \bar{F}_T = (1+c)\Lambda_T^c \left(\eta Q + \bar{Q} + \lambda^{-1}(J_0 - \bar{\alpha}_0)\right)^c (1-\eta)Q.$$

Proof. Proposition 2.6.14 applies:

$$\begin{aligned}
&\mathbb{E}\left[Y_T - Y_0\right] \\
&\quad = \frac{1}{\lambda}\mathbb{E}\left[\frac{1}{1+c}\left|J_0 + \bar{\alpha}_0\right|^{1+c} - \frac{1}{1+c}\left|J_T + \bar{\alpha}_T\right|^{1+c} - \int_0^T \beta\left(J_t + \bar{\alpha}_t\right)^c J_t dt\right].
\end{aligned}$$

The last term in the expectation simplifies:

$$\begin{aligned}
\int_0^T \beta\left(J_t + \bar{\alpha}_t\right)^c J_t dt &= \int_0^T \beta\Lambda_T^{1+c}\left(\lambda^{-1}(J_0 - \alpha_0) - \eta Q - \bar{Q}\right)^c (\lambda^{-1}J_0 - \eta Q)dt \\
&= \beta T\Lambda_T^{1+c}\left|\lambda^{-1}(J_0 - \alpha_0) - \eta Q - \bar{Q}\right|^{1+c} \\
&\quad + \beta T\Lambda_T^{1+c}(\lambda^{-1}\alpha_0 + \bar{Q})\left(\lambda^{-1}(J_0 - \alpha_0) - \eta Q - \bar{Q}\right)^c.
\end{aligned}$$

The two equalities lead to

$$\begin{aligned}
\lambda\mathbb{E}\left[Y_T - Y_0\right] = &\ \mathbb{E}\left[\frac{1}{1+c}\left|J_0 + \bar{\alpha}_0\right|^{1+c} - \Lambda_T^c\left|\lambda^{-1}(J_0 - \alpha_0) - \eta Q - \bar{Q}\right|^{1+c}\right] \\
&- \mathbb{E}\left[\beta T\Lambda_T^{1+c}(\lambda^{-1}\alpha_0 + \bar{Q})\left(\lambda^{-1}(J_0 - \alpha_0) - \eta Q - \bar{Q}\right)^c\right].
\end{aligned}$$

□

Figure 4.10 simulates the stress test described in Proposition 4.3.16 across hedging and crowding scenarios. Three broad observations follow.

(a) Even without hedging, crowding affects the portfolio through external impact.

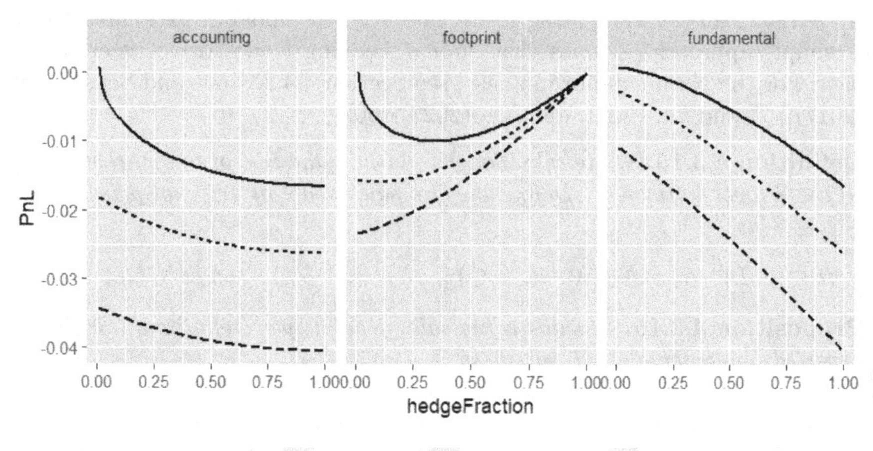

FIGURE 4.10
Simulated fire sale P&L across hedging fractions η and crowding scenarios. See Proposition 4.3.16.

(b) In the simple fire sale model, the trader considers their price impact but not the impact of other traders. Therefore, the trader is not entirely strategic. Section 4.4 studies Nash equilibria in a non-distressed setting.

(c) For $\eta > 0.5$ the accounting P&L reaches a plateau as trading footprint and fundamental P&L balance. This balancing act raises the question of whether the manager wants to lock in current losses or expects price impact to revert before other risks materialize.

4.4 Combining Two Portfolios' Trading

Section 1.3.4 in Chapter 1 motivates why senior management may compare the economics of two portfolios trading jointly or separately. This section takes the point of view of a trading team tasked to quantify the portfolio consolidation's economic benefit. Game theory plays an essential role in the analysis, and traders must model how the two portfolios share information. Consider two scenarios and compare them against a collaborative trading system.

(a) The two portfolios share no information.

Both portfolios solve independent execution problems. However, side-by-side trading links their *impact costs*. See Figure 4.11.

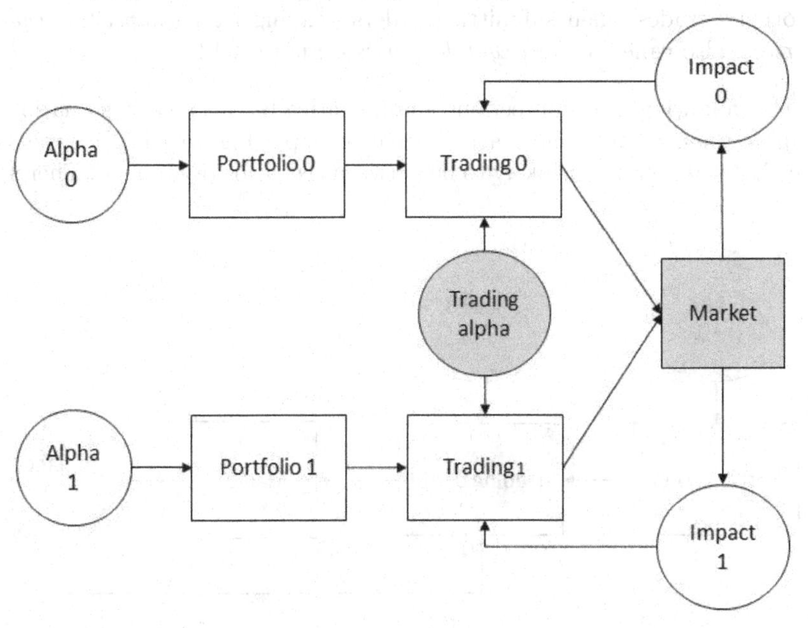

FIGURE 4.11
No mutual information case: independent trading model.

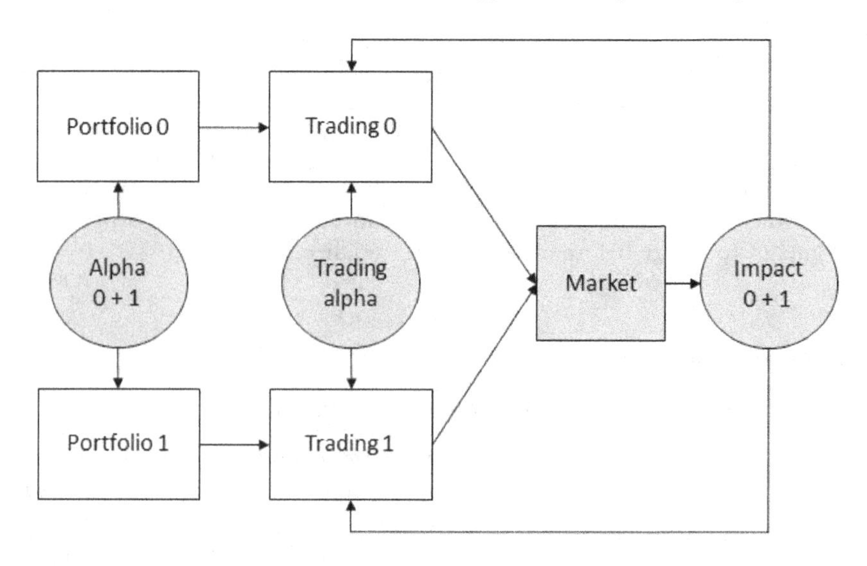

FIGURE 4.12
Mutual information case: competitive trading model.

(b) The two portfolios fully share their information, including their alphas.

Both portfolios consider the price impact and the implied alpha of each other's trades when submitting orders, leading to a *competitive equilibrium*, also called a *Nash equilibrium*. See Figure 4.12.

The following two subsections mathematically cover each scenario and compare them to the collaborative scenario, as per Figure 4.13. However, the teams' actual behavior is likely to lie between scenarios (a) and (b). Therefore,

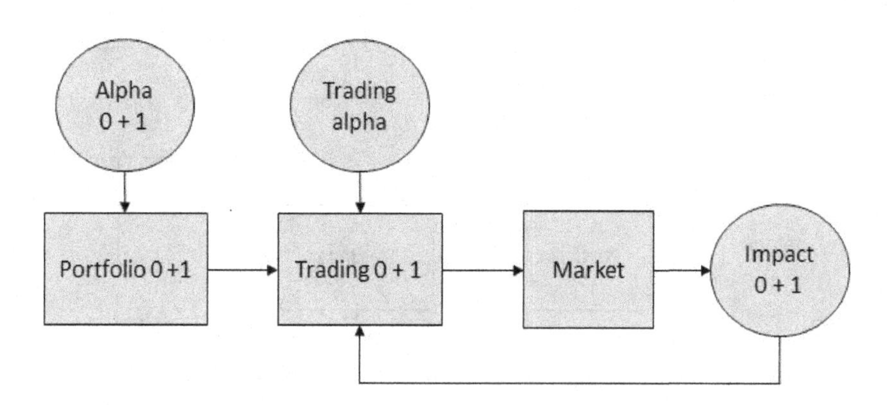

FIGURE 4.13
Mutual information case: collaborative trading model.

the third subsection introduces an alternative empirical method. In addition, two papers study the hybrid scenario where multiple portfolios have joint information on each other's trades and price impact but distinct alpha signals: they agree to disagree.

(a) Casgrain and Jaimungal (2019)[66] "restrict the flow of information to the agents by allowing agent j to only have access to the information of the price process" (p. 5). In this section's language, the agents trade with knowledge of other agents' price impact but their alpha signals do not agree. Casgrain and Jaimungal solve the game's mean-field limit and provide numerical simulations.

(b) Donnelly and Lorig (2020)[86] simulate and compare two information-sharing scenarios.

> "Two cases of the subjective views of agents are considered, one in which they all share the same information, and one in which they all have an individual signal correlated with price innovations" (p. 1).

4.4.1 Theory in the case without mutual information

To focus the discussion on portfolio interaction effects, assume that liquidity fluctuations are second-order: this assumption allows the reader to leverage the original OW model. While one can generalize the approach to the generalized OW model or the globally concave AFS model, the solutions are out of scope for this section.

Definition 4.4.1 (Independent trading). *Consider two traders $i \in \{0;1\}$, controlling trading processes $Q^i \in \mathcal{S}$ under separate filtrations $\mathcal{F}^i$. Assume both traders*

(a) trade under the original OW model I with parameters β, λ. Denote the price impact of the trading process Q^i by I^i and assume $I^i_0 = 0$.

(b) trade based on the overnight alpha $\tilde{\alpha}^i = \mathbb{E}\left[\tilde{\alpha}^r \middle| \mathcal{F}^i_0\right]$. Trader i only has access to $\tilde{\alpha}^i \in \mathcal{S}(\mathcal{F}^i)$, where $\mathcal{S}(\mathcal{F}^i)$ is the space of bounded $\mathcal{F}^i$-semimartingales.

(c) assume the external impact, including the impact from each other, is zero.

(d) share the trading alpha α_t for $t \in (0, T)$ satisfying the deterministic trading alpha assumption from Corollary 2.4.1 in Chapter 2.

Mathematically, each trader solves the stochastic control problem

$$\sup_{Q^i \in \mathcal{S}(\mathcal{F}^i)} \mathbb{E}\left[\tilde{\alpha}^i Q^i_T + \int_0^T \left(\alpha_t - I^i_t\right) dQ^i_t - \frac{1}{2}[I^i, Q^i]_T + [\alpha, Q^i]_T\right]$$

with

$$dI_t^i = -\beta I_t^i dt + \lambda dQ_t^i.$$

The two portfolios trade side-by-side without considering each other's price impact but use the same execution algorithm, including trading alphas. The execution from Section 3.2 describes each trader's trading behavior and yields Corollary 4.4.2.

Corollary 4.4.2 (Independent trading). *Consider the independent trading model from Definition 4.4.1. The optimal trading strategies satisfy the following equations.*

- *Each trader's impact state reflects their overnight alpha and the common trading alpha*

$$\forall t \in (0,T) \quad I_t^i = \frac{1}{2}\left(\tilde{\alpha}^i + \alpha_t - \beta^{-1}\alpha_t'\right); \quad I_T^i = \tilde{\alpha}^i.$$

- *The expected combined order size is*

$$\mathbb{E}\left[Q_T^0 + Q_T^1\right] = \frac{\tilde{\alpha}^0 + \tilde{\alpha}^1}{\Lambda_T} + \frac{\beta}{\lambda}\mathbb{E}\left[\int_0^T \left(\alpha_t - \beta^{-1}\alpha_t'\right) dt\right]$$

where

$$\Lambda_T = \frac{2\lambda}{2+\beta T}.$$

Proof. Corollary 4.4.2 is a consequence of Proposition 3.2.3 in Chapter 3. □

A trader combining the two executions derives a combined overnight alpha $\tilde{\alpha} = \mathbb{E}\left[\tilde{\alpha}^r \middle| \bar{F}_0\right]$ under the combined filtration $\bar{F}$. In practice, a trader fits the combined signal using alpha profiling, which Section 3.2.1 in Chapter 3 describes. Denote by Q^* the optimal strategy for this single trader and by Q^0 and Q^1 the strategies of each individual trader. Define similarly the target impact states I^*, I^0, I^1 for the single and individual traders. Apply Corollary 2.5.9 from Chapter 2 to estimate the opportunity cost of trading separately:

$$\mathbb{E}\left[Y_T(Q^0+Q^1)\right] = \mathbb{E}\left[Y_T(Q^*)\right] - \mathbb{E}\left[\int_0^T \frac{\beta}{\lambda}\left(I_t^* - I_t^0 - I_t^1\right)^2 dt\right].$$

Therefore, one compares I^* and $I^0 + I^1$ to decompose the gains missed when trading independently. For $t \in (0,T)$,

$$I_t^* = \frac{1}{2}\left(\tilde{\alpha} + \alpha_t - \beta^{-1}\alpha_t'\right)$$

and

$$I_t^0 + I_t^1 = \frac{1}{2}\left(\tilde{\alpha}^0 + \tilde{\alpha}^1\right) + \alpha_t - \beta^{-1}\alpha_t'$$

Hence,

$$I_t^* - I_t^0 - I_t^1 = \underbrace{\frac{1}{2}\left(\tilde{\alpha} - \tilde{\alpha}^0 - \tilde{\alpha}^1\right)}_{overnight\ alpha\ calibration} - \underbrace{\frac{1}{2}\left(\alpha_t - \beta^{-1}\alpha_t'\right)}_{double\ counting\ of\ trading\ alpha}$$

(a) One best formulates the first loss as a statistical question: are the overnight alphas additive? The combined alpha is generally sub-linear, and the independent orders sum larger than optimal.[3] In trade space, non-additive implied alphas cause the two portfolios to oversize their orders: they do not consider each other's price impact.

(b) The second loss stems from a mechanical double-counting of the trading alpha. In theory, the team can fix this without combining the two portfolios: simply divide the alpha among the portfolios to avoid double-counting.[4]

4.4.2 Theory in the case with mutual information

Section 4.4.1 highlights the value of the common information set of two portfolios. This value stems from two sources: sizing the order based on an improved overnight alpha and not double-counting the trading alpha.

However, sharing information between the two portfolio teams is not enough: Proposition 4.4.6 shows that, even if both portfolios trade on the joint information set, competition effects diminish their P&L. In this case, sub-optimal information does not cause the loss, and, instead, the two teams miss gains due to an *arms race*: a standard instance of a competitive equilibrium under-performing a collaborative equilibrium.

Many papers solve the Nash equilibrium with joint information. For example,

- Brunnermeier and Pedersen (2005)[41] solve a particular case with two agent types: traders liquidating a distressed position and strategic traders aware of this flow. Price impact follows the Almgren and Chriss model. The Nash equilibrium includes *predatory* behavior: the strategic traders cause additional price impact during the distressed liquidation before later unwinding their positions.

- Carlin, Lobo, and Viswanatha (2005)[57] and Schied and Schöneborn (2008)[204] extend the game to include more scenarios between the agents. These scenarios yield two Nash equilibria: *predatory* behavior and *liquidity provision.* The agents' strategy depends on price impact parameters in

[3] This behavior is in line with Donnelly and Lorig (2020)[86], who combine partial information-sharing with distinct alpha signals. The authors "study how the overall price impact on the asset depends on how much information is shared between agents as dictated by the correlation between their signals" (p. 2).

[4] In practice, a trading team may face organizational challenges when allocating their alpha across competing portfolio teams.

the Almgren and Chriss model and the number of distressed and strategic traders.

- Schied, Strehle, and Zhang (2017)[205] solve the two-player game for the original OW model with instantaneous costs. Shied, Strehle, and Zhang show that the existence of Nash equilibria depends on the ratio between price impact and immediate costs.
- Strehle (2018)[217] extends the solution for the original OW model from two to an arbitrary number of players and derives a closed-form formula.
- Voß (2019)[228] solves a two-player game in the Almgren and Chriss model. Again, the existence of Nash equilibria depends on the ratio between price impact and instantaneous costs.
- Neuman and Voß (2021)[176] characterize the solution to a game with an arbitrary number of players under the original OW model using coupled solutions of FBSDEs. They prove the existence and uniqueness of the Nash equilibria and mean-field limit.
- Fu, Horst, and Xia (2022)[100] derive the mean-field game solution for the generalized OW model.

Definition 4.4.3 (Competitive trading). *Consider two traders $i \in \{0;1\}$ controlling trading processes $Q^i \in \mathcal{S}$ under the shared filtration $\mathcal{F}$. Assume both traders*

(a) trade under the original OW model I with parameters β, λ. Denote the price impact of the trading process Q^i by I^i and assume $I_0^i = 0$.

(b) share the overnight alpha $\tilde{\alpha} \in \mathbb{R}$.

(c) consider each other's impact as an external impact on their execution.

(d) share the trading alpha α_t for $t \in (0,T)$ satisfying the deterministic trading alpha assumption from Corollary 2.4.1 in Chapter 2.

Given candidate trading processes $\left(Q^i\right)_{i\in\{0,1\}}$, or equivalently target impact states $\left(I^i\right)_{i\in\{0,1\}}$, for both traders, define a game where each trader's expected P&L is

$$J_i\left(I^i, I^{1-i}\right) =$$
$$\mathbb{E}\left[\tilde{\alpha} Q_T^i + \int_0^T Q_t^i dS_t - \int_0^T \left(I_t^i + I_t^{1-i}\right) dQ_t^i - \frac{1}{2}\left[I^i, Q^i\right]_T - \left[I^{1-i}, Q^i\right]_T\right].$$

Define an open-loop Nash equilibrium for the game as a pair of $\left(I^i\right)_{i\in\{0,1\}} \in \mathcal{S}_2$ such that for $i \in \{0,1\}$

$$\forall I \in \mathcal{S} \quad J_i\left(I, I^{1-i}\right) \leq J_i\left(I^i, I^{1-i}\right).$$

Definition 4.4.3 is one possible competitive equilibrium. Variants introduce heterogeneity between the traders, further information-sharing, and other advanced features. Remarks 4.4.4 and 4.4.5 provide examples.

Remark 4.4.4 (Heterogeneous traders). *To the author's best knowledge, Fu, Horst, and Xia (2022)[100] solve the broadest trading equilibrium under the generalized OW model. Fu, Horst, and Xia cover scenarios between heterogeneous traders by adding initial positions and terminal constraints. For instance, Fu, Horst, and Xia show that a trader without alpha on a stock finds profitable round-trip trades based on another portfolio's price impact. This adverse behavior remains* even if the other portfolio is aware of the strategy*:*

> *Full trading transparency between portfolio managers within a firm does not lead to the best firm outcome.*

Remark 4.4.5 (Closed-loop Nash equilibrium). *One challenging extension arises when two teams do not just share information but share their strategies' implementation. This extension leads to a* closed-loop Nash equilibrium. *One refers to Carmona and Delarue (2016)[58, 59] for the mathematical definition of a closed-loop Nash equilibrium. Under an Almgren and Chriss price impact model, Carmona and Yang (2008)[62] solve and numerically simulate a* closed-loop Nash equilibrium *in a two-player game. Carmona and Yang apply their model to simulate profitable round-trip trades based on the impact of another trader's execution strategy.*

Micheli, Muhle-Karbe, and Neuman (2021)[167] extend the closed-loop equilibrium of Carmona and Yang (2008)[62] to N players with alpha signals.

> *"Compared to the corresponding open-loop Nash equilibrium, both the agents' optimal trading rates and their performance move towards the central-planner solution, in that excessive trading due to lack of coordination is reduced" (p. 1, [167]).*

The model's mean-field limit remains an open question.

> *"An intriguing theoretical question is whether the difference [between open- and closed-loop equilibrium] vanishes completely in the "mean-field limit" of many small agents" (p. 15, [167]).*

Proposition 4.4.6 (Nash equilibrium example). *The pair of strategies satisfying the backward ODE for $t \in (0, T]$*

$$I'_t = \beta \left(3I_t - \tilde{\alpha} - \alpha_t + \beta^{-1}\alpha'_t\right)$$

with terminal condition $I_T = \frac{1}{2}\tilde{\alpha}$ defines an open-loop Nash equilibrium for the game as per Definition 4.4.3.

Proof. Mapping $J_i(I, I^{1-i})$ into impact space yields:

$$J_i(I, I^{1-i}) = \mathbb{E}\left[\int_0^T \frac{\beta}{\lambda}\left(\left(\tilde{\alpha} + \alpha_t - I_t^{1-i}\right) I_t - I_t^2\right) dt\right]$$
$$+ \mathbb{E}\left[\int_0^T \frac{I_t}{\lambda} d\left(\alpha_t - I_t^{1-i}\right) + \frac{1}{\lambda}\left(\left(\tilde{\alpha} - I_T^{1-i}\right) I_T - \frac{1}{2} I_T^2\right)\right].$$

For the candidate solution, I^{1-i} exhibits a jump at $t = 0$ and is differentiable for $t > 0$. The candidate solution leads to

$$J_i(I, I^{1-i}) = \mathbb{E}\left[\int_0^T \frac{\beta}{\lambda}\left(\left(\tilde{\alpha} + \alpha_t - I_t^{1-i} - \beta^{-1}\left(\alpha_t' - (I_t^{1-i})'\right)\right) I_t - I_t^2\right) dt\right]$$
$$+ \mathbb{E}\left[\frac{1}{\lambda}\left(\left(\tilde{\alpha} - I_T^{1-i}\right) I_T - \frac{1}{2} I_T^2\right)\right].$$

Therefore, the partial equilibrium for I^{1-i} is

$$\forall t \in (0, T) \quad I_t = \frac{1}{2}\left(\tilde{\alpha} + \alpha_t - I_t^{1-i} - \beta^{-1}\left(\alpha_t' - (I_t^{1-i})'\right)\right)$$

and

$$I_T = \tilde{\alpha} - I_T^{1-i}.$$

Looking for a symmetric equilibrium reduces the problem to the backward ODE for $t \in (0, T]$

$$I_t' = \beta\left(3I_t - \tilde{\alpha} - \alpha_t + \beta^{-1}\alpha_t'\right)$$

with terminal condition $I_T = \frac{1}{2}\tilde{\alpha}$. This backward ODE has a solution satisfying the Nash condition. □

Remark 4.4.7 (Estimating the loss). *As in the scenario without information-sharing, a trader estimates the competitive equilibrium's opportunity cost by comparing the total price impact in the competitive case* $2I_t$ *and the impact in the collaborative case* I_t^*. *The dynamics of* I_t *depend on the signal* α. *Two general observations follow.*

(a) *For* $t \in (0, T)$, *the competitive impact* $2I_t$ *mean-reverts around* $\frac{4}{3}I_t^*$: *even with a combined information set, two competitive traders overpay their total impact by a third. This equilibrium is an example of the* tragedy of the commons: *competitive agents over-trade on shared information. For instance, see Subrahmanyam (1998)[219]:*

> *"we develop a model that captures a key element of the real-world competition between informed agents, namely, the race to uncover information earlier than others" (p. 83).*

(b) *The total impact reaches the same final state as in the collaborative scenario*

$$2I_T = I_T^*.$$

Remark 4.4.8 (Illustrating the arms race). *Assume $\alpha_t = 0$ for $t \in (0, T)$ and $T \gg \beta^{-1}$. The two teams perform block trades at $t = 0$, and the total impact state jumps to*

$$2I_{0+} \approx \frac{4}{3} I_{0+}^*$$

before monotonically increasing to

$$2I_T = I_T^*.$$

In words, two competitive teams

(a) *rush to trade the shared alpha,*

(b) *trade an elevated impact state, leading to increased execution speed for $t > 0$, and*

(c) *do not perform a block trade at $t = T$.*

The lack of jump at $t = T$ and the sizable jump at $t = 0$ stem from an arms race*: each team accelerates to not trade after the other's impact. The Nash equilibrium leads this arms race to a logical conclusion with a sizable jump at $t = 0$ and no jump at $t = T$. Figure 4.14 illustrates the arms race.*

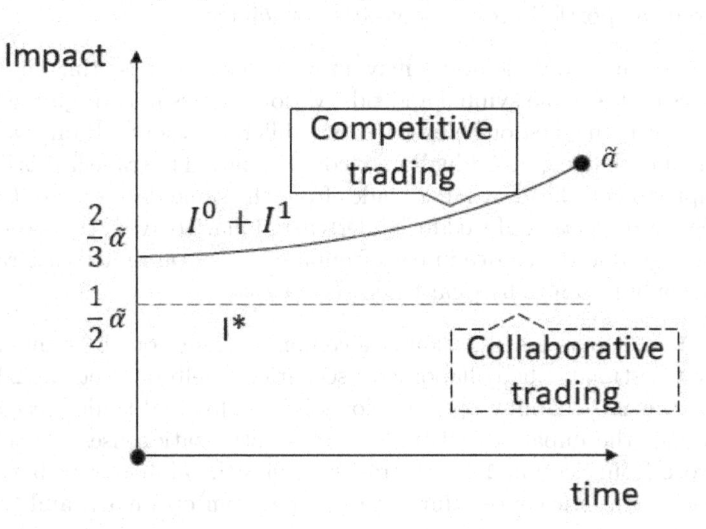

FIGURE 4.14
Illustration of the impact state for competitive and collaborative trading without trading alpha. See Remark 4.4.8.

4.4.3 Empirical simulation approach

Sections 4.4.1 and 4.4.2 illustrate three ways combining two portfolios' trading adds value.

(a) If the two portfolios trade independently, then their orders over-trade. Therefore, combining the execution adds value by correctly sizing the combined order.

(b) If the two portfolios trade independently, they will double-count trading alphas. Therefore, combining the trading adds value by not double-counting the trading alpha.

(c) If the two portfolios compete with the same information set, they will enter an arms race and trade aggressively. Therefore, combining the trading adds value by reducing transaction costs.

Unfortunately, these results are illustrative toy models: real-life data is unlikely to make the information-sharing regime between trading teams explicit. Therefore, the results provide intuition and assess the robustness of simulations. However, one does not need to model the information-sharing to simulate the two portfolios trading separately. Indeed, an empirical approach uses historical data from the two teams at face value, regardless of whether they stem from an independent or competitive scenario. For example, in Waelbroeck et al. (2012)[231], portfolio teams trade through a single trading desk. Waelbroeck and Gomes study whether the trading desk should analyze the data from the portfolio teams separately or jointly.

> "One client (trading desk) may provide data representing multiple trades on the same symbol and side; various trades may originate from different managers, or be executed via different traders. If one were to look at each trade individually, a recommendation to trade fast is likely, simply to get ahead of other trades from the same desk - impact from other trades gets confused for underlying alpha. To avoid this problem, it is desirable to remove impact from a client's overall activity, rather than only for an individual trade" (p. 56).

One implements a simulation of a combined trader on the same historical window to estimate the collaborative scenario. Waelbroeck et al. (2012)[231] describe this as "enabling optimization of the individual trade piece after accounting for the impact of all trades". This optimization uses the simulation framework from Section 4.2.3. A trader implies the alphas from both portfolios, runs a statistical procedure to combine them optimally, and solves the combined execution problem.

4.5 Summary of Results

This section summarizes price impact results for portfolio applications. Let Q be a trading strategy's position, P the observed mid-price, I a price impact model, and $S = P - I$ the unperturbed price.

4.5.1 Alpha research

> One trader's impact is another trader's alpha.

This observation allows a trader to turn any *predictive* flow model into an alpha signal using a price impact model, such as the Order Flow Imbalance (OFI) model of Cont, Cucuringu, and Zhang (2021)[73].

In general, alpha signals predict unperturbed returns

$$\begin{aligned}\alpha_t &= \mathbb{E}\left[S_T - S_t \middle| \mathcal{F}_t\right] \\ &= \mathbb{E}\left[P_T - I_T - P_t + I_t \middle| \mathcal{F}_t\right].\end{aligned}$$

One derives unperturbed returns from observed returns by adjusting for price impact.

Alpha researchers predict returns to maximize their portfolio's P&L. Therefore, practitioners must align their statistical loss function to their algorithm's expected P&L:

> A loss in an alpha signal's predictive power maps to a P&L regret for the trading strategy.

For the generalized OW model, an alpha signal α_t maximizes a risk-neutral trader's fundamental P&L by minimizing weighted least-squares in *impact* space. Hence, the weighted least squares loss function to minimize is

$$\min_{\alpha} \mathbb{E}\left[\int_0^T \frac{2\beta_t + \gamma_t'}{2\lambda_t} \left(I_t(\alpha) - I_t(\alpha^r)\right)^2 dt\right]$$

where $I(\alpha)$ is the optimal impact given an alpha signal α, and α^r is the realized alpha the signal predicts.

4.5.2 Market simulator

Assume given the following inputs.

(a) Let Q^r, P^r be the realized trading process and prices.

(b) Let Q be a trading process representing novel trades. The trading process Q simulates an alternative scenario to the realized trading process Q^r.

(c) Let I be a price impact model.

A market simulator outputs an unperturbed price

$$S_t = P_t^r - I_t(Q^r)$$

and simulated market prices

$$P_t(Q) = P_t^r - I_t(Q^r) + I_t(Q)$$

under the novel trading strategy Q.

4.5.3 Liquidity risk management

A position or trade's actual value is its expected *closing P&L*, defined as

$$\mathbb{E}[X_\tau | \mathcal{F}_t]$$

where τ is the position's future closing time.

Assuming no further alpha opportunities, a position's expected closing P&L is equal to its fundamental P&L, minus liquidation costs:

$$\mathbb{E}[X_\tau | \mathcal{F}_t] = Y_t - \frac{\Lambda}{2} Q_t^2$$

where $\frac{\Lambda}{2} Q_t^2$ is a pre-trade cost model.

Therefore, a position's *accounting* P&L, also called its *mark-to-market* P&L, is a biased estimator of its fundamental value. A team assesses this bias using

(a) the trading footprint

$$X_t - Y_t = Q_t I_t$$

which has a positive expectation and reverts to zero when the position closes.

(b) the liquidation costs, as per a pre-trade cost model.

A risk-neutral trader values position-entering and -exiting trades symmetrically: they have equal arrival slippage and fundamental P&L. However, position-entering and -exiting trades have *asymmetric* trading footprints. Therefore, they have asymmetric accounting P&L, and a portfolio's trading footprint assesses this P&L asymmetry.

> Managing liquidity risk is synonymous with understanding price impact's effect on mark-to-market P&L, which the trading footprint encapsulates.

4.5.4 Combining two portfolios' trading

Two portfolios can trade in three separate ways:

(a) Independently

Two portfolios trading independently double-count shared alphas. If the two portfolios are correlated, they incur high impact costs.

(b) Competitively

Two portfolios trading competitively based on the same information enter an arms race: each has the incentive to trade ahead of the other team's impact.

(c) Collaboratively

The collaborative case merges the two teams' alphas and executes a single, optimal order. This strategy avoids double-counted alphas and the arms race but requires coordination between the portfolios.

A trading team can simulate, quantify, and compare the three scenarios using back tests.

4.6 Exercises

Exercise 11 Price simulation

This exercise implements the price simulation introduced by Waelbroeck et al. (2012) [231] and tests the original optimal execution strategy by Obizhaeva and Wang (2013)[180]. Fix the parameters

$$N = 10000; \quad \sigma = 0.05; \quad \beta = 6.5; \quad \lambda = 0.1 \cdot \sigma.$$

1. Simulate a random walk for S_n over $n = 1...N$. Assume Gaussian increments with mean zero and variance $\frac{\sigma^2}{N}$.
2. Implement the function $I(Q)$ for the original OW model.
3. Plot the price impact, fundamental P&L, and accounting P&L for a TWAP execution over $[0, \frac{N}{2}]$, a TWAP execution over $[0, \frac{N}{4}]$, and the optimal execution strategy.
4. Assume in this question that the simulated price includes the price impact from a previous TWAP execution over $[0, \frac{N}{4}]$. Then, re-simulate the price path by replacing this execution's price impact with that of the slower TWAP over $[0, \frac{N}{2}]$.

Exercise 12 Alpha simulation

This exercise builds upon the previous exercise and adds an alpha signal to the simulation.

Synthetic alpha signals I

Consider a synthetic signal

$$\alpha_t = a\left(S_T - S_t\right) + b\sigma\left(W_T - W_t\right)$$

for parameters a, b, and an independent random walk W. Section 1.2.4 in Chapter 1 defines $S_T - S_t$ as the signal's *realized alpha*. Therefore, one constructs the synthetic alpha by adding noise to the realized alpha.

1. Compute the correlation ρ between α_t and $S_T - S_t$ for given a, b.
2. Pick values of a, b such that $\alpha_t = \mathbb{E}\left[S_T - S_t \middle| \alpha_t\right]$ and $\rho = 0.05$.

Simulated trading II

Simulate a rolling prediction horizon $h = T - t$ where $h = 100$ for α_t and $h = 1$ for α'_t.

1. What are the strategy's simulated profits? How do trading metrics scale with ρ?

Exercise 13 Footprint simulation

This exercise studies position inflation. The reader replicates the footprint simulation from Example 4.3.9 on a Markovitz portfolio by building upon the previous two exercises. Fix the following parameters:

$$N = 1000; \quad M = 500; \quad \tilde{\sigma} = 0.05; \quad \beta = 1; \quad \lambda = 0.3\sigma$$

where $n = 1...N$ now represent trading days and M the number of stocks in a portfolio. Each stock i has variance σ^i. Sample σ^i from the absolute value of a Gaussian with mean zero and variance $\tilde{\sigma}^2$. In this setting, the signal is a *rolling* alpha predicting returns over the next $n = 20$ days. This leads to

$$\alpha_t = a\left(S_{t+20} - S_t\right) + b\left(W_{t+20} - W_t\right)$$

for the synthetic alpha.

1. Calibrate a, b such that the alpha prediction has correlation $\rho = 0.05$ with returns.

   Assume that at the end of each day, the portfolio implements a simplified Markovitz portfolio

   $$Q^i_n = \frac{\alpha^i_n}{\left(\sigma^i\right)^2}.$$

2. Assume the trader executes order $Q^i_n - Q^i_{n-1}$ using a TWAP algorithm and compute the daily value of I.

3. Assume the trader executes order $Q^i_n - Q^i_{n-1}$ using the original OW execution model and compute the daily value of I.
4. In both cases, simulate and plot the time-series of the portfolio's trading footprint and fundamental P&L.

Exercise 14 N-player Nash equilibrium

1. Generalize the 2-player Nash equilibrium from Proposition 4.4.6 to an N-player game.
2. Describe the total impact state $\sum_k I_k$ at equilibrium as $N \to \infty$.

Part III

Measuring Price Impact

Part III focuses on *measuring* price impact and emphasizes building price impact models considering common trading biases. The reader learns to

(a) define a causal model, detect causal biases, and account for them.

(b) disentangle an order's alpha from its impact.

(c) measure crowding and leakage in market trades.

(d) consider microstructure when measuring price impact.

(e) adjust price impact measurements for past orders.

(f) fit and compare price impact models on public trading data.

(g) build a cross-impact model.

Chapter 5 is technical and lays the mathematical foundation for the applications in Chapters 6 and 7. Therefore, the reader can skip this chapter in their first study. Furthermore, Section 5.4 provides a concise reference to mathematical results.

Chapter 6 works through four ubiquitous trading biases. Chapter 7 statistically compares benchmark price impact models. Both chapters end with sections summarizing concrete models, formulas, and statistics for practitioners.

5

An Introduction to the Mathematics of Causal Inference

> *"Fortunately, the days of 'statistics can only tell us about associations, and association is not causation' seem to be permanently over."*
> – Rubin (2004)[202]

The study of causality has recently undergone a transformation: previously relegated to footnotes and phrased in terms of vague interpretations, causal inference is now a precise mathematical theory. Indeed, causal inference extends Bayesian statistics and is an active area of research in machine learning. Knight (2019)[130] interviewed Yoshua Bengio on causal inference applied to machine learning, also called causal machine learning. Yoshua Bengio won the Turing Award for his contribution to deep learning.

> "It's a big thing to integrate [causality] into AI," Bengio says. "Current approaches to machine learning assume that the trained AI system will be applied on the same kind of data as the training data. In real life it is often not the case."

Schölkopf et al. (2021)[207] tackle the increasing intersection of machine learning and causal inference.

> "[G]eneralizing well outside the i.i.d setting requires learning not mere statistical associations between variables, but an underlying *causal model*" (p. 1).

Technology and healthcare companies already actively apply causal machine learning to their businesses.

(a) The Microsoft Research Summit of 2021[194] had over a dozen talks within its causal machine learning track, one of its seven science tracks.

> "This track focuses on emerging causal machine learning technologies and the opportunities for practical impact at the intersection of academia and industry, with contributions from researchers at Microsoft and the broader academic and industrial research communities."

(b) Netflix (2018)[174] made live experiments and causal machine learning "the driving force behind the pace of our innovations."

DOI: 10.1201/9781003316923-5

> "Netflix consistently employs a simple but powerful approach to product innovation: we ask our members, through online experiments, which of several possible experiences resonate with them."
>
> "We use controlled A/B experiments to test nearly all proposed changes to our product, including new recommendation algorithms, user interface (UI) features, content promotion tactics, originals launch strategies, streaming algorithms, the new member signup process, and payment methods."

Netflix organized a week-long *Causal Inference and Experimentation summit* (2022)[175].

> "Netflix scientists recently came together for an internal Causal Inference and Experimentation Summit. The weeklong conference brought speakers from across the content, product, and member experience teams to learn about methodological developments and applications in estimating causal effects."

(c) Prosperi et al. (2020)[190] published a paper titled *Causal inference and counterfactual prediction in machine learning for actionable healthcare.* According to Prosperi et al. machine learning solves practical business problems in healthcare when combined with causal inference.

> "When pursuing intervention modelling, the bio-health informatics community needs to employ causal approaches and learn causal structures" (p. 2).

Trading teams also rely, explicitly or implicitly, on causal inference to solve problems in algorithmic trading. For example,

- Chapter 6 highlights and solves common causal biases traders face when fitting price impact models.
- Section 3.3 of Chapter 3 highlights the role of A-B testing and live trading experiments in TCA.

More generally, Pearl (2004)[186] highlights the need to document assumptions when drawing conclusions from models.

> "Since such assumptions are based primarily on human judgment, it is important to formally assess to what degree the target conclusions are sensitive to those assumptions or, conversely, to what degree the conclusions are robust to violations of those assumptions" (p. 1).

Causal graphs visually communicate model premises to stakeholders, such as portfolio managers or technology teams. Such documentation aids novel methods' adoption, such as machine learning, by visualizing assumptions without constraining models to simplistic parametric forms.

Today, investment professionals are beginning to explore applications of causal inference to finance. For instance, Lopez de Prado (2022)[83] introduces "Causal factor investing" (p. 1) and argues for the use of causal graphs, causal mathematics, and controlled live experiments to "rebuild factor investing on the more solid scientific foundations of causal inference" (p. 5). In addition, the Abu Dhabi Investment Authority (ADIA), a sovereign wealth fund managing an estimated $790 billion (SWFI (2022)[220]), created the "ADIA Lab Award for Causal Research in Investments" (ADIA (2022)[3]). The award "aims to promote the use of the formal language of causal inference to express findings in the areas of finance", cites Lopez de Prado (2022)[83], and provides further potential applications of causal inference to finance.

> "For example, papers may propose causal graphs to model the effects of various monetary policies, use Bayesian networks to implement a coherent stress testing framework, propose randomized controlled trials designed to assess the efficacy of broker algo-wheels, etc."

This chapter provides a technical introduction to the mathematical theory of causal inference. The impatient reader can skip this chapter in their first reading and directly study the applications presented in Chapters 6 and 7. Section 5.2 covers the theory's basic building blocks and follows the book *Causality* by Pearl (2009)[187]. Section 5.3 focuses on regression methods for live trading experiments. Finally, Section 5.4 summarizes causal inference's essential mathematical definitions and theorems.

5.1 A Pedagogical Example

For traders, this chapter's essential takeaway is that drawing conclusions from trading data requires causal modeling. For teaching purposes, this section provides a simple example where practitioners draw opposite conclusions from the same trading data based on their understanding of cause and effect.

Algorithm	Order Size	Sample Size	Slippage	Slippage
aggressive	small	40k	-10bps	-16bps
aggressive	large	10k	-40bps	
passive	small	10k	-5bps	-21bps
passive	large	40k	-25bps	

FIGURE 5.1
Example TCA table comparing the arrival slippage of an aggressive and a passive algorithm across order sizes.

A trader determines whether the aggressive or passive algorithm performs better considering the TCA Table 5.1. Then, the trader compares three metrics.

(a) The algorithms' overall arrival slippages,

$$-16 > -21.$$

The aggressive algorithm has better overall slippage than the passive algorithm.

(b) The algorithms' arrival slippages for small orders,

$$-10 < -5.$$

The aggressive algorithm has worse slippage than the passive algorithm for small orders.

(c) The algorithms' arrival slippages for large orders,

$$-40 < -25.$$

The aggressive algorithm has worse slippage than the passive algorithm for large orders.

A trader can draw two opposite conclusions *from the same data:*

The aggressive algorithm has lower expected slippage than the passive algorithm, *but* it has higher expected slippage for small orders and higher expected slippage for large orders.

The reader checks that the size-based slippages do not contradict the overall slippages considering the *sample sizes.* On average, the aggressive algorithm executed smaller orders. Therefore, the apparent contradiction is not due to an *algebra mistake.* Instead, these counter-intuitive metrics are an example of *Simpson's paradox.* Section 5.2.3 formalizes *Simpson's paradox* and its application to ranking trading algorithms.

A trader resolves Simpson's paradox by stating whether order size causes the algorithm choice or whether the algorithm causes the order's size.

(a) For example, assume the trader assigns small orders with a higher probability to the aggressive algorithm. In that case, the aggressive algorithm did not *decide* to have smaller orders. Conversely, the passive algorithm did not *decide* to have larger orders. Because larger orders mechanically exhibit a larger price impact, comparing overall slippages is not fair. Therefore, the trader draws the correct conclusion using size-based arrival slippage.

(b) On the other hand, assume the trader sends the same alpha to both algorithms, and the aggressive algorithm *decides* to execute a smaller size than the passive algorithm. Then, the aggressive algorithm's actions cause the overall slippage. Therefore, the trader draws the correct conclusion using overall arrival slippage.

5.2 A Technical Primer on Causal Inference

The following quote by Pearl (2009)[187] motivates this technical primer on the mathematics of causal inference. Emphasis was added.

> "In the last decade, owing partly to advances in graphical models, causality has undergone a major transformation: from a concept shrouded in mystery into a mathematical object with well-defined semantics and well-founded logic. Paradoxes and controversies have been resolved, slippery concepts have been explicated, and practical problems relying on causal information that long were regarded as either metaphysical or unmanageable can now be solved using elementary mathematics. **Put simply, causality has been mathematized**" (p. xv).

The mathematical theory of causal inference is an extension of Bayesian statistics. Any model in probability theory starts with the line

Let $(\Omega, \mathcal{F}, \mathbb{P})$ be a probability space...

to mathematically ground definitions, results, and proofs. Ω determines the space of possible outcomes, $\mathcal{F}$ the collection of measurable events, and $\mathbb{P}$ a probability measure over these events. Causal inference adds a further object to the probability space: *a causal structure* $\mathcal{G}$.

A causal structure encodes conditional independence assumptions through functional relationships between random variables. Adding a causal structure to a probability space resolves statistical paradoxes and biases found, for example, in trading. Causal inference also extends Bayes' formula into a broader calculus, named do-calculus. Section 5.2.2 covers do-calculus.

Notation 5.2.1 (Conditional distributions). *Given a probability space $(\Omega, \mathcal{F}, \mathbb{P})$ with random variables X, Y, and Z, one is interested in expressing statements about conditional distributions.*

One understands equalities such as

$$\mathbb{P}(X|Y,Z) = \mathbb{P}(X|Z)$$

in the following way:

(a) For discrete variables, the equality

$$\mathbb{P}(X = x|\,Y = y, Z = z) = \mathbb{P}(X = x|\,Z = z)$$

holds for all valid outcomes x, y, and z for the random variables X, Y, and Z.

(b) For continuous variables, the equality

$$p(X = x|\,Y = y, Z = z) = p(X = x|\,Z = z)$$

holds, where p is the associated conditional probability density function.

5.2.1 Causal structures

First, recall the standard definition of conditional independence in probability.

Definition 5.2.2 (Conditional independence). *Let $(\Omega, \mathcal{F}, \mathbb{P})$ be a probability space and X, Y, and Z be three random variables. Define X and Y to be conditionally independent given Z if*

$$\mathbb{P}(X \mid Y, Z) = \mathbb{P}(X \mid Z).$$

The definition extends to three sets of random variables X, Y, and Z.

Conditional independence plays a crucial role in statistical models. Statisticians may prove conditional independence from other model properties or assume conditional independence as a core model property. In the latter case, a *conditional independence assumption* constrains the class of statistical models under consideration. A statistician motivates such assumptions with a fundamental understanding of the underlying mechanism generating the data.

Example 5.2.3 (Causal interpretation of an independence assumption). *Imagine the following two dice-throwing experiments.*

(a) Flip one fair coin per hand. For each coin, if it lands on heads, roll an eight-sided dice; otherwise, roll a six-sided dice with the corresponding hand.

(b) Flip a single fair coin. If it lands on heads, roll an eight-sided dice with your left hand and a six-sided dice with your right hand. On tails, roll a six-sided dice with your left hand and an eight-sided dice with your right hand.

In experiment (a), the outcomes of the two dice rolls are independent. For example, if one observes the value rolled on the left hand, this observation provides no information on the value rolled on the right hand. In experiment (b), the outcomes are dependent. For example, if one observes a roll of eight on the left hand, one determines that the right hand is rolling a six-sided dice and rules out the values seven and eight for the right hand.

Intuitively, the initial coin flip in experiment (b) assigns the dice to each hand and causes the dependency between the dice rolls. Therefore, a statistician recovers independence between the two dice rolls by conditioning on the initial coin flip's value.

A *causal structure* formally represents such conditional independence assumptions and visually documents them in graphical form.

Definition 5.2.4 (Causal structure). *A* causal structure *of a set of variables V is a directed acyclical graph (DAG) in which each node corresponds to a distinct element of V.*

The graph's directed links represent direct functional relationships between the corresponding variables.

A statistician promotes the causal structure to a *causal model*, as per Definition 5.2.7, by instantiating these links with specific functional relationships and adding a probability measure to the variables.

Example 5.2.5 (Simpson's graph). *Figure 5.2 provides an example of a causal structure $\mathcal{S}$ expressed as a directed acyclical graph and refers to it as* Simpson's graph *in relation to* Simpson's paradox. *Section 5.2.3 covers Simpson's paradox on a trading example. The paradox refers to a standard statistical bias when estimating the effect of a treatment variable X on an outcome variable Y in the presence of a confounding variable Z.*

The linked variables X, Y, and Z have the following causal interpretations.

(a) Z has no parents and is an exogenous *variable with a direct effect on children X and Y.*

(b) X directly affects child Y and is an outcome of parent Z.

(c) Y has no descendants and is an outcome *of parents X and Z.*

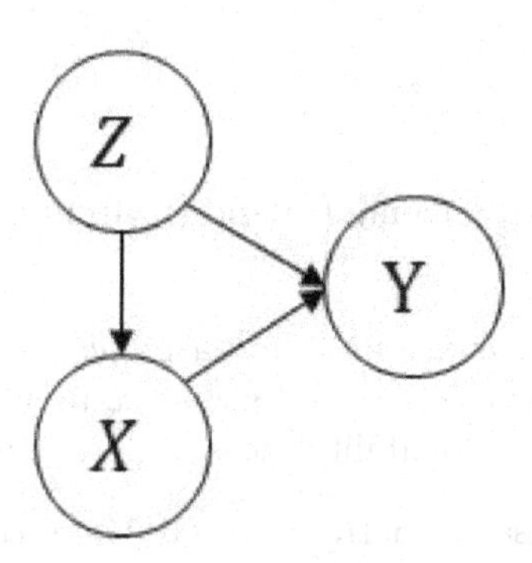

FIGURE 5.2
Causal structure $\mathcal{S}$ for three variables X, Y, and Z from Example 5.2.5.

In the language of causal inference, one describes Z as a confounding variable *for estimating the effect of X on Y leading to a causal* bias. *Section 5.2.4 formalizes this description in the language of* do-calculus *and defines* causal biases.

Example 5.2.6 (Causal graph for Example 5.2.3). *A statistician distinguishes the two experiments from Example 5.2.3 by their causal graphs, illustrated in Figure 5.3.*

(a) Causal graph $\mathcal{D}_a$ represents experiment (a). For $x \in \{l, r\}$ representing the left- and right-hand indexes, C_x and D_x are the outcomes of the corresponding coin flip and dice roll. The lack of links between the left and right

sides of the causal graph visually represents the independence between the left-hand and right-hand variables.

(b) *Causal graph $\mathcal{D}_b$ represents experiment (b). A single coin flip C affects both dice rolls D_l and D_r. This creates a dependency between the observed outcomes of D_l and D_r. Intuitively, observing the common cause C separates nodes D_l and D_r: the dice rolls D_l and D_r are independent conditional on the initial coin flip C.*

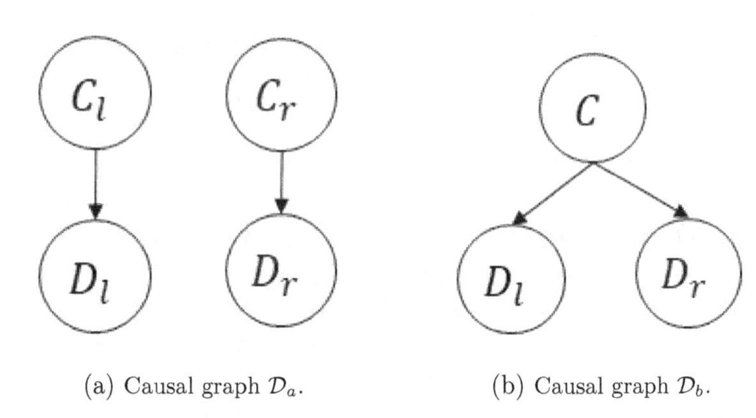

(a) Causal graph $\mathcal{D}_a$. (b) Causal graph $\mathcal{D}_b$.

FIGURE 5.3
Causal structures for experiments (a) and (b) from Example 5.2.3.

The terms *causal structure* and *causal graph* are interchangeable. Moreover, a statistician promotes a *causal structure* to a *causal model* by adding functional links f_i and a probability measure $\mathbb{P}$, as per Definition 5.2.7.

Definition 5.2.7 (Causal model)**.** *A* causal model *consists of*

(a) *a causal structure, as per Definition 5.2.4.*

(b) *a set of functions f_i compatible with the causal structure,*

$$f_i : (parents(x_i), \epsilon_i) \mapsto f_i(parents(x_i), \epsilon_i)$$

where $parents(x_i)$ are potential outcomes of the parent variables *of X_i and ϵ_i is a noise term idiosyncratic to X_i.*

(c) *a probability space $(\Omega, \mathcal{F}, \mathbb{P})$ that assigns probabilities to all the ϵ_i, with each ϵ_i being independent.*

Example 5.2.8 (Causal model for Simpson's graph)**.** *Consider the causal structure $\mathcal{S}$ introduced in Example 5.2.5. A statistician can promote the causal structure $\mathcal{S}$ to a linear causal model $\mathcal{M}$. The model assigns the indexes $i = 1, 2, 3$ to the nodes X, Y, and Z.*

The linear formulas

$$X = \alpha Z + \epsilon_1$$
$$Y = \beta X + \gamma Z + \epsilon_2$$
$$Z = \epsilon_3$$

for parameters α, β, and γ specify the functions f_i. The noise terms ϵ_i are independent Gaussians with mean zero and variance σ_i under $\mathbb{P}$. Figure 5.4 illustrates the causal model $\mathcal{M}$.

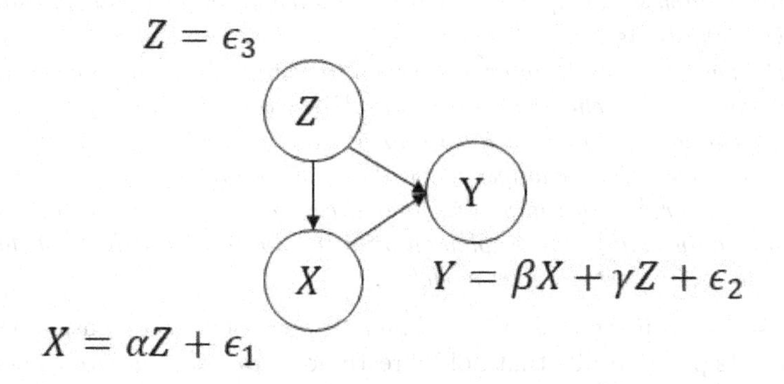

FIGURE 5.4
Causal model $\mathcal{M}$ for three variables X, Y, and Z from Example 5.2.8.

Causal models represent a *constructivist* point of view on causal inference and enable the reader to implement complex simulations. Remark 5.2.9 expands on the constructivist point of view and highlights how a causal model is also a *data generation model* for $\mathbb{P}$.

Remark 5.2.9 (Causal models as data generation models). *Computer scientists use data generation models to simulate random processes, such as in a Monte Carlo simulation. However, such data generation models are complex and challenging to manage, modify, and reproduce. Practitioners in High-Performance Computing (HPC) rely on computational DAGs to articulate a simulation's logical components and control its complexity. For example, Bosilca et al. (2012)[30] present "A generic distributed DAG engine for High Performance Computing" (p. 1).*

A causal model naturally fits into this computational framework:

(a) A causal model $\mathcal{M}$ begins by assigning probabilities to the ϵ_i.

(b) These probabilities then propagate along the causal graph $\mathcal{G}$ and assign a joint probability distribution $\mathbb{P}$ over all the variables X_i.

The model's causal graph is a natural template for the simulation's computational graph.

Example 5.2.10 (A simple computational graph). *To highlight the robustness of a causal graph approach to implementing a simulation, consider Example 5.2.8. The causal model naturally articulates the simulation's components and the order in which the simulation can execute the components.*

(a) The simulation first generates node Z, which has no parents.

(b) The simulation generates X after generating its parent Z.

(c) The simulation generates Y after generating its parents Z and X.

The causal graph's lack of cycles guarantees such an order exists by ruling out recursive definitions.

Furthermore, a practitioner who wants to change the underlying model will likely make minimal changes to the causal structure. For example, a statistician can change the distribution ϵ_3 to be positive and X to depend on the square root of Z. Such changes leave the model's underlying DAG structure untouched. Therefore, a computer scientist can re-use the same computational graph and only modify the implementation of nodes Z and X to update the simulation and reflect the new model.

Causal structures represent a *probabilistic* point of view on causal inference. Probabilists prove results that hold directly for a *causal structure* rather than a specific causal model where possible. Such results are robust to particular choices of the f_i and ϵ_i and allow the application of non-parametric fitting methods, such as machine learning. Definition 5.2.11 grounds the probabilistic point of view.

Definition 5.2.11 (Consistency). *Define a probability distribution $\mathbb{Q}$ on the set of variables V* as consistent *with the causal structure $\mathcal{G}$ if there exists a causal model $\mathcal{M}$ based on the causal structure $\mathcal{G}$ that generates $\mathbb{Q}$ for the variables in V.*

Define statements that hold for *any* probability measure consistent with a *causal structure* $\mathcal{G}$ as *causal statements* under $\mathcal{G}$. Causal inference is a deep theory: other notions related to causal structures and statements go beyond this chapter's scope. However, Remark 5.2.12 provides examples and references for such causal concepts.

Remark 5.2.12 (Informal definitions of model preferences). *A given causal model is not the* only *representation of either the probability model or the data generation model. Pearl (2009; pp. 45–49)[187] and the references within provide a mathematical treatment of model preferences, which explore causal models' potential* uniqueness *properties.*

(a) Minimality

There exists an incomplete ordering over the space of causal structures. The ordering leads to a notion of minimality over causal graphs. Because the ordering is incomplete, there does not necessarily exist a single minimal causal structure.

(b) Stability

A class of causal models is stable when one rules out distributions for the ϵ_i that invalidate *dependency links. Stability guarantees that the causal structure $\mathcal{G}$ accurately represents the dependencies between variables regardless of the specific choice of $\mathbb{P}$ made in the causal model $\mathcal{M}$.*

The power of causal structures stems from their ability to visually encode dependencies *for any $\mathbb{P}$ consistent with the causal structure.* The reader will quickly build up intuition and be able to prove conditional independence assumptions from a causal graph's visual properties. The first step is understanding the three *paths* that create dependencies between variables.

Definition 5.2.13 (DAG paths). *Let $\mathcal{G}$ be a DAG. Define a* path *connecting nodes X and Y as a sequence of consecutive edges,* regardless of their directionality, *leading from X to Y.*

Furthermore, define the following three-node paths for three nodes i, m, and j.

(a) A chain

$$i \to m \to j$$

(b) A fork

$$i \leftarrow m \to j$$

(c) A collider

$$i \to m \leftarrow j$$

Definition 5.2.14 (d-separation). *Let $\mathcal{G}$ be a DAG. Define a path p to be d-separated (or blocked) by a set of nodes Z if and only if*

(a) the path p contains a chain $i \to m \to j$ or a fork $i \leftarrow m \to j$ such that m is in Z.

(b) the path p contains a collider $i \to m \leftarrow j$ such that neither m nor its descendants are in Z.

Define two sets of nodes X and Y to be d-separated by the set Z if Z d-separates every path from a node in X to a node in Y.

Chains and *forks* relate to one of the two nodes' *parents:* observing m prevents dependencies from traveling along chains and forks. Remark 5.2.17 addresses *colliders.* Theorem 5.2.15 links the visual d-separation of nodes on a causal graph with the probabilistic independence of the associated random variables.

Theorem 5.2.15 (Probabilistic implications of d-separation). *Let $\mathcal{G}$ be a causal structure and X, Y, and Z be three sets of variables on $\mathcal{G}$.*

(a) *If Z d-separates X and Y, for every probability measure $\mathbb{P}$ consistent with $\mathcal{G}$, X and Y are conditionally independent given Z.*

(b) *Conversely, if Z does not d-separate X and Y, then there exists at least one probability measure $\mathbb{P}$ consistent with $\mathcal{G}$ for which X and Y are* not *conditionally independent given Z.*

Proof. Theorem 5.2.15 is exactly Theorem 1.2.4 from Pearl (2009, p. 18)[187]. □

The terms *d*-separation and *causal independence* under a causal structure $\mathcal{G}$ are interchangeable. Theorem 5.2.15 proves that causal independence is equivalent to statistical independence across all probability measures consistent with a given causal structure $\mathcal{G}$.

Example 5.2.16 (Establishing conditional independence from a causal graph). *Consider the causal graph $\mathcal{S}'$ from Figure 5.5. The goal is to find a set of variables to obtain conditional independence of X and Y given those variables. Consider the following three options.*

(a) *The empty set $\emptyset$ does* not *d-separate X and Y.*

The fork $X \leftarrow Z \rightarrow Y$ prevents causal independence between X and Y, as the empty set does not control Z.

(b) *The singleton $\{Z\}$ d-separates X and Y.*

Z blocks the fork $X \leftarrow Z \rightarrow Y$. While there is a collider $X \rightarrow W \leftarrow Y$, W is not part of the conditioning variables and has no descendants.

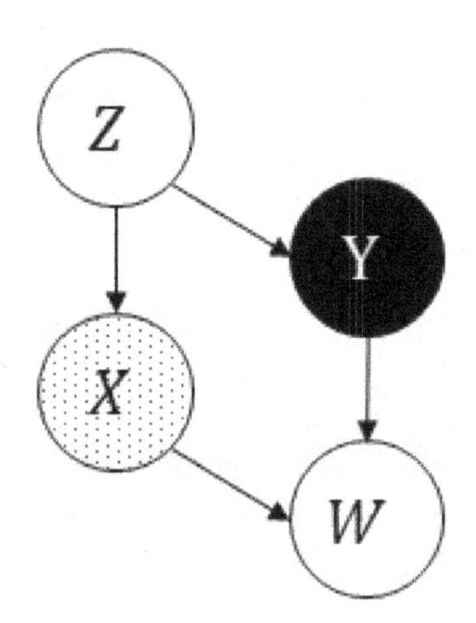

FIGURE 5.5
Causal graph $\mathcal{S}'$ from Example 5.2.16.

(c) *The set* $\{Z, W\}$ *does* not *d-separate* X *and* Y.

The collider $X \to W \leftarrow Y$ *creates a dependence between* X *and* Y *when conditioning on* W.

Remark 5.2.17 (Intuition on the collider condition). *The following quote by Pearl (2009)[187] summarizes the intuition on colliders. Emphasis was added.*

> *"At first glance, readers might find it a bit odd that conditioning on a node not lying on a blocked path may unblock the path. However, this corresponds to a general pattern of causal relationships:* ***observations on a common consequence of two independent causes tend to render those causes dependent****, because information about one of the causes tends to make the other more or less likely, given that the consequence has occured. This pattern is known as* selection bias *or* Berkson's paradox *in the statistical literature (Berkson 1946) and as the* explaining away effect *in artificial intelligence (Kim and Pearl 1983)" (p. 17).*

Example 5.2.18 (The Monty Hall problem). *The Monty Hall problem is a classic probability puzzle that illustrates the power of Bayes' formula. This example views the Monty Hall problem within a causal model to explain Remark 5.2.17 on colliders.*

The Monty Hall problem, named after the original host of the American television game show "Let's make a deal," involves a game with the following rules:

(a) *The host presents three doors to a game participant. Behind one of the three doors, chosen by the host before the start of the game, stands a valuable prize, such as a car. The participant wins the prize if they guess the door the prize stands behind. A goat stands behind the other two doors and is considered worthless.*

Mathematically, let H *be the variable denoting the door choice made by the host.* H *takes value in* $\{1, 2, 3\}$.

(b) *The participant chooses one of the three doors. The host does not reveal the chosen door at this stage.*

Denote by P *this choice made by the participant.* P *takes value in* $\{1, 2, 3\}$ *and is independent of* H.

(c) *The host then* reveals *one of the remaining worthless doors.*

This step is always feasible: if the participant's chosen door does not contain the valuable prize, then one of the two other doors does, and the host reveals the third door as worthless. On the other hand, if the participant's chosen door contains the valuable prize, both other doors are worthless, and the host can pick one at random.

Denote by R *the revealed door.* R *takes value in the set difference* $\{1, 2, 3\} \setminus \{P, H\}$ *given knowledge of* P *and* H.

(d) Given the revealed worthless door R, the participant has the option to reveal the door P chosen initially or to pick the third door. Their final door choice determines their prize.

The causal structure $\mathcal{H}$ illustrated in Figure 5.6 summarizes the relationship between H, P, and R: the game host and participant independently reach their decisions H, P. The revealed door R is a common cause of H and P.

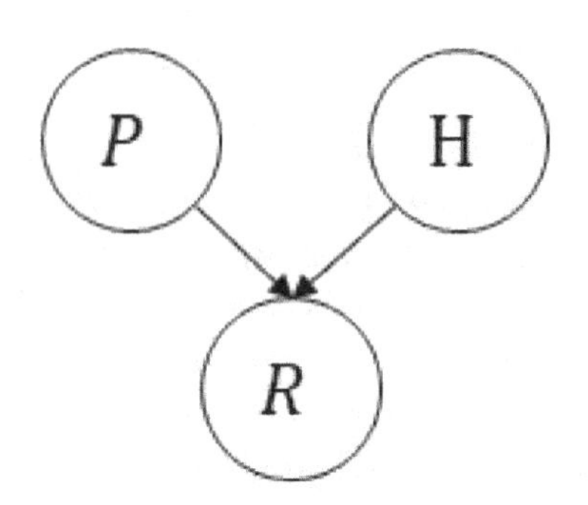

FIGURE 5.6
Causal graph $\mathcal{H}$ for the Monty Hall problem from Example 5.2.18. P and H represent the participant and host's independent choices. R is the revealed door, which depends on both P and H.

The game's last step corresponds to the estimation of the conditional probability distribution

$$\mathbb{P}(H|\,P,R)\,.$$

Without the information from the revealed door, the participant picks the right door with probability $\frac{1}{3}$. Indeed, H and P are independent and

$$\mathbb{P}(H|\,P) = \mathbb{P}(H)\,.$$

However, Bayesian calculations show

$$\mathbb{P}(H|\,P,R) \neq \mathbb{P}(H|\,R)\,.$$

Indeed,

$$\begin{aligned}\mathbb{P}(H=3|\,P=1,R=2) &= \frac{\mathbb{P}(H=3,P=1,R=2)}{\mathbb{P}(P=1,R=2)}\\ &= \frac{\mathbb{P}(H=3,P=1)}{\mathbb{P}(P=1,R=2,H=1)+\mathbb{P}(P=1,R=2,H=3}\\ &= \frac{(1/3)^2}{(1/2)\cdot(1/3)^2+(1/3)^2}\\ &= 2/3\end{aligned}$$

while

$$\begin{aligned}\mathbb{P}(H=3|\, R=2) &= \frac{\mathbb{P}(H=3, R=2)}{\mathbb{P}(R=2)} \\ &= \frac{\mathbb{P}(H=3, R=2, P=1) + \mathbb{P}(H=3, R=2, P=3)}{1/3} \\ &= \frac{(1/3)^2 + (1/2)\cdot(1/3)^2}{1/3} \\ &= 1/2.\end{aligned}$$

> *The host and participant's choices H and P are not independent conditional on the revealed door R.*

Indeed, the third *door, the one not picked by the participant and not revealed by the host, has a probability $\frac{2}{3}$ of containing the valuable prize.*

> *The participant maximizes their odds of winning the valuable prize by considering the revealed door and switching their choice in the Monty Hall problem's last step.*[1]

The statistician cannot quantify the probabilities associated with the decision without promoting the causal structure $\mathcal{H}$ to a causal model. However, the causal structure $\mathcal{H}$ establishes the negative result

$$\mathbb{P}(H|\, P, R) \neq \mathbb{P}(H|\, R)$$

due to the collider path $P \to R \leftarrow H$. Conditional on observing the revealed door, the decisions made by the host and the participant become dependent. To repeat Pearl (2009)[187]:

> *"observations on a common consequence of two independent causes tend to render those causes dependent" (p. 17).*

A common thread in this chapter is that *causal structures* function as *fundamental constraints* on candidate probability models. Probabilists leverage these constraints to prove *causal statements* for probability measures consistent with the causal structure. Likewise, statisticians leverage these constraints to simplify the calibration of non-parametric models.

---

[1] Burns and Wieth (2003)[46] study how people react to the Monty Hall problem. Burns and Wieth cite multiple other studies and contribute their own, all indicating that the public has difficulties understanding the optimal strategy for the Monty Hall problem. The correct answers to the Monty Hall problem in the studies range from 9 to 12%. However, when the researchers presented the Monty Hall problem with clear causal links, up to 51% of the public identified the optimal strategy:

> "participants were more likely to solve the problem correctly when presented with versions that made the causal structure of the problem easier to understand" (p. 201).

5.2.2 Do-calculus

Do-calculus is a set of mathematical rules used to prove *causal statements.* Traders ask causal questions, also called *what-if* questions, in the informal language of *counterfactuals.* A *counterfactual* answers the question:

What if I had done X?

Counterfactuals describe in words an *alternative scenario* to the observed data. This section attaches mathematical definitions and statistical estimates to such counterfactuals. Definition 5.2.19 formalizes the counterfactual

What if I had done X?

by defining the action do(X) as a change of probability measure.

Definition 5.2.19 (The do() operator). *Given two disjoint sets of variables X and Y on a causal model $\mathcal{M}$, define the do(X) action*

$$\mathbb{P}\,(Y|\,do(X = x)) = \tilde{\mathbb{P}}\,(Y)$$

where one obtains $\tilde{\mathbb{P}}$ by replacing all the functions f_i from Definition 5.2.7 that pertain to variables $X_i \in X$ with the constant function $X_i = x_i$.

Definition 5.2.19 defines the action do($X = x$) for x constant. Causal inference refers to such actions as *constant interventions.* Definition 5.2.19 extends to *randomized interventions* where x is an *independent* random variable. Traders can express randomized, controlled experiments as randomized interventions, as seen in Example 5.2.20.

Example 5.2.20 (Do actions in a linear causal model). *Consider the linear causal model from Example 5.2.8. The constant intervention $do(X = x)$ for $x \in \mathbb{R}$ leads to a new probability measure $\tilde{\mathbb{P}}$ generated by the truncated causal model*

$$\begin{aligned} X &= \cancel{\alpha Z + \epsilon_1} \\ &= x \\ Y &= \beta X + \gamma Z + \epsilon_2 \\ Z &= \epsilon_3. \end{aligned}$$

Similarly, the randomized intervention $do(X = \epsilon)$ where

$$\mathbb{P}(\epsilon = 1) = \mathbb{P}(\epsilon = -1) = \frac{1}{2}$$

leads to a new probability measure $\tilde{\mathbb{P}}$ generated by the truncated causal model

$$\begin{aligned} X &= \epsilon \\ Y &= \beta X + \gamma Z + \epsilon_2 \\ Z &= \epsilon_3. \end{aligned}$$

Remark 5.2.21 (Relationship with Bayesian conditioning). *The choice to use a similar notation for Bayesian conditioning and do actions is not by accident. Under certain conditions, the so-called* naive *estimation formula*

$$\mathbb{P}(Y|\, do(X)) = \mathbb{P}(Y|X)$$

holds.

However, naive estimation does not always *hold: a common cause of both X and Y may confound the effect of X on Y. When naive estimation fails, one states that the causal structure presents a* causal bias *for the identification of X's effect on Y.*

Causal inference establishes purely Bayesian *formulas for* action *estimates. Section 5.2.4 studies this* identifiability problem *in detail. Establishing a* causal structure *is vital to linking counterfactuals to Bayesian statistics, as Schölkopf et al. (2021)[207] articulate.*

> *"Once a causal model is available, either by external human knowledge or a learning process,* causal reasoning *allows to draw conclusions on the effect of interventions, counterfactuals and potential outcomes. In contrast, statistical models only allow to reason about the outcome of i.i.d. experiments" (p. 6).*

Definition 5.2.22 (Mixing do() with Bayesian conditioning). *Given three disjoint sets of variables X, Y, and Z on a causal model $\mathcal{M}$, one defines*

$$\mathbb{P}(Y|\, do(X=x), Z) = \frac{\mathbb{P}(Y, Z|\, do(X=x))}{\mathbb{P}(Z|\, do(X=x))}.$$

Definition 5.2.19 follows a *constructivist* point of view. Definition 5.2.23 and Proposition 5.2.24 outline the *probabilistic* point of view on the do-operator based on the concept of *link erasure* and *trimmed causal graphs.* The *probabilistic* point of view leads to Theorem 5.2.26, which outlines the rules of do-calculus.

Definition 5.2.23 (Trimming causal graphs). *Let X and Y be two sets of variables on a DAG $\mathcal{G}$. Define the following* link erasures.

(a) Define erasing the parents *of X as the graph $\mathcal{G}_{\overline{X}}$ where one erases all links pointing to nodes in X from $\mathcal{G}$.*

(b) Define erasing the children *of X as the graph $\mathcal{G}_{\underline{X}}$ where one erases all links emerging from nodes in X from $\mathcal{G}$.*

(c) One can combine multiple erasure operations by appending graph indexes. For example, $\mathcal{G}_{\overline{X}\underline{Y}}$ erases the parents of X and the children of Y.

Proposition 5.2.24 (Do action). *Let $\mathcal{G}$ be a causal structure and $\mathbb{P}$ a probability measure consistent with $\mathcal{G}$. Then the probability measure $\tilde{\mathbb{P}}$ obtained from the do-action do(X) is consistent with the causal structure $\mathcal{G}_{\overline{X}}$.*

The probabilistic point of view interprets the do(X) action as *erasing the parents* of X in the causal model. Figure 5.7 visually represents the do(X) action on the causal graph $\mathcal{G}_{\overline{X}}$.[2]

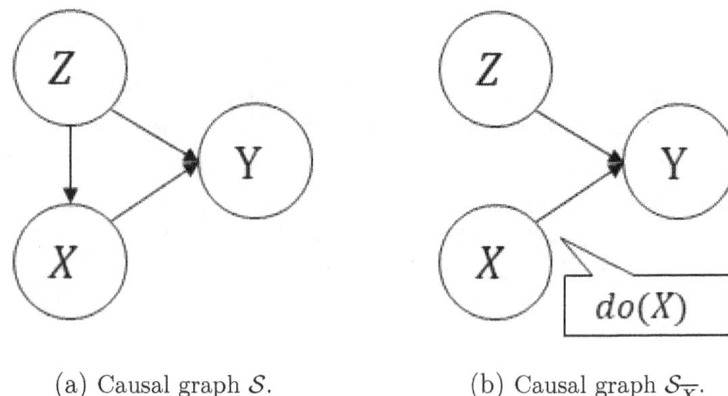

(a) Causal graph $\mathcal{S}$. (b) Causal graph $\mathcal{S}_{\overline{X}}$.

FIGURE 5.7
Visual representation of the action do(X) on the causal graph $\mathcal{S}$ as per Definition 5.2.23.

Remark 5.2.25 (Generating interventional data). *The trimmed causal model reflects the* counterfactual *associated with the do(X) action: the* intervention *on node X changes the data generation model to reflect the new probability measure $\tilde{\mathbb{P}}$.* Observational data *refers to data generated under the original causal model and measure $\mathbb{P}$.*

A statistician can estimate any Bayesian formula under $\mathbb{P}$ given enough observational data. However, there is no guarantee that observational data recover counterfactuals, *as they, by definition, correspond to a new probability measure $\tilde{\mathbb{P}}$ generated by a* different *DAG.*

Interventional data *refers to data explicitly generated from the counterfactual scenario and probability measure $\tilde{\mathbb{P}}$. Statisticians can estimate naive Bayesian formulas for do()-actions on interventional data as the data directly reflects $\tilde{\mathbb{P}}$. Randomized interventions and counterfactuals are vital concepts in A-B testing, and the do(X) operator formalizes A-B testing. Section 5.3.1 applies interventional measurement to compare trading strategies' quality.*

[2]Lattimore and Rohde (2019)[144] highlight the effectiveness of do-calculus and the ability for more involved Bayesian models to recover the same results by explicitly modeling interventions.

> "[D]o-calculus and Bayesian inference can both be used in order to make causal estimates, although this is not to suggest that each approach does not have its strengths and weaknesses" (p. 1).

Other methods besides do-calculus exist to establish counterfactuals, including the potential outcomes framework by Rubin (2004)[202].

Do-calculus follows three basic rules. Shipster and Pearl (2006)[214] and Huang and Valtorta (2006)[118] have shown the rules of do-calculus to be *complete* for the derivation of all identifiable causal effects. Section 5.2.4 introduces a direct method, the back-door criterion, to establish a causal formula's *identifiability*.

Theorem 5.2.26 (Rules of do-calculus). *Let $\mathcal{M}$ be a causal model with causal structure $\mathcal{G}$ and probability measure $\mathbb{P}$. For any disjoint subsets of variables X, Y, Z, and W, one has the following rules.*

(a) Rule 1 *(Insertion/deletion of observations):*

$$\mathbb{P}\left(Y \mid do(X), Z, W\right) = \mathbb{P}\left(Y \mid do(X), W\right)$$

if Y and Z are conditionally independent given X and W on the graph $\mathcal{G}_{\overline{X}}$.

(b) Rule 2 *(Action/observation exchange):*

$$\mathbb{P}\left(Y \mid do(X), do(Z), W\right) = \mathbb{P}\left(Y \mid do(X), Z, W\right)$$

if Y and Z are conditionally independent given X and W on the graph $\mathcal{G}_{\overline{X}\underline{Z}}$.

(c) Rule 3 *(Insertion/deletion of actions):*

$$\mathbb{P}\left(Y \mid do(X), do(Z), W\right) = \mathbb{P}\left(Y \mid do(X), W\right)$$

if Y and Z are conditionally independent given X and W on the graph $\mathcal{G}_{\overline{X}\overline{Z(W)}}$, where $Z(W)$ is the subset of Z that are not ancestors of nodes in W in $\mathcal{G}_{\overline{X}}$.

Proof. Theorem 5.2.26 is exactly Theorem 3.4.1 from Pearl (2009, p. 85)[187]. □

5.2.3 Simpson's paradox

> "Simpson's paradox provides an excellent case study for demonstrating that raw data cannot be used for inferring causality without further assumptions" (Lattimore and Rohde (2019, p. 4)[144]).

Like the Monty Hall problem, Simpson's paradox is a counter-intuitive statistical result that can catch traders off-guard. This section illustrates Simpson's paradox on a trading example and applies do-calculus to solve the associated trading problem. It revisits the pedagogical example from Section 5.1 in a formal setting.

Example 5.2.27 (Measuring two trading algorithms' performance). *Traders deploy multiple trading algorithms across their orders. A generic problem in trading is the* performance measurement *of trading algorithms: traders compare algorithms and deploy more orders to the trading algorithms that perform better.*

For example, assume a trader can deploy two trading algorithms: a passive algorithm and an aggressive algorithm. The trader wants to quantify the effect of switching over to one or the other algorithm. Let A denote a variable representing the choice of algorithm. A takes value in $\{p, a\}$, where p stands for the passive algorithm and a for the aggressive algorithm. The arrival slippage Y *measures, in basis points, an order's performance. For example, an order with $Y = -20$ underperformed the arrival price by 20bps. Section 1.2.7 of Chapter 1 defines arrival price and arrival slippage. The counterfactual of interest is*

What if I always use algorithm A?

The action $do(A)$ captures the counterfactual, and the trader estimates the causal formula

$$\mathbb{E}\left[Y \middle| do(A)\right]$$

to quantify the counterfactual.

In Example 5.2.27, the trader considers three scenarios for the observed trading data.

(a) A fair coin toss determines the choice of trading algorithm A.

(b) The order size determines the choice of trading algorithm A.

Let S be a discrete random variable capturing the order size. S takes values in $\{s, l\}$, where s stands for small and l for large orders.

(c) A fair coin toss determines the choice of trading algorithm A and the trading speed depends on algorithm A.

Let $\bar{S}$ be a discrete variable capturing the order's trading speed. $\bar{S}$ takes values in $\{\bar{s}, \bar{l}\}$, where $\bar{s}$ stands for small and $\bar{l}$ for large trading speeds.

Under scenario (a), the naive estimator

$$\mathbb{E}\left[Y \middle| A\right]$$

is the only available and correct estimator for the counterfactual of interest. The trader directly compares

$$\mathbb{E}\left[Y \middle| A = a\right]; \quad \mathbb{E}\left[Y \middle| A = p\right]$$

to decide whether algorithm a or p performs better.

Under scenario (b), a choice appears due to the trading variable S. The trader could directly estimate each algorithm's arrival slippage

$$\mathbb{E}\left[Y \middle| A\right],$$

or the trader could estimate each algorithm's arrival slippage for large and small orders separately:

$$\mathbb{E}\left[Y \mid A, S\right].$$

The trader then estimates which algorithm performs better separately for small and large orders.

At first glance, one would expect both approaches to yield identical results. However, Simpson's paradox states that not only do those two methods differ, but probability distributions exist for which the two conclusions are opposite! Example 5.2.28 provides one such probability distribution.

Example 5.2.28 (A counter-intuitive example)**.** *Consider scenario (b) from Example 5.2.27 with the expected slippage numbers from Figure 5.8. Basic Bayesian calculus provides the conditional expectation*

$$\begin{aligned}
\mathbb{E}\left[Y \mid A = a\right] &= \mathbb{E}\left[\mathbb{E}\left[Y \mid A = a, S\right] \mid A = a\right] \\
&= \mathbb{E}\left[Y \mid A = a, S = s\right] \cdot P\left(S = s \mid A = a\right) \\
&\quad + \mathbb{E}\left[Y \mid A = a, S = l\right] \cdot P\left(S = l \mid A = a\right) \\
&= (-10) \cdot (0.8) + (-40) \cdot (0.2) \\
&= -16
\end{aligned}$$

where the first equality holds by the tower property of conditional expectations. Similarly, one estimates

$$\begin{aligned}
\mathbb{E}\left[Y \mid A = p\right] &= (-5) \cdot (0.2) + (-25) \cdot (0.8) \\
&= -21.
\end{aligned}$$

When studying the aggregate statistics, the trader concludes that the aggressive algorithm $A = a$ has less negative arrival slippage and performs better than the passive algorithm. Indeed, the aggressive algorithm outperforms the passive algorithm by 5bps.

A	S	Sample size	$E[Y \mid A, S]$
a	s	40k	-10bps
a	l	10k	-40bps
p	s	10k	-5bps
p	l	40k	-25bps

FIGURE 5.8
Expected arrival slippages across orders from scenario (b) from Example 5.2.27.

> *The conclusion when studying the aggregate statistics contradicts the conclusion when analyzing* the same data *for small and large orders separately.*

Indeed, the aggressive algorithm underperforms the passive algorithm by 5*bps for small orders and* 15*bps for large orders!*

$$\mathbb{E}[Y|A=a,S=s]-\mathbb{E}[Y|A=p,S=s]=-5$$
$$\mathbb{E}[Y|A=a,S=l]-\mathbb{E}[Y|A=p,S=l]=-15$$

Example 5.2.28 provides a probability distribution for which

$$\mathbb{E}[Y|A]$$

and

$$\mathbb{E}[Y|A,S]$$

lead to contradictory results on the effect of A on Y. Both expressions are valid Bayesian estimators: the ambiguity arises when attaching either estimator to the *counterfactual*

> What if I always use algorithm A?

The probability distribution from Figure 5.8 drives the ambiguity in Example 5.2.28. For instance, a trader could observe identical statistics for scenario (c) by replacing S, s, and l with $\bar{S}$, $\bar{s}$, and $\bar{l}$, leading to the same ambiguity.

The trader lifts the ambiguity by adding a causal structure to the probability distribution and formalizing their counterfactual of interest. The trader then applies do-calculus, as defined in Theorem 5.2.26, to identify the correct formula for their counterfactual.

Example 5.2.29 applies do-calculus to resolve Simpson's paradox from first principles. In addition, Section 5.2.4 introduces the backdoor criterion as a user-friendly alternative.

Example 5.2.29 (A basic application of do-calculus). *Consider scenarios (a), (b), and (c) from Example 5.2.27 under causal structures $\mathcal{S}^a$, $\mathcal{S}^b$, and $\mathcal{S}^c$, as defined in Figure 5.9.*

(a) The graph $\mathcal{S}^a_{\underline{A}}$ separates A from Y. Indeed, the graph $\mathcal{S}^a_{\underline{A}}$ removes the links originating from A: with no links going from or to A, there are no paths from A to Y. Applying rule 2 from Theorem 5.2.26 leads to

$$\mathbb{P}(Y|\,do(A))=\mathbb{P}(Y|A)\,.$$

as Y and A are independent under $\mathcal{S}^a_{\underline{A}}$. It follows that

$$\mathbb{E}[Y|\,do(A)]=\mathbb{E}[Y|A]\,.$$

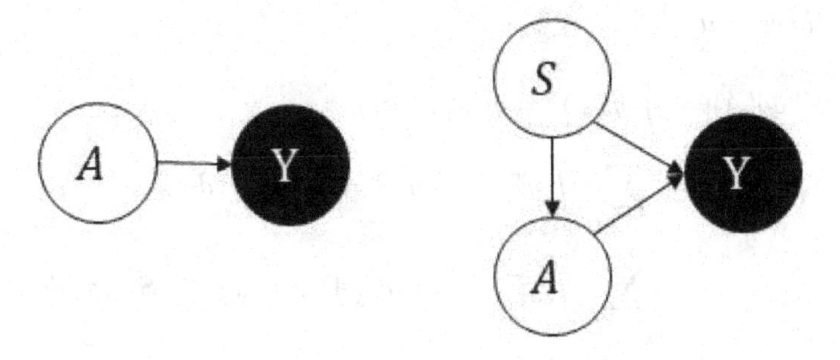

(a) Causal graph $\mathcal{S}^a$ for scenario (a).

(b) Causal graph $\mathcal{S}^b$ for scenario (b).

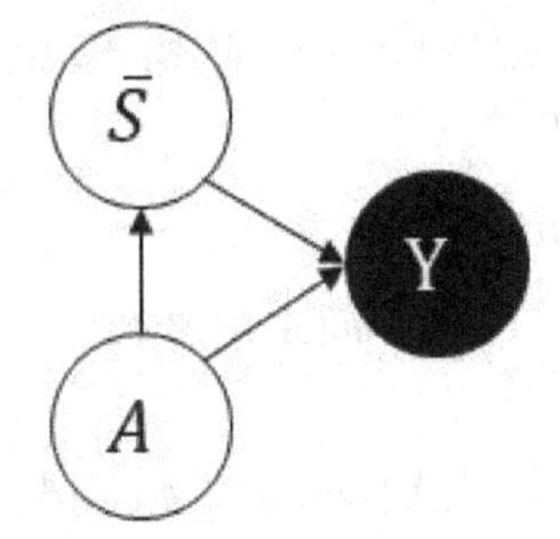

(c) Causal graph $\mathcal{S}^c$ for scenario (c).

FIGURE 5.9
Causal structures for scenarios (a), (b), and (c) from Example 5.2.27.

(b) *Under the graph $\mathcal{S}^b_{\underline{A}}$, A and Y are d-separated by S. Indeed, the graph $\mathcal{S}^b_{\underline{A}}$ removes the links originating* from *A but keeps the link $S \to A$ pointing to A. Therefore, A and Y are not independent under $\mathcal{S}^b_{\underline{A}}$, but they are independent* conditional *on S: all paths between A and Y go through S under $\mathcal{S}^b_{\underline{A}}$. Applying rule 2 from Theorem 5.2.26 leads to*

$$\mathbb{P}(Y|\, do(A), S) = \mathbb{P}(Y|\, A, S)\,.$$

Morever, S, A are d-separated on $\mathcal{S}^b_{\overline{A}}$. Hence, rule 3 from Theorem 5.2.26 yields

$$\mathbb{P}(S|\, do(A)) = \mathbb{P}(S)\,.$$

It follows that

$$\begin{aligned}
\mathbb{E}[Y|\, do(A)] &= \int y\mathbb{P}(Y = y|\, do(A))\, dy \\
&= \sum_{x\in\{s,l\}} \int y\mathbb{P}(Y = y, S = x|\, do(A))\, dy \\
&= \sum_{x\in\{s,l\}} \int y\mathbb{P}(Y = y|\, do(A), S = x) \cdot \mathbb{P}(S = x|\, do(A))\, dy \\
&= \sum_{x\in\{s,l\}} \int y\mathbb{P}(Y = y|\, A, S = x) \cdot \mathbb{P}(S = x)\, dy \\
&= \sum_{x\in\{s,l\}} \mathbb{E}[Y|\, A, S = x] \cdot \mathbb{P}(S = x).
\end{aligned}$$

One also has the negative result

$$\mathbb{P}(Y|\, do(A)) \neq \mathbb{P}(Y|\, A)$$

and hence,

$$\mathbb{E}[Y|\, do(A)] \neq \mathbb{E}[Y|\, A]$$

since A and Y are not *causally independent on $\mathcal{A}^b_{\underline{A}}$. The fork $A \leftarrow S \rightarrow Y$ requires conditioning on S to d-separate A and Y.*

(c) *Under the graph $\mathcal{S}^c_{\underline{A}}$, A and Y are separate. Indeed, the graph $\mathcal{S}^b_{\underline{A}}$ removes the links originating* from *A: this includes the link $A \rightarrow \bar{S}$ and leaves A isolated. Applying rule 2 from Theorem 5.2.26 leads to*

$$\mathbb{E}[Y|\, do(A)] = \mathbb{E}[Y|\, A].$$

However, because $\bar{S}$ is a descendant of A, one has the negative result

$$\mathbb{E}[Y|\, do(A)] \neq \sum_{x\in\{\bar{s},\bar{l}\}} \mathbb{E}\left[Y|\, A, \bar{S} = x\right] \cdot \mathbb{P}\left(\bar{S} = x\right).$$

Conditioning on the common cause $\bar{S}$ re-introduces a dependence between Y and A: one cannot identify the effect of A on Y by conditioning on the common cause $\bar{S}$.

Example 5.2.29 shows that, under the causal structure $\mathcal{S}^a$, scenario (a) generates interventional data for the action do(A). The trader identifies the effect of A on Y with a naive Bayesian formula.

Under the causal model $\mathcal{S}^b$, the trader must condition by S to estimate the counterfactual of interest because S is a common cause of A and Y. S is a *confounding variable* for the estimation of the effect of A on Y.

Scenario (c) provides an example of the opposite result: one can include too many conditioning variables. Scenarios (b) and (c) are near identical, with the distinction that $\bar{S} \leftarrow A$ replaces the direction $S \rightarrow A$. Therefore, the trader lifts the ambiguity in Simpson's paradox differently depending on the causal link's orientation.

Remark 5.2.30 (Interpretation and trader intuition). *Assume a trader observes the expected slippage numbers from Figure 5.8 for either scenario (b) or (c).*

- *In scenario (b), S is a choice outside the algorithms' control. Intuitively, the trading team allocated less costly orders to the aggressive algorithm and the allocation, rather than the algorithm's merit, caused the lower slippage.*

  Traders demand fair *data and comparisons to evaluate trading strategies. For straightforward trading infrastructures and strategies, a trader's intuition can be enough to establish an analysis's fairness. However, a causal structure documents more sophisticated trading infrastructures and demonstrates fairness for analyzing trading strategies at scale.*

- *In scenario (c), the algorithm controls $\bar{S}$. Hence, the aggressive algorithm trading more orders in the $\bar{S} = \bar{s}$ category is due to its merit rather than an external choice. In conclusion, traders should be wary of over-using control variables: selecting a consequence rather than a cause introduces rather than removes confounding in the analysis.*

Simpson's paradox is an example of non-trivial identifiability. Moreover, it illustrates the pitfalls of using Bayesian estimates to study counterfactuals.

5.2.4 Identifiability of causal formulas

Section 5.2.2 introduces the do() operator and the associated rules of do-calculus. The do() operator mathematically formalizes the notion of *counterfactual* and explores changes to the data generating process. The do() operator shares a common notation with Bayesian conditioning but is mathematically distinct. The goal of causal inference is to formulate and estimate counterfactuals.

(a) A trader formulates the counterfactual of interest by phrasing the counterfactual as a do() action.

(b) A statistician estimates the counterfactual of interest using interventional data or applying the rules of do-calculus to simplify a causal into a purely Bayesian expression.

Definition 5.2.31 introduces *identifiability.*[3] Theorem 5.2.34, called the back-door criterion, provides a condition to simplify causal expressions using do-calculus.

[3] This definition is a mathematical shortcut: it corresponds to Corollary 3.4.2 by Pearl (2009, p. 86)[187], which proves its equivalence to Definition 3.2.4 (p. 77).

Definition 5.2.31 (Identifiability). *Define a causal formula*

$$\mathbb{P}(Y|\, do(X), Z)$$

to be identifiable *given a causal structure $\mathcal{G}$ and a set of observable variables if there exists a sequence of do-calculus operations from Theorem 5.2.26 that leads to a purely Bayesian formula (without do() operator) involving only observable variables.*

Define the formula to be naively identifiable *if one can prove the following equality via do-calculus*

$$\mathbb{P}(Y|\, do(X), Z) = \mathbb{P}(Y|\, X, Z)\,.$$

By Definition 5.2.22, the identifiability of

$$\mathbb{P}(Y|\, \mathrm{do}(X), Z)$$

is equivalent to the identifiability of

$$\mathbb{P}(Y, Z|\, \mathrm{do}(X))\,; \quad \mathbb{P}(Z|\, \mathrm{do}(X))\,.$$

Without loss of generality, one can state results for the latter, simpler form.

Corollary 5.2.32 is a straightforward, common sufficient condition for establishing naive identifiability of causal expressions.

Corollary 5.2.32 (A sufficient naive identifiability condition). *Let $\mathcal{M}$ be a causal model with causal structure $\mathcal{G}$. Let X and Y be two variables such that X has no parents. Then the effect of X on Y is naively identifiable:*

$$\mathbb{P}(Y|\, do(X)) = \mathbb{P}(Y|\, X)\,.$$

Proof. X has no parents on $\mathcal{G}$. Therefore, X is an entirely isolated node on $\mathcal{G}_{\underline{X}}$. One applies rule 2 of Theorem 5.2.26: Y and X are independent on $\mathcal{G}_{\underline{X}}$ and

$$\mathbb{P}(Y|\, \mathrm{do}(X)) = \mathbb{P}(Y|\, X)\,.$$

For clarity, the variable X in this corollary does not match the variable X in Theorem 5.2.26 but instead matches variable Z. □

The intuition behind Corollary 5.2.32 is that, with no parents for X, only direct paths from X to Y can cause correlation between X and Y. Definition 5.2.33 and Theorem 5.2.34 introduce the notion of back-door to generalize this insight.

Definition 5.2.33 (Back-door). *Let X and Y be two variables in a causal structure $\mathcal{G}$. Define the set of variables Z to satisfy the* back-door *criterion* from X to Y *if:*

(a) Z d-separates every path between X and Y with a link pointing into X.

(b) No descendant of X is in Z.

The back-door *criterion extends to sets of variables X and Y by requiring Z to satisfy the back-door criterion for every* pair *formed from X and Y.*

Definition 5.2.33 generalizes the naive situation from Corollary 5.2.32: conditioning on the correct variables leaves node X with only direct paths to Y.

Theorem 5.2.34 (Back-door criterion). *Let $\mathcal{M}$ be a causal model with causal structure $\mathcal{G}$. Assume the set Z satisfies the back-door criterion* from X to Y. *Then controlling for Z identifies the effect of action do(X) on Y.*

For discrete variables, this leads to

$$\mathbb{P}(Y|\, do(X)) = \sum_{z} \mathbb{P}(Y|\, X, Z = z)\, \mathbb{P}(Z = z).$$

For continuous variables, this leads to

$$p(Y|\, do(X)) = \int p(Y|\, X, Z = z)\, p(Z = z) dz.$$

Proof. Theorem 5.2.34 is exactly Theorem 3.3.2 from Pearl (2009, p. 79)[187]. □

The back-door criterion 5.2.34 provides a straightforward, graphical way to choose control variables that identify a given causal formula. In words, the back-door criterion selects a set of variables that separates the action from the effect and does not contain the action's descendants.

Example 5.2.35 (Revisiting Simpson's paradox). *Consider Simpson's graph $\mathcal{S}$ illustrated in Figure 5.10. The back-door criterion 5.2.34 applies: a statistician can measure the effect of X on Y by controlling for Z, as $\{Z\}$ satisfies the back-door condition from Definition 5.2.33. In addition, the following negative results also hold:*

(a) One cannot measure the effect of X on Y without controlling for Z due to the back-door $X \leftarrow Z \rightarrow Y$.

(b) One cannot measure the effect of Z on Y by controlling for X because X is a descendant of Z.

Identification is not unique: multiple sets of conditioning variables can identify a causal effect on a given causal graph. Furthermore, more complex graphs may require a *sequence* of conditioning variables, as outlined in the *sequential back-door criterion* of Theorem 5.2.36.

Theorem 5.2.36 (Sequential back-door criterion). *Let Y and $(X_i)_{i=1...n}$ be variables in a causal structure $\mathcal{G}$. Assume given a sequence of sets of observable variables Z_k such that*

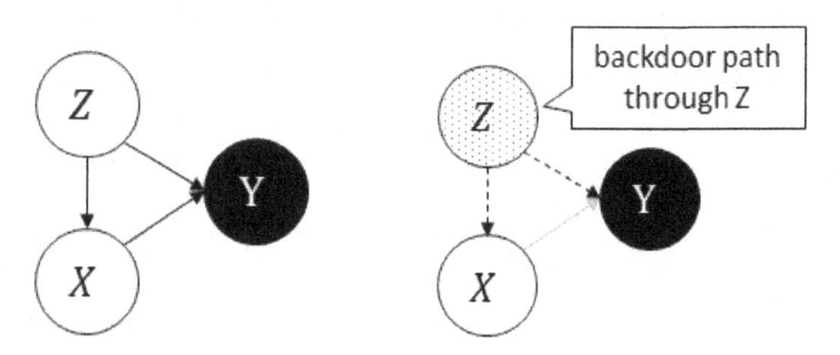

(a) Simpson's causal graph $\mathcal{S}$.

(b) Back-door for estimating $X \to Y$.

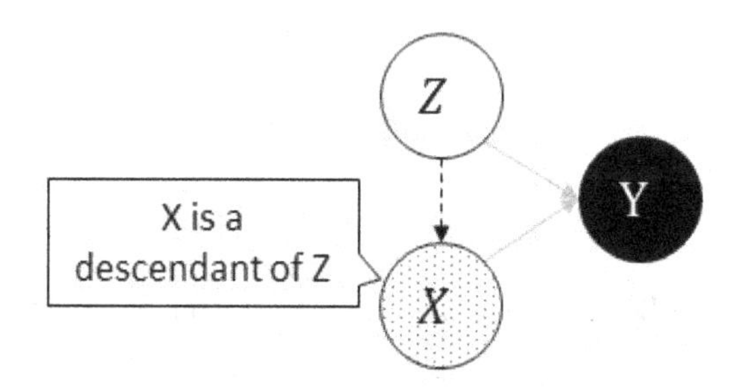

(c) Negative result for estimating $Z \to Y$ conditioning on X.

FIGURE 5.10
Identifiability results on Simpson's causal graph $\mathcal{S}$ from Example 5.2.35.

(a) each element of Z_k is not *a descendant of $\{X_k, ..., X_n\}$.*

(b) Y and X_k are conditionally independent given $X_1, ..., X_{k-1}, Z_1, ..., Z_k$ on the graph $\mathcal{G}_{\underline{X_k}\,\overline{X}_{k+1}\cdots\overline{X}_n}$

Then one has the following identifiability equation for discrete variables

$$\mathbb{P}(Y|\, do(X_1), ..., do(X_n)) = \sum_{z_1,...,z_n} \mathbb{P}(Y|\, Z_1 = z_1, ..., Z_n = z_n, X_1, ..., X_n)$$
$$\times \prod_{k=1}^{n} \mathbb{P}(Z_k = z_k|\, Z_1 = z_1, ..., Z_{k-1} = z_{k-1}, X_1, ..., X$$

and the equivalent formula for continuous variables.

Proof. Theorem 5.2.36 is exactly Theorem 4.4.1 from Pearl (2009, p. 121)[187]. □

For completeness's sake, this section ends with the *front-door criterion*, another straightforward method to apply the rules of do-calculus to identify causal effects.

Definition 5.2.37 (Front-door). *Let X and Y be two variables in a causal structure $\mathcal{G}$. Define the set of variables Z to satisfy the* front-door *criterion* from X to Y *if:*

(a) Z blocks every directed *path* from X to Y,

(b) there is no unblocked back-door path from X to Z, and

(c) X blocks all back-door paths from Z to Y.

Intuitively, the criterion allows directed paths of the form $Z \to \cdots \to Y$ to proxy for paths $X \to \cdots \to Y$ even in the presence of a backdoor confounder between X and Y. Indeed, the criterion rules out any dependence between such a hidden confounder and Z and guarantees that $X \to \cdots \to Y$ must pass through Z. Theorem 5.2.38 provides the corresponding identification formula.

Theorem 5.2.38 (Front-door criterion). *Let $\mathcal{M}$ be a causal model with causal structure $\mathcal{G}$. Assume the set Z satisfies the front-door criterion* from X to Y. *Then one has the following identifiability equation for discrete variables*

$$\mathbb{P}(Y = y|\, do(X = x)) = \sum_z \mathbb{P}(Z = z|\, X = x) \sum_{x'} \mathbb{P}(Y = y|\, X = x', Z = z)\, \mathbb{P}(X = x')$$

and the equivalent formula for continuous variables.

Proof. Theorem 5.2.38 is exactly Theorem 3.3.4 from Pearl (2009, p. 83)[187]. □

5.3 Methods to Reduce Causal Biases

Example 5.2.27 from Section 5.2.3 is not a contrived brainteaser for trading interviews: trading desks face variants of Simpson's paradox daily. Causal biases are a *core challenge* in quantitative trading when evaluating strategies and algorithms.

Traders implement algorithms that trigger based on signals and order characteristics.

Unfortunately, the signals and the order characteristics predict price moves regardless of the trading strategy used. Therefore, these triggers *confound* the evaluation of trading strategies. Trading teams solve confounding in trading using a combination of two approaches with advantages and disadvantages. The first approach, covered by point (a), uses *interventions* to design bias-free data. The second approach, covered by point (b), uses *observational data* and removes the bias by estimating the correct identifiability equation using appropriate control variables.

(a) Section 1.2.8 of Chapter 1 defines live trading experiments, such as A-B tests.

    Live trading experiments randomize trading decisions to remove causal biases. However, the trading experiment introduces a cost: the random decision is unlikely to be optimal. One concrete example is order allocation to trading algorithms, which leads to a classic learning problem: should one allocate the orders to the algorithm that has performed best so far or randomize the decision to get a better estimate of each algorithm? The machine learning literature describes this problem as the *exploration versus exploitation trade-off*. For instance, see Bubeck and Cesa-Bianchi (2012)[42]:

    > "Multi-armed bandit problems are the most basic examples of sequential decision problems with an exploration–exploitation trade-off. This is the balance between staying with the option that gave highest payoffs in the past and exploring new options that might give higher payoffs in the future" (p. 3).

(b) Section 5.2.4 defines identifiability and describes how to control for confounding variables in observational data.

    Due to opportunity costs, *interventional data* generated from a live trading experiment tends to be a small fraction of the team's overall trading data. Therefore, the second approach's advantage is the extensive observational data available to the trading team. However, because the observational data is not bias-free, the statistical estimators involved are more complicated and not robust to omitted confounders.

Remark 5.3.1 (A warning). *In principle, practitioners can implement both options without explicitly proposing a causal graph.*

(a) For interventional data, one can randomize attributes of the trading strategy and hope that the confounding variable has been randomized.

(b) For observational data, one hopes the trader controls for the correct variables in their statistical estimation. However, as illustrated in Example 5.2.29 and Remark 5.2.30, it is possible to simultaneously under- and over-specify the control variables in estimating a causal formula.

Both are dangerous practices: while there is always a risk that the causal graph leaves out an unknown confounding variable, a trading team should propose a causal graph to guide causal inference. ***Traders play a leading role by contributing their domain knowledge to the causal graph.*** *At the very least, this allows the trading team to document and communicate assumptions explicitly. Section 5.3.1 walks through an A-B test example.*

Finally, Section 5.3.2 introduces a machine learning method: causal regularization. Causal regularization leverages observational and interventional data for their respective strengths.

5.3.1 Standard A-B testing

This section illustrates best practices when designing and drawing conclusions from live trading experiments. Example 5.3.2 describes a common trading scenario extending Example 5.2.27 from Section 5.2.3.

Example 5.3.2 (Evaluating trading algorithms). *Consider a more general version of Example 5.2.27 introduced in Section 5.2.3 to illustrate Simpson's paradox.*

(a) For a given order, A represents the choice of a trading algorithm from a set $\{A_1, ..., A_k\}$ of available algorithms.

(b) Y measures the order's arrival slippage *in basis points and serves as the performance metric.*

(c) The trading team observes both the order size S and trading speed $\bar{S}$.

Like in Example 5.2.27, assume that the trading algorithm controls $\bar{S}$ but does not control S. The causal graph $\mathcal{A}$ from Definition 5.3.3 captures these assumptions. Unlike in Example 5.2.27, let both S and $\bar{S}$ take values over $(0, \infty)$.

The trading team evaluates its trading algorithms and improves the team's performance with two actions. First, the team can increase the allocation to better-performing algorithms. Second, the team can tune an underperforming algorithm's trading parameters. In this example, the trading speed $\bar{S}$ is a parameter under the algorithm's control that the team can adjust. The trading team defines two counterfactuals of interest:

(a) Algorithm switch counterfactual.

What if I switch to algorithm A?

(b) Algorithm acceleration counterfactual.

What if I switch to a sped-up version of algorithm A?

Definition 5.3.3 mathematically describes the causal assumptions made in Example 5.3.2.

Definition 5.3.3 (Causal structure for trading algorithms). *Assume given the following causal structure $\mathcal{A}$.*

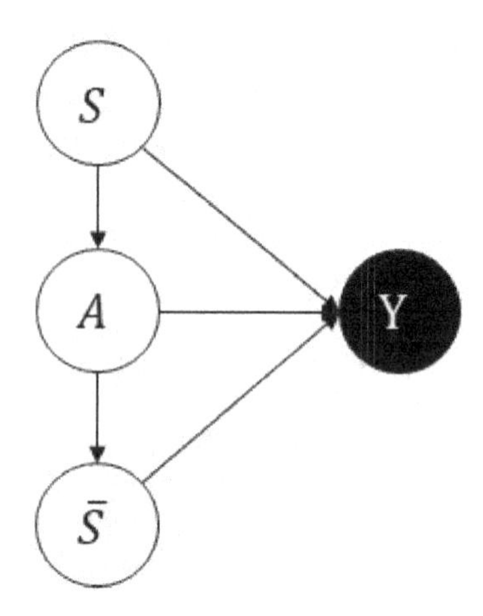

Definition 5.3.4 formalizes the counterfactuals of interest on the causal structure $\mathcal{A}$ using the do()-action.

Definition 5.3.4 (Actions to measure). *The actions the trader measures on the causal structure $\mathcal{A}$ are:*

(a) $E\left[Y|\, do(A)\right]$, *corresponding to the* algorithm switch counterfactual.

(b) $E\left[Y|\, do(\bar{S}), do(A)\right]$, *corresponding to the* algorithm acceleration counterfactual.

Proposition 5.3.5 identifies the counterfactuals of interest using do-calculus.

Proposition 5.3.5 (Identifiability of actions). *Under the causal structure $\mathcal{A}$, the following identifiability equations hold.*

(a) Controlling for S identifies the action associated with the algorithm switch counterfactual

$$E\left[Y|\, do(A)\right] = \int E\left[Y|\, A, S = s\right] p(S = s) ds.$$

(b) Controlling for S identifies the action associated with the algorithm acceleration counterfactual

$$E\left[Y|\, do(\bar{S}), do(A)\right] = \int E\left[Y|\, \bar{S}, A, S = s\right] p(S = s) ds.$$

Proof. (a) The conditioning variable $Z = \{S\}$ satisfies the back-door condition from Theorem 5.2.34.

(b) The conditioning sequence $Z_1 = \{S\}$, $Z_2 = \emptyset$ for the sequence $X_1 = A$, $X_2 = \bar{S}$ satisfies the sequential back-door condition from Theorem 5.2.36. □

Following Proposition 5.3.5, the trading team must understand how order size affects order allocation in the observational data to evaluate their algorithms under causal structure $\mathcal{A}$. The trading team requires interventional data if they cannot establish this relationship or if the trading team wants to complement the analysis with bias-free data. Definition 5.3.6 defines live trading experiments as causal graphs trimmed by specific do()-actions. These experiments are well-designed: naive identification formulas hold for the counterfactuals of interest.

Definition 5.3.6 (Experiments). *Consider the following two experiments under the causal model $\mathcal{S}$:*

(a) Assign the trading algorithm randomly, $do(A = \mu)$, where μ is a random variable over $\{A_1, ..., A_n\}$ independent of all other variables.

(b) Assign the trading algorithm and the trading speed randomly, $do(A = \mu)$, $do(\bar{S} = \nu)$.

μ and ν are random variables independent of all other variables and each other. μ takes value over $\{A_1, ..., A_n\}$ and ν over $(0, \infty)$.

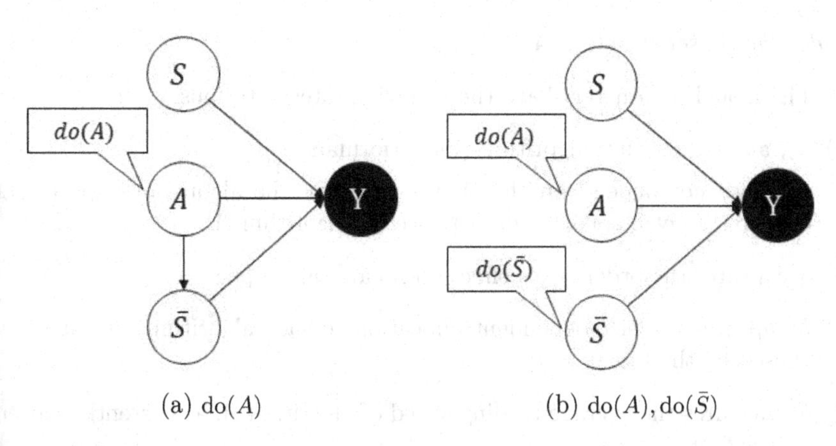

FIGURE 5.11
Graph representations of the counterfactuals from Definition 5.3.6 on the causal model $\mathcal{A}$.

Proposition 5.3.5 proves causal statements for evaluating trading algorithms. These statements hold for all probability distributions $\mathbb{P}$ consistent with the causal structure $\mathcal{A}$. One of causal inference's strengths is the ability to state such results for a large class of candidate statistical models. A statistician implements the derived formulas either using non-parametric methods such as machine learning or specifying a parametric model for the distribution $\mathbb{P}$. Example 5.3.7 promotes the causal structure $\mathcal{A}$ to a causal model $\mathcal{M}$ and provides a parametric example for $\mathbb{P}$.

Example 5.3.7 (Causal model for trading algorithms). *Let* $(\Omega, \mathcal{F}, \mathbb{P})$ *be a probability space and*

(a) α, β, *and* δ *three vectors indexed over* $\{A_1, ..., A_k\}$,

(b) γ *a real,*

(c) f *and* g *two real functions,*

(d) h *a vector-valued function that takes value over the set of probability measures over* $\{A_1, ..., A_k\}$, *and*

(e) ϵ, ν, η *three independent Gaussians with zero mean.*

Define the causal model $\mathcal{M}$ *via the functional relationships*

$$\begin{aligned} Y &= \alpha_A + \beta_A f\left(\bar{S}\right) + \gamma g\left(S\right) + \epsilon \\ \mathbb{P}\left(A = A_i | S\right) &= h_i(S) \\ \bar{S} &= \delta_A + \nu \\ S &= \eta \end{aligned}$$

under the causal structure $\mathcal{A}$.

The model parameters have the following interpretations.

(a) α_A and β_A capture algorithm A's performance.

α_A does not depend on the choice made for the algorithm's speed. β_A captures a given speed's performance for algorithm A.

(b) γ captures the order size's effect on its arrival slippage.

(c) h captures the size-dependent allocation to each algorithm, as currently chosen by the team.

(d) δ_A captures the average trading speed of algorithm A, as currently chosen by the team.

While useful from a communication perspective, the model parameters' interpretations do not directly translate into actions for the trading team. *Counterfactuals*, which are not tied to a specific functional form, directly translate into actions for the trading team. Under the causal model $\mathcal{M}$, a trader identifies the two counterfactuals of interest with the following formulas.

The formula

$$E\left[Y|\,\mathrm{do}(A=a)\right]=\alpha_a+\beta_a\mathbb{E}\left[f\left(\delta_a+\nu\right)\right]+\gamma\mathbb{E}\left[g\left(S\right)\right],$$

where the last term is a constant offset across all algorithms, estimates the *algorithm switch counterfactual.* The estimator's variance scales in

$$\frac{1}{n_i^{\mathrm{int}}},$$

where n_i^{int} is the size of the interventional data for algorithm A_i. A statistician estimates the counterfactual with two methods. They compute the empirical expectation of Y over values of A on the interventional data or regress

$$Y=\sum_i\theta_{A_i}\mathbb{1}_{A=A_i}+\gamma g\left(S\right)+\epsilon$$

on the observational data. The regression coefficients θ_{A_i} satisfy

$$\theta_{A_i}-\theta_{A_j}=E\left[Y|\,\mathrm{do}(A=A_i)\right]-E\left[Y|\,\mathrm{do}(A=A_j)\right].$$

The regression must control for S to estimate the correct causal expression due to the causal link $S\to A$. This dependence increases the estimator's variance, which scales in

$$\frac{1}{(1-\rho_i^2)n_i^{\mathrm{obs}}}.$$

n_i^{obs} is the size of the observational data for algorithm A_i and ρ^i the correlation between $\mathbb{1}_{A=A_i}$ and $g\left(S\right)$.

The formula

$$E\left[Y|\,\mathrm{do}(\bar{S}=\bar{s}),\mathrm{do}(A=a)\right]=\alpha_a+\beta_a f\left(\bar{s}\right)+\gamma\mathbb{E}\left[g\left(S\right)\right]$$

estimates the *algorithm acceleration counterfactual.* The same observations hold regarding this counterfactual's interventional and observational data estimates.

Remark 5.3.8 (How much interventional data does one need?). *A statistician can quantify the trade-off between using interventional and observational data under a given causal model.*

a) *In general, the variance of the estimator based on interventional data scales in $\frac{1}{n^{int}}$, where n^{int} is the intervention size. The interventional data's size typically precludes models with high dimension, such as neural networks.*

b) *The variance of the estimator based on observational data depends on the strength in the observational data of the dependence between the do-action variable and the controlling variable present. The* variance-inflation factor *quantifies this effect for linear models.*

For example, consider a scenario where 95% of the time, an order's size determines which algorithm trades it. As a rule of thumb, the statistician needs

$$\frac{1}{1-0.95^2} \approx 10$$

times as much observational as interventional data to quantify the algorithm's advantage with the same confidence.

5.3.2 Causal regularization

Remark 5.3.8 contrasts observational data with interventional data: observational data is extensive and allows non-parametric models. Interventional data is bias-free and allows straightforward formulas. This section introduces a hybrid approach, *causal regularization*, which benefits from both data.

The reader may already be familiar with ridge, lasso, and other regularization penalties for *predictive* models. For example, Tibshirani (1996)[222] compares lasso favorably to two other regression techniques:

> "It produces interpretable models like subset selection and exhibits the stability of ridge regression" (p. 1).

Ghojogh and Crowley (2019)[107] provide a comprehensive tutorial covering "The Theory Behind Overfitting, Cross Validation, Regularization, Bagging, and Boosting" (p. 1). The reader studies Section 2.4.11 "Smoothing and regularization" (p. 80) by Isichenko (2021)[121] for a finance-centric guide.

Belloni, Chernozhukov, and Hansen (2014)[22] provide a stark warning on the *naive* use of regularization for causal inference.

> "Part of the difficulty in drawing inferences after regularization or model selection is that these procedures are designed for forecasting, not for inference about model parameters" (p. 33).

Regularization fails in the presence of a confounding variable because *the same causal bias* is present in the testing data. Janzing (2019)[126] generalizes regularization for use in causal inference. The central insight from Janzing (2019)[126] is that statisticians can apply regularization's benefit for predictions -the reduction of in-sample confirmation bias via Bayesian validation- to *other* biased estimators for which *bias-free testing data* is available. Live trading experiments, such as TCA experiments introduced in Section 3.3 from Chapter 3 provide bias-free testing data for specific causal biases.

5.3.2.1 Regularization in the predictive case

Example 5.3.9 describes how evaluating a model on the same data used for training leads to a *confirmation bias*. The confirmation bias is due to the data's finite nature: the bias goes to zero as the sample size goes to infinity. For a limited sample size, the model detects spurious correlations between features and overestimates model parameters' magnitude when using the same data for training and testing.

Example 5.3.9 (Bias in the case of finite data). *Consider the model*

$$Y = \beta X + \epsilon$$

where β is of dimension n. Let $\hat{X}$, $\hat{Y}$, and $\hat{\epsilon}$ be the random variables' in-sample realizations. Then the least-squares estimator satisfies

$$\hat{\beta} = \left(\hat{X}^t\hat{X}\right)^{-1} \hat{X}^t\hat{Y} = \beta + \left(\hat{X}^t\hat{X}\right)^{-1} \hat{X}^t\hat{\epsilon}.$$

The finite data correlations between X and ϵ bias the in-sample beta parameter's magnitude:

$$\mathbb{E}\left[\left|\left|\hat{\beta}\right|\right|^2\right] = ||\beta||^2 + \mathbb{E}\left[\left|\left|\left(\hat{X}^t\hat{X}\right)^{-1} \hat{X}^t\hat{\epsilon}\right|\right|^2\right] > ||\beta||^2.$$

As $n \to \infty$, $\mathbb{E}\left[\left|\left|\left(\hat{X}^t\hat{X}\right)^{-1} \hat{X}^t\hat{\epsilon}\right|\right|^2\right] \to 0$, and the bias goes to zero.

Remark 5.3.10 describes regularization as a general *recipe* statisticians use to avoid overfitting a model on a given sample. One can formalize the recipe and prove that regularization removes the finite data bias from Example 5.3.9.

Remark 5.3.10 (Regularization recipe for predictive models). *For predictive models, regularization's practical use involves four steps.*

(a) Split the data into training and testing data.

A statistician can split the data in many ways. K-fold cross-validation is standard outside of finance. Because traders deal with time-series data, they use rolling samples for testing.[4]

(b) Establish a regularization penalty.

As shown in Remark 5.3.11, the regularization penalty expresses a Bayesian prior on the model parameters' distribution.

(c) Fit a family of models in-sample on the training data. A meta-parameter θ parametrizes the family.

A statistician fits most of the model parameters using the training data, leaving only a low-dimensional degree of freedom called the meta-parameter. Therefore, all the candidate models live in a low-dimensional manifold defined by the meta-parameter.

[4] Isichenko (2021)[121] motivates this best practice in finance: traditional cross-validation techniques "are not directly applicable due to the overlap of features and targets: predictors of future returns often include past returns" (p. 68). On the other hand, "rolling cross-validation is occasionally termed leave-future-out (LFO) CV, a method especially suited for online learning" (p. 69).

(d) Use the testing data to calibrate the meta-parameter θ out-of-sample.

The training step fits the model's shape *on the training data and parametrizes the shape by θ. The testing step, also called the tuning step, explores the low-dimensional space of candidate models on the testing data and calibrates the Bayesian prior. Finally, the* tuned *meta-parameter θ narrows the models down to one.*

Remark 5.3.11 (Bayesian interpretation of regularization). *Regularization aims to reduce the bias on the magnitude of β at the cost of in-sample explanatory power. The general idea is to consider β as a random variable and express a prior on the magnitude of β. For instance, assume a Gaussian prior for β:*

$$\beta \sim N\left(0, \eta^2\right).$$

Furthermore, assume that ϵ is Gaussian, i.i.d., and independent of β and X,

$$\epsilon \sim N\left(0, \sigma^2\right).$$

Then the conditional density function p for Y is

$$p\left(y|\, x, \beta\right) = \frac{1}{\sigma\sqrt{2\pi}} \exp\left(-\frac{1}{2\sigma^2}||y - \beta x||^2\right).$$

By Bayes, the posterior probability density function satisfies

$$\log p\left(\beta|\, x, y\right) = C - \frac{1}{2\sigma^2}||y - \beta x||^2 - \frac{1}{2\eta^2}||\beta||^2$$

for a constant C.

The ratio $\theta = \frac{\sigma^2}{\eta^2}$ summarizes the Bayesian prior's strength. Given a prior strength θ, a statistician maximizes the posterior probability by minimizing the least-squares objective function with ridge penalty θ:

$$\beta_\theta = argmax_\beta ||\hat{Y} - \beta\hat{X}||^2 + \theta||\beta||^2.$$

In practice, both σ and η are unknown, and the statistician calibrates the meta-parameter θ using the testing data. Remark 5.3.12 outlines how to calibrate θ.

Remark 5.3.12 (Bias-free testing data). *The testing data is essential in regularization because, by construction, it does not contain the* same *spurious finite data correlations as the training data. A statistician views the testing data as bias-free when applied to the in-sample model.*

The testing data's bias-free nature justifies its use to calibrate *the prior by scanning over values of the meta-parameter θ. Next, a statistician compares performance metrics, for example, the model's R^2, across values of θ and training and testing data. This comparison quantifies the effect of a specific prior on β on the model's quality in the training and testing data.*

The performance on the training data decreases with θ: by definition, as one increases θ, the prior on the size of β becomes stronger. The performance on the testing data initially typically increases with θ: regularization reduces overfitting. As θ becomes large enough, the model over-regularizes β, and the performance on the testing data decreases. Figure 5.12 illustrates an example of such a regularization curve. A trader toggles θ to move the estimate of β closer to its in-sample value, given by $\hat{\beta}$, or closer to the prior that β is small and of mean zero.

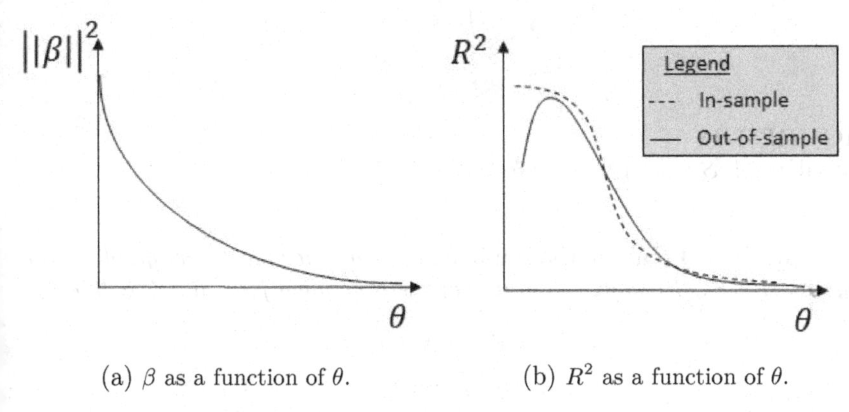

(a) β as a function of θ. (b) R^2 as a function of θ.

FIGURE 5.12
Regularization's effect on $||\beta||^2$ and R^2 as per Remark 5.3.12.

5.3.2.2 Regularization in the causal case

Under a specific causal model, Janzing (2019)[126] proves a statistical correspondence between the error introduced when overfitting finite data and the error introduced in a causal model with confounding.

> "The idea of the present paper is simply that standard regularization techniques do not care about the *origin* of this error term. Therefore, they can temper the impact of confounding in the same way as they help avoiding to overfit finite data. Such a strong statement, for course, relies heavily on our highly idealized generating model for the confounding term. We therefore ask the reader not to quote it without also mentioning the strong assumptions" (p. 2).

Example 5.3.13 provides a correspondence between formulas for a finite-data bias and a causal bias on the same example as Janzing (2019)[126]. Regularization requires no specific assumptions on the nature of the bias to remove, simply that the testing data does not present the same bias as the training data

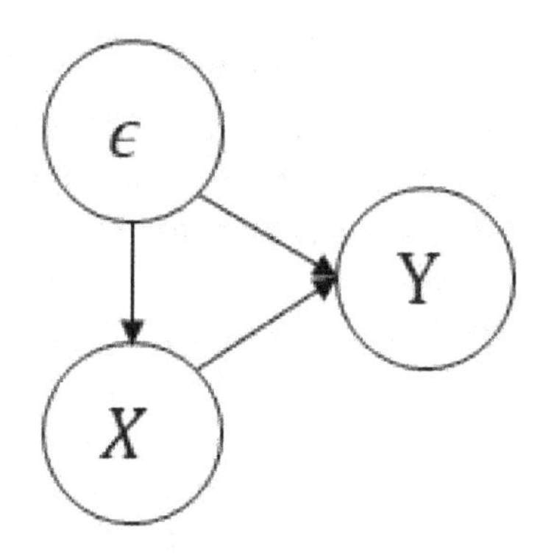

FIGURE 5.13
Causal graph $\mathcal{S}$ from Example 5.3.13.

Example 5.3.13 (Bias in the causal case). *Consider the causal graph $\mathcal{S}$ from Figure 5.13. Furthermore, assume that the functional form for Y is linear,*

$$Y = \beta X + \epsilon$$
$$X = \gamma \epsilon + \epsilon_0$$

where β, γ are of dimension n and ϵ and ϵ_0 are independent Gaussians. Let $\hat{X}$, $\hat{Y}$, and $\hat{\epsilon}$ be the in-sample realizations of the random variables. Then the least-squares estimator still satisfies

$$\hat{\beta} = \left(\hat{X}^t \hat{X}\right)^{-1} \hat{X}^t \hat{Y} = \beta + \left(\hat{X}^t \hat{X}\right)^{-1} \hat{X}^t \hat{\epsilon},$$

and one still has the in-sample bias

$$\mathbb{E}\left[\left|\left|\hat{\beta}\right|\right|^2\right] = ||\beta||^2 + \mathbb{E}\left[\left|\left|\left(\hat{X}^t \hat{X}\right)^{-1} \hat{X}^t \hat{\epsilon}\right|\right|^2\right] > ||\beta||^2.$$

Unlike in the previous no-confounding case from Remark 5.3.9, the bias does not vanish when $n \to \infty$: X and ϵ remain correlated. Indeed, in the causal case, β relates to

$$\mathbb{E}\left[Y \mid do(X)\right] = \beta X$$

while $\hat{\beta}$ relates to

$$\hat{\beta} X \to \mathbb{E}\left[Y \mid X\right] = \beta X + \underbrace{\mathbb{E}\left[\epsilon \mid X\right]}_{\neq 0}$$

as $n \to \infty$. The above negative result is a linear case of the negative causa statement

$$\mathbb{E}\left[Y \mid do(X)\right] \neq \mathbb{E}\left[Y \mid X\right]$$

on the causal graph $\mathcal{S}$ due to the fork $X \leftarrow \epsilon \to Y$.

Based on the insight from Example 5.3.13, Remark 5.3.14 generalizes the regularization recipe from the predictive case to the causal case.

Remark 5.3.14 (Regularization recipe for causal models). *For causal models, regularization's practical use involves the following steps.*

(a) Split the data into training and testing data.

The training data includes any valid observational data. On the other hand, testing data only includes interventional data that removes the targeted causal bias.

(b) Establish a regularization penalty.

This step is identical to the predictive case.

(c) Fit a family of models in-sample on the training data. A meta-parameter θ parametrizes the family.

This step is identical to the predictive case. All the candidate models live in a low-dimensional manifold.

(d) Use the testing data to calibrate the meta-parameter θ on the bias-free data.

The interventional data calibrates the prior on the strength of the confounding. The stronger the prior, the more the model shrinks the confounding variable's effect.

Causal regularization blends the practical advantages of observational *and* interventional data. Furthermore, in trading applications, observational data is plentiful but biased. Interventional data, on the other hand, is scarce but bias-free. Figure 5.14 illustrates the advantage of using *both* data when estimating trading models in the presence of a causal bias.

(a) Using the observational data gives a *biased estimator with a small variance.* The first panel illustrates the estimator on observational data, represented by grey squares.

(b) Using the interventional data produces an *unbiased estimator with a large variance.* The second panel illustrates the estimator on interventional data, represented by black circles.

(c) Using causal regularization achieves the best of both worlds: an *unbiased estimator with a small variance.* The third panel illustrates the estimator that leverages both data for their respective strengths.

In trading applications, the interventional data is rarely extensive enough to fit a high-dimensional model, such as a neural network. However, a statistician can fit a high-dimensional model on observational data, then *de-bias* the

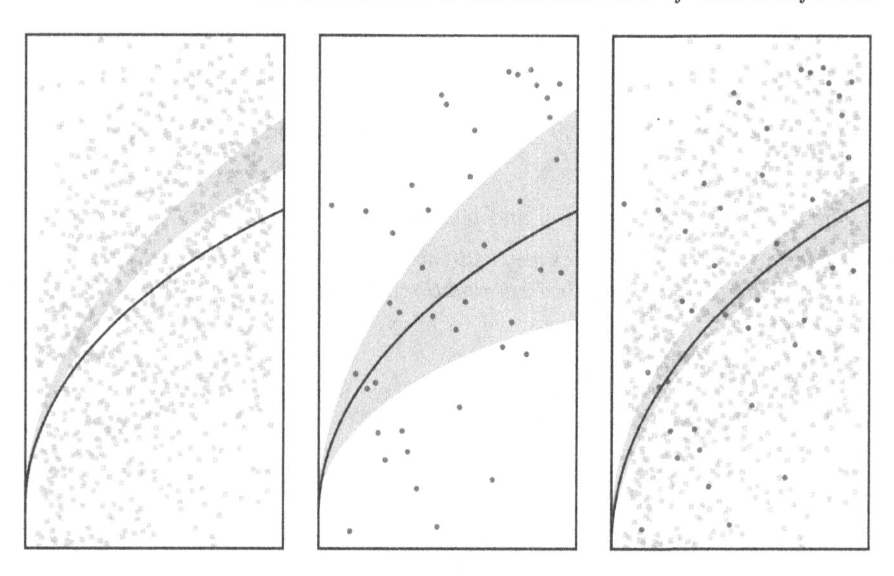

FIGURE 5.14
Illustration of the bias and variance trade-off between observational data (grey squares) and interventional data (black circles). The last panel combines both data through *causal regularization*.

model using interventional data.[5] Section 6.3.4 from Chapter 6 applies causal regularization to resolve the so-called *prediction bias* and estimate price impact in the presence of an unobserved alpha signal.

5.4 Summary of Results

This section summarizes mathematical results in causal inference.

5.4.1 Causal structures and models

Definition 5.4.1 (Causal structure). *A* causal structure *of a set of variables V is a directed acyclical graph (DAG) in which each node corresponds to a distinct element of V.*

[5]Rothenhäusler et al. (2021)[201] deal with the case where bias-free data is unavailable for testing. Rothenhäusler et al. propose a method that leverages many heterogeneous interventions, none of which are entirely unbiased. The method avoids the cost of implementing a new, bias-free trading experiment by leveraging past, related, but imperfect experiments.

Definition 5.4.2 (Causal model). *A causal model consists of*

(a) a causal structure.

(b) a set of functions f_i compatible with the causal structure,

$$f_i : (parents(x_i), \epsilon_i) \mapsto f_i(parents(x_i), \epsilon_i)$$

where $parents(x_i)$ are potential outcomes of the parent variables *of X_i and ϵ_i is a noise term idiosyncratic to X_i.*

(c) a probability space $(\Omega, \mathcal{F}, \mathbb{P})$ that assigns probabilities to all the ϵ_i, with each ϵ_i being independent.

Definition 5.4.3 (Consistency). *Define a probability distribution $\mathbb{Q}$ on the set of variables V as* consistent *with the causal structure $\mathcal{G}$ if there exists a causal model $\mathcal{M}$ based on the causal structure $\mathcal{G}$ that generates $\mathbb{Q}$ for the variables in V.*

Definition 5.4.4 (d-separation). *Let $\mathcal{G}$ be a DAG. Define a path p to be d-separated (or blocked) by a set of nodes Z if and only if*

(a) the path p contains a chain $i \to m \to j$ or a fork $i \leftarrow m \to j$ such that m is in Z.

(b) the path p contains a collider $i \to m \leftarrow j$ such that neither m nor its descendants are in Z.

Theorem 5.4.5 (Probabilistic implications of d-separation). *Let $\mathcal{G}$ be a causal structure and X, Y, and Z be three sets of variables on $\mathcal{G}$.*

(a) If Z d-separates X and Y, for every probability measure $\mathbb{P}$ consistent with $\mathcal{G}$, X and Y are conditionally independent given Z.

(b) Conversely, if Z does not d-separate X and Y, then there exists at least one probability measure $\mathbb{P}$ consistent with $\mathcal{G}$ for which X and Y are not *conditionally independent given Z.*

5.4.2 Do-calculus

Definition 5.4.6 (The do() operator). *Given two disjoint sets of variables X and Y on a causal model $\mathcal{M}$, define the $do(X)$ action*

$$\mathbb{P}(Y|\, do(X = x)) = \tilde{\mathbb{P}}(Y)$$

where one obtains $\tilde{\mathbb{P}}$ by replacing all the functions f_i that pertain to variables $X_i \in X$ with the constant function $X_i = x_i$.

Definition 5.4.7 (Mixing do() with Bayesian conditioning). *Given three disjoint sets of variables X, Y, and Z on a causal model $\mathcal{M}$, one defines*

$$\mathbb{P}(Y|\, do(X = x), Z) = \frac{\mathbb{P}(Y, Z|\, do(X = x))}{\mathbb{P}(Z|\, do(X = x))}.$$

Corollary 5.4.8 (A sufficient naive identifiability condition). *Let $\mathcal{M}$ be a causal model with causal structure $\mathcal{G}$. Let X and Y be two variables such that X has no parents. Then the effect of X on Y is naively identifiable:*

$$\mathbb{P}(Y|\, do(X)) = \mathbb{P}(Y|\, X).$$

Definition 5.4.9 (Back-door). *Let X and Y be two variables in a causal structure $\mathcal{G}$. Define the set of variables Z to satisfy the* back-door *criterion* from X to Y *if:*

(a) Z d-separates every path between X and Y with a link pointing into X.

(b) No descendant of X is in Z.

Theorem 5.4.10 (Back-door criterion). *Let $\mathcal{M}$ be a causal model with causal structure $\mathcal{G}$. Assume the set Z satisfies the back-door criterion* from X to Y. *Then controlling for Z identifies the effect of action $do(X)$ on Y.*

For discrete variables, this leads to

$$\mathbb{P}(Y|\, do(X)) = \sum_z \mathbb{P}(Y|\, X, Z = z)\, \mathbb{P}(Z = z).$$

For continuous variables, this leads to

$$p(Y|\, do(X)) = \int p(Y|\, X, Z = z)\, p(Z = z) dz.$$

Definition 5.4.11 (Front-door). *Let X and Y be two variables in a causal structure $\mathcal{G}$. Define the set of variables Z to satisfy the* front-door *criterion* from X to Y *if:*

(a) Z blocks every directed *path* from X to Y,

(b) there is no unblocked back-door path from X to Z, and

(c) X blocks all back-door paths from Z to Y.

Theorem 5.4.12 (Front-door criterion). *Let $\mathcal{M}$ be a causal model with causal structure $\mathcal{G}$. Assume the set Z satisfies the front-door criterion* from X to Y. *Then one has the following identifiability equation for discrete variables*

$$\mathbb{P}(Y = y|\, do(X = x)) = \sum_z \mathbb{P}(Z = z|\, X = x) \sum_{x'} \mathbb{P}(Y = y|\, X = x', Z = z)\, \mathbb{P}(X = x')$$

and the equivalent formula for continuous variables.

5.4.3 A-B testing

5.4.3.1 Interventional data

The do(X) action generates data under a new probability measure $\tilde{\mathbb{P}}$. *Observational data* refers to data generated under the original measure $\mathbb{P}$.

A statistician can estimate any Bayesian formula under $\mathbb{P}$ given enough observational data. However, there is no guarantee that observational data recover *counterfactuals*, as they, by definition, correspond to a new probability measure $\tilde{\mathbb{P}}$.

Interventional data refers to data explicitly generated from the counterfactual scenario. Statisticians can estimate naive Bayesian formulas for do()-actions on interventional data as the data directly reflects $\tilde{\mathbb{P}}$. Randomized interventions and counterfactuals are vital concepts in A-B testing, and the do(X) operator formalizes A-B testing.

5.4.3.2 Causal regularization

For *causal models*, regularization's practical use involves four steps.

(a) Split the data into training and testing data. The training data includes any valid observational data. On the other hand, testing data only includes interventional data that removes the targeted causal bias.

(b) Establish a regularization penalty.

(c) Fit a family of models in-sample on the training data. A meta-parameter θ parametrizes the family.

(d) Use the testing data to calibrate the meta-parameter θ on the bias-free data. The interventional data calibrates the prior on the confounding's strength. The stronger the prior, the more the model shrinks the explanatory variable to consider the confounding variable's effect.

5.5 Exercises

Exercise 15 Examples of d-separation

Consider the causal structure outlined in Figure 5.15. From a trading perspective, this is an extension of the causal structure $\mathcal{A}$ from Definition 5.3.3 in Section 5.3.1. This exercise shows which trading variables are independent of each other and which require further conditioning to be independent. The two nodes added are:

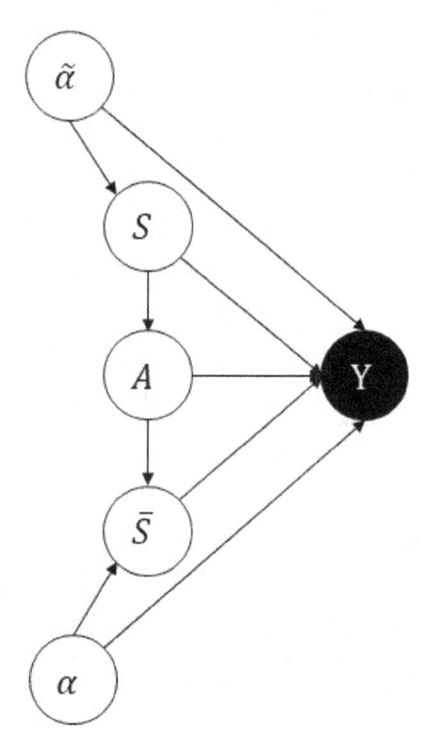

FIGURE 5.15
Causal graph for exercises 15 and 16.

(a) A long-term alpha $\tilde{\alpha}$ that predicts long-term returns contributing to Y, and is one of the variables determining the order size S.

(b) A short-term alpha α that predicts short-term returns contributing to Y, and is one of the variables determining the trading speed $\bar{S}$.

Assume that the two alphas are independent and that the trading infrastructure enforces the above causal graph, in particular ruling out causal paths such as $\tilde{\alpha} \rightarrow A$, $\tilde{\alpha} \rightarrow \bar{S}$, $\alpha \rightarrow A$, and $\alpha \rightarrow S$.[6]

1. Prove that α and $\tilde{\alpha}$ are independent.
2. Prove that α and $\tilde{\alpha}$ are dependent conditional on $\bar{S}$. Comment on the dependency's nature.
3. Prove that S and $\bar{S}$ are dependent. Comment on the dependency's nature.
4. Prove that S and $\bar{S}$ are independent conditional on A.
5. Prove that A and α are independent.

[6]From an organizational perspective, these causal assumptions simplify the performance attribution of algorithms and alpha signals.

6. Prove that A and α are dependent conditional on $\bar{S}$.
7. Prove that A and $\tilde{\alpha}$ are dependent.
8. Prove that A and $\tilde{\alpha}$ are independent conditional on S.
9. Comment on the distinction between the dependency structure of A with α and A with $\tilde{\alpha}$

Exercise 16 Examples of do-calculus

Consider the causal structure outlined in Figure 5.15, with the same interpretation as in the previous exercise. This exercise defines counterfactuals of interest and establishes identifiability equations. Consider the following counterfactuals:

(a) *What if I had submitted a larger order?*

$$\mathbb{E}\left[Y|\,\mathrm{do}(S)\right]$$

(b) *What if I had chosen algorithm A?*

$$\mathbb{E}\left[Y|\,\mathrm{do}(A)\right]$$

(c) *What if I had traded algorithm A faster?*

$$\mathbb{E}\left[Y|\,\mathrm{do}(A),\mathrm{do}(\bar{S})\right]$$

1. Identify each counterfactual succinctly.

Exercise 17 Example of causal regularization

This exercise numerically implements causal regularization for the model outlined in Example 5.3.13. As a reminder,

$$Y = \beta X + \epsilon$$
$$X = \gamma\epsilon + \epsilon_0$$

where β, γ are of dimension n and ϵ and ϵ_0 are independent Gaussians. Assume ϵ, ϵ_0 are unobserved. Fix the following parameters:

$$\beta = 1; \quad \gamma = 0.1; \quad \sigma_\epsilon = 10; \quad \sigma_{\epsilon_0} = 1.$$

1. Generate observational data of size $n_{\mathrm{obs}} = 1e6$. Estimate β using observational data only.
2. Generate a hundred independent interventions of size $n_1 = 1e2$, $n_2 = 1e3$, and $n_3 = 1e4$.
3. For each intervention, estimate β using interventional data only. Comment on the distribution of the estimated β.

4. For each intervention, estimate β using causal regularization. Comment on the distribution of the estimated β.

Exercise 18 Variance inflation

Consider the regression model

$$\Delta P = \beta\alpha + \lambda\Delta I + \epsilon$$

which cofits an observed alpha signal α with an impact model ΔI to predict observed returns ΔP. Assume that α and ΔI are correlated Gaussians with variances $\sigma_\alpha^2, \sigma_I^2$, and correlation ρ. Let σ^2 be the variance of ϵ and assume all samples to be i.i.d.

1. Prove the variance inflation factor on β and λ from first principles.
2. Plot the minimum amount of interventional data needed to outperform observational data as a function of the correlation in the observational data.

6

Dealing with Biases When Fitting Price Impact Models

This chapter builds upon the mathematical framework from Chapter 5 and applies causal inference to trading. The focus is on *identifying, communicating, and accounting for causal assumptions* related to price impact. *Four common trading biases* motivate this chapter. Bouchaud et al. (2018)[37] list the biases and their root causes.

(a) **Prediction bias:**

> Trades triggered by or considering an alpha signal exhibit bias when estimating their price impact.

(b) **Synchronization bias:**

> "The impact of a metaorder can change according to whether or not other traders are seeking to execute similar metaorders at the same time" (p. 238).

(c) **Implementation bias:**

> Tactical deviations from the strategic trading trajectory introduce biases.

(d) **Issuer bias:**

> "Another bias may occur if a trader submits several dependent metaorders successively" (p. 239).

The chapter begins with a roadmap to situate the reader in what may be unfamiliar mathematics.

(a) Subsection 6.2.1 provides a brief non-technical primer on causal inference and summarizes the essential terms from Chapter 5.

(b) Subsection 6.2.2 lists the practical goals and trading applications of causal inference. These extend beyond price impact research and interest other stakeholders, such as traders and developers.

(c) Subsection 6.2.3 provides a template for researchers to identify, communicate and account for causal biases. Each section applies the template to one of the four trading biases listed by Bouchaud et al. (2018)[37].

DOI: 10.1201/9781003316923-6

While the mathematics of causal inference roots this chapter, the reader can skip proofs and focus on applications in their first study. Section 6.8 summarizes these applications.

6.1 A Pedagogical Example

For traders, this chapter's essential takeaway is that one must consider causal biases inherent to trading when estimating price impact. For teaching purposes, this section outlines a simple *prediction bias* example.

In an interview with the *Hedge Fund Journal* (2020)[158], Robert Almgren describes prediction bias for a broker analyzing a single order.

> "Impact cannot actually be disentangled from alpha due to the endogeneity issue – it is not possible to work out if the order caused the move or the reverse. The drift could in fact be alpha"

Consider a broker executing orders on their clients' behalf. The broker estimates a price impact model from their clients' orders to obtain the best execution for their clients. For instance, this involves a regression

$$\Delta P = \lambda f(Q) + \epsilon \tag{6.1}$$

where ΔP is an order's arrival slippage; Q, the order size; f, a function; λ, the regression's parameter; and ϵ, the regression's residuals.

However, the broker *knows* that the client likely submitted the trade based on a price predictor. For example, consider a client trading based on an alpha signal α. Then the correct model is

$$\Delta P = \alpha + \lambda f(Q) + \epsilon. \tag{6.2}$$

However, the broker does not observe α: the client is unlikely to share their trading signal. To quote Bacidore (2020)[12],

> "traders are understandably reluctant to pass their alphas over to a broker due to the potential for lost intellectual property, front-running, etc." (p. 51).

Prediction bias distinguishes between regressions 6.1 and 6.2: fitting an intercept is insufficient. Indeed, because clients trade larger orders when they have stronger signals, α and Q are dependent: for example, in algorithmic trading, a sizable α *mechanically causes* a sizable Q. As a result, regression 6.1 will estimate a greater λ than regression 6.2 to compensate for the unobserved, correlated α. Unfortunately, this leads the broker to overestimate price impact and propose a slower execution strategy than warranted.

Section 6.3 solves prediction bias using the tools of causal inference: causal graphs and do-calculus. Furthermore, the chapter outlines and solves other recurrent trading biases.

6.2 Chapter Roadmap

6.2.1 A non-technical primer on causal inference

Concisely, causal inference studies *counterfactuals*. A *counterfactual* answers a *what-if* question,

What if I had done X?

Example 6.2.1 (Alpha and impact counterfactuals). *Counterfactuals recast the causality challenge defined in Section 1.1.3 of Chapter 1 as* what-if *scenarios. A trade's realized alpha answers the question*

What would the price have been if I had not traded?

Price impact answers the question

What would the price have been if I had traded differently?

Counterfactuals describe *alternative scenarios* to the observed data. Causal inference attaches mathematical definitions and statistical estimates to counterfactuals. These estimates enable traders or trading algorithms to act upon hypothetical scenarios in live trading, as seen in Chapters 3 and 4. Counterfactuals relate to actions and differ from passive observation. For example, consider two scenarios:

(a) A trader always trades at a faster speed when alpha is higher.

(b) A trader randomly selects trades to execute faster, regardless of their alpha.

In scenario (a), the trader *observes* a correlation between trading speeds and returns. In scenario (b), the trader *intervenes* on the trades using a live experiment, as defined in Section 1.2.8 of Chapter 1. The *intervention* breaks the correlation between alpha and trading speed. In the language of causal inference, scenario (b) is the *interventional data.* It emulates the *price impact counterfactual* from Example 6.2.1. Scenario (a) is the *observational data* and reflects the baseline scenario the counterfactual replaces.

Remark 6.2.2 (Causal biases). *Trading biases arise when estimating counterfactuals from observational data. Such biases are* causal *and require a* causal graph *to understand and remove. Concretely, naively implementing Bayesian statistics to evaluate counterfactuals leads to costly errors. Under conditions, a statistician removes causal biases from observational data. In other cases, traders require interventional data to lift the bias.*

The primary tool to visualize and prove results on counterfactuals is a *causal graph*, also called a *causal structure.* A causal structure is *distinct* from

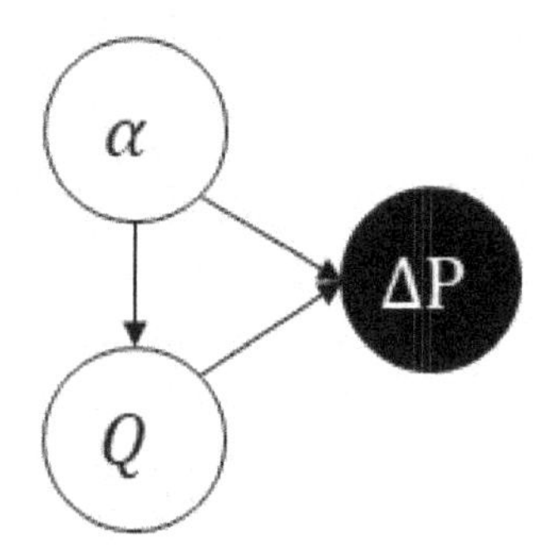

FIGURE 6.1
The graph represents the causal relationships between the features of the alpha model α, the trading process Q, and the observed returns ΔP.

a statistical model: one establishes it *before* proposing a statistical model. A causal graph comprises nodes and links, as illustrated in Figure 6.1. Links point from causes to effects. A causal structure represents assumptions made on the underlying real-life process. These premises take two broad shapes in trading:

(a) Mechanical assumptions describe the trading system's *internal logic.*

For example, link $\alpha \to Q$ in Figure 6.1 captures the execution algorithm's internal logic: trading algorithms react to signals.

(b) Economic assumptions describe *how markets function.*

For example, link $Q \to \Delta P$ in Figure 6.1 reflects price impact: trades cause price moves. Similarly, investors assume their signal features reflect fundamental drivers of stock prices.

There are three advantages to stating assumptions through a *causal structure.*

(a) The graph represents *intuitive, visual documentation of the trading system's underlying premises.* The team shares these assumptions with stakeholders, e.g., clients or developers, to further their business understanding. For instance, the reader studies "Causal Forecasting at Lyft" (2022)[195].

> A causal graph "represents our data and assumptions and is valid for forecasting and planning. [...] A read of this DAG communicates the function of our business."

For another successful example of a causal inference ecosystem in the technology industry, the reader studies "Supercharging A/B Testing at Uber"

(2022)[108]. Uber's causal inference system achieves a crucial objective for its stakeholders:

> "The results are universally trusted, allowing teams to align on the ground truth quickly and act on it without endlessly reinvestigating surprising findings, rerunning faulty experiments, and second-guessing decisions."

(b) The graph serves as a basis to express *counterfactuals* and *trading actions.* Traders visualize counterfactuals by erasing causal links the counterfactual negates. See Figure 6.2 for an example.

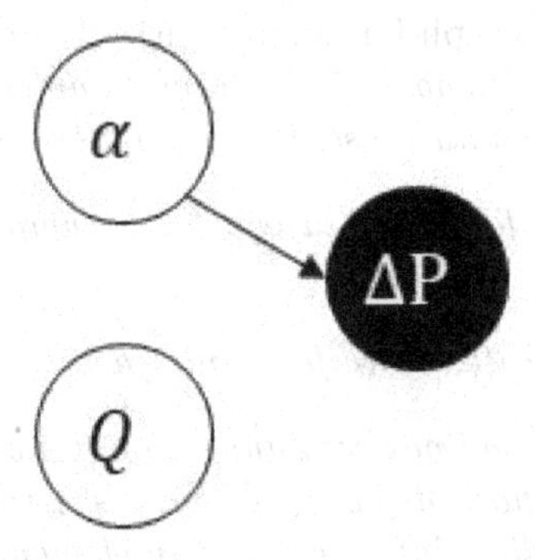

FIGURE 6.2
Graph representation of the counterfactual "What would the price have been if I had not traded?" obtained by erasing links from and to the trading process Q.

(c) The assumptions expressed in a causal structure apply to *all* statistical models consistent with the graph, including machine learning models. Therefore, the causal structure is a *contract* between the impact researcher and stakeholders.

The last term this section introduces is *causal regularization. Causal regularization* uses *observational data* as training data and *interventional data* as testing data. The following trade-off in trading motivates this machine learning technique:

- Observational data is plentiful and comes at no cost: teams generate it as part of ongoing trading.
- Interventional data is scarce and expensive: teams generate it with live trading experiments that purposefully deviate from optimum.

Causal regularization leverages both data for their strengths, leading to more *robust* statistical models.

6.2.2 Applying causal inference to trading

One applies causal inference in four steps.

(a) Document *causal assumptions* in the trading system.

(b) Articulate *counterfactuals.*

(c) Estimate counterfactuals, given the causal assumptions.

(d) *Communicate* causal assumptions and counterfactuals to stakeholders.

Remark 6.2.3 outlines a second trading application of causal inference.

Remark 6.2.3 (Causal graph for simulation)**.** *A causal model is also a* data generation process. *See Remark 5.2.9 from Chapter 5 for details. Using a causal model to simulate data presents two vital benefits:*

(a) The simulation data faithfully satisfy the assumptions that the team documents.

(b) The simulation replicates real-life counterfactuals and experiments.

This graph approach streamlines communication between traders, researchers, and developers and anchors discussions on a shared graph. Furthermore, a simulation and live trading infrastructure implemented with a shared graph improve the system's reproducibility.

6.2.3 A template for dealing with causal biases

The chapter repeats a template to solve four biases identified by Bouchaud et al. (2018)[37]. This template applies to other trading biases and formalizes causal inference's four steps.

(a) Document *causal assumptions* in the trading system.

- Define the causal graph.

(b) Articulate *counterfactuals.*

- Express counterfactuals using *do-calculus.*

(c) Estimate counterfactuals, given the causal assumptions.

- Prove or disprove the identifiability of counterfactuals on the causal structure.
- Express trading experiments using *do-calculus.* Prove the intervention removes the causal bias.
- Estimate the counterfactual using causal regularization.

(d) *Communicate* causal assumptions and counterfactuals to stakeholders.

- Provide the graph associated with the causal structure.
- Provide the counterfactuals and illustrate them on the causal graph.
- Illustrate the trading experiments on the causal graph.
- Present the counterfactual estimates.

6.3 Prediction Bias

Prediction bias affects the separation of alpha and impact when predicting returns. This section presents two related applications.

(a) *Estimate an impact model when alpha is unknown.* For example, a client does not provide the alpha, or the historical signal is not reproducible.

(b) *Estimate an alpha model when impact is unknown.* For example, an alpha researcher adjusts a signal for impact without relying on broker inputs.

6.3.1 Definitions

Bouchaud et al. (2018)[37] define two prediction biases.

> "**Prediction bias 1:** The larger the volume Q of a metaorder, the more likely it is to originate from a stronger prediction signal" (p. 238).

> "**Prediction bias 2:** Traders with strong short-term price-prediction signals may choose to execute their metaorders particularly quickly" (p. 238).

The distinction Bouchaud et al. (2018)[37] make is causal: in the first case, the *portfolio team's* decision to size the order based on their alpha biases the data. In the second case, the *execution team's* decision to accelerate the execution based on their alpha biases the data. The section defines *prediction bias* more generally.

> **Prediction bias:** Trades triggered by or considering an alpha signal exhibit bias when estimating their price impact.

This definition ignores the communication between the portfolio and execution teams. Section 3.2.1 of Chapter 3 covers communication between portfolio and execution teams.

Definition 6.3.1 (Simpson's trading graph). *Define the causal structure $\mathcal{S}$:*

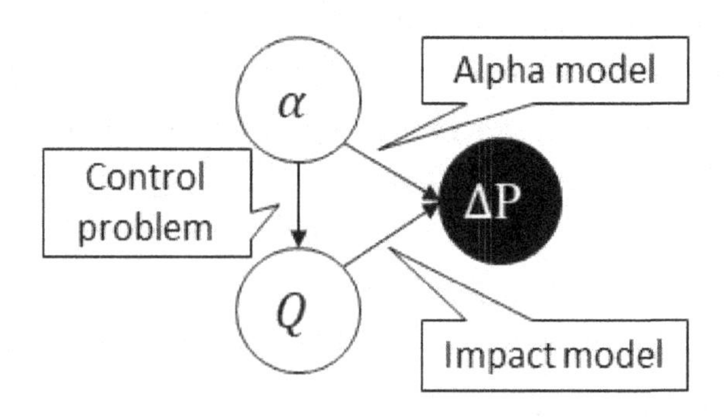

Name $\mathcal{S}$ Simpson's trading graph *in reference to* Simpson's paradox.[1]

Simpson's trading graph consists of three causal links.

(a) $\alpha \to \Delta P$ represents the *alpha model*: alpha features drive prices.

(b) $\alpha \to Q$ represents the *control problem*: trading reacts to signals.

(c) $Q \to \Delta P$ represents *price impact*: trades cause price moves.

Of the three causal links in S, $\alpha \to Q$ is the most directly verifiable: the link reflects the trading system's internal logic. However, the more complex a trading infrastructure, the less traders can verify the implementation of $\alpha \to Q$. This is one reason quantitative trading teams emphasize their trading infrastructure's *reproducibility.*

Remark 6.3.2 (Causal assumptions). *A trader reads the causal assumptions from $\mathcal{S}$ by listing the* missing *links and variables. The following statement summarizes the missing links.*

The trading system allows no feedback loops.

For instance, a feedback loop between Q and α introduces causal biases for the signal that are challenging to identify. Furthermore, feedback loops in production create instabilities that research does not reproduce. Therefore, trading teams should avoid such feedback loops and guarantee exogenous alpha signals, as per Section 1.2.3 of Chapter 1.

Proposition 6.3.3 formalizes the pedagogical example from Section 6.1. It proves that the naive Bayesian price impact estimator is incorrect on Simpson's trading graph: price impact exhibits a causal bias due to the prediction α.

---

[1] Kissell (2021)[129] links transaction cost estimation to Simpson's paradox.

"[T]he statistical results may fall victim to Simpson's paradox (e.g., dangers that arise from drawing conclusions from aggregate samples)" (p. 96).

Section 5.2.3 of Chapter 5 covers Simpson's paradox.

Proposition 6.3.3 (Prediction bias). *Under the causal structure $\mathcal{S}$, the prediction bias*

$$\mathbb{E}[\Delta P|\, do(Q)] \neq \mathbb{E}[\Delta P|\, Q]$$

holds.

Proof. The fork $Q \leftarrow \alpha \rightarrow \Delta P$ prevents identification without controlling for α. □

6.3.2 Actions and counterfactuals

Under Simpson's trading graph, traders state two practical counterfactuals:

(a) *Alpha counterfactual*

How much alpha am I missing when not trading?

(b) *Impact counterfactual*

What would the price have been if I had traded differently?

Definition 6.3.4 (Actions). *Under the causal structure $\mathcal{S}$,*

(a) $p(\Delta P|\, do(Q=0), \alpha)$ *defines the* alpha counterfactual.

(b) $p(\Delta P|\, do(Q))$ *defines the* impact counterfactual.

Proposition 6.3.5 (Identifiability of the counterfactuals). *The following identifiability equations hold under the causal structure $\mathcal{S}$:*

(a) One naively identifies the alpha counterfactual*:*

$$p(\Delta P|\, do(Q=0), \alpha) = p(\Delta P|\, Q=0, \alpha).$$

(b) Conditioning on alpha identifies the impact counterfactual*:*

$$p(\Delta P|\, do(Q)) = \int p(\Delta P|\, Q, \alpha)\, p(\alpha)\, d\alpha.$$

Proof. (a) follows from rule 2 of Theorem 5.2.26 in Chapter 5: ΔP and Q are conditionally independent given α on $\mathcal{S}_{\underline{Q}}$.

(b) The conditioning variable $Z = \{\alpha\}$ satisfies the back-door condition from Theorem 5.2.34 in Chapter 5.

□

6.3.2.1 Why is impact research complex?

Given the causality challenge, or endogeneity issue, the reader is likely unsurprised that identifying price impact is more challenging than identifying alpha. A researcher estimates the impact from observational data by conditioning on α. Furthermore, in practice, α is the *point-in-time* signal deployed in the live process Q, *not the alpha model's current version.*

For instance, consider a return prediction

$$\Delta I(\lambda, Q)$$

from an impact model I with parameter λ to calibrate on data. The two statistical regressions in equation 6.3 yield biased estimators of λ.

$$\underbrace{\Delta P = \Delta I(\lambda, Q) + \epsilon;}_{\text{prediction bias}} \quad \underbrace{\Delta P = \Delta I(\lambda, Q) + \beta\alpha_{\text{latest}} + \epsilon}_{\text{prediction bias}} \tag{6.3}$$

The statistical regression 6.4 yields a consistent estimator of λ.

$$\underbrace{\Delta P = \Delta I(\lambda, Q) + \beta\alpha_{\text{historical}} + \epsilon}_{\text{correct}} \tag{6.4}$$

Frequent signal improvements hamper this crucial point and pressure impact researchers to demand historical signals' *reproducibility.* Bouchaud (2022)[31] outlines this best practice and highlights CFM's ability to reproduce past trading decisions.

> "One of the recurrent criticism is that metaorders are not exogenous, and possibly conditioned on trading signals [...]. CFM's proprietary data allows one to eliminate many of these biases, since the strength of the trading signal is known and can be factored in the regression" (p. 4).

Example 6.3.6 (Impact-adjusted alpha research). *Consider a trader who can reproduce historical alphas but fits a novel model going forward. Let $\alpha_{historical}$ and α_{new} be the historical and new alpha signals. Then, a two-step regression resolves the ambiguity in a bias-free way.*

(a) The historical regression

$$\Delta P = \Delta I(\lambda, Q) + \beta^h \alpha_{historical} + \epsilon$$

yields the correct λ for impact.

(b) Given the impact model $I(\lambda, Q)$, one computes unperturbed *returns ΔS via $\Delta S = \Delta P - \Delta I(\lambda, Q)$. Finally, the regression*

$$\Delta S = \beta^n \alpha_{new} + \epsilon'$$

provides the correct β^n for the novel alpha model.

If the trader does not observe historical α, for instance, if a broker executes orders on a client's behalf, α becomes a hidden confounding variable. Under that premise, observational data cannot identify the *impact counterfactual.* Hence, the trading team must generate *interventional data* through live trading experiments to identify price impact.

6.3.3 Experiments and regularization

Under Simpson's trading graph, two actions provide interventional data to identify models when traders do not observe historical alpha. Furthermore, using *causal regularization*, interventions complement observational data.

Definition 6.3.7 (Expressing two experiments in do-calculus)**.** *Under the causal structure S, consider two experiments:*

(a) Submit random trades, do(Q).

Bouchaud et al. (2004)[34] outline this experiment.

> *"[Impact] could be in principle measured empirically by launching on the market a sequence of real trades of totally random signs, and averaging the impact over this sample of trades (a potentially costly experiment!)" (p. 14).*

Bouchaud (2022)[31] implements *this experiment and assesses "purely random trades that were studied at CFM during a specifically designed experimental campaign in 2010-2011." (p. 4)*

(b) Randomly do not submit trades, do($Q = 0$).

Before measuring alpha, *a trader sets aside a random subset of stocks to not trade.*

Each experiment naively identifies a counterfactual.

a) *Random* trades naively estimate price impact:

$$p\,(\Delta P|\,\mathrm{do}(Q))\,.$$

b) *Randomly* unsubmitted trades naively estimate alpha signals:

$$p\,(\Delta P|\,\mathrm{do}(Q=0),\alpha)\,.$$

Both experiments erase confounding causal paths, such as $\alpha \to Q \to \Delta P$. This approach corresponds to the A-B testing framework from Section 5.3.1 of Chapter 5. The regressions

$$\underbrace{\Delta P = \Delta I(\lambda, Q) + \epsilon|\,\mathrm{do}(Q);}_{\text{random trades}} \quad \underbrace{\Delta P = \beta\alpha + \epsilon|\,\mathrm{do}(Q=0)}_{\text{randomly unsubmitted trades}}$$

yield consistent price impact and alpha estimates on their respective interventional data.

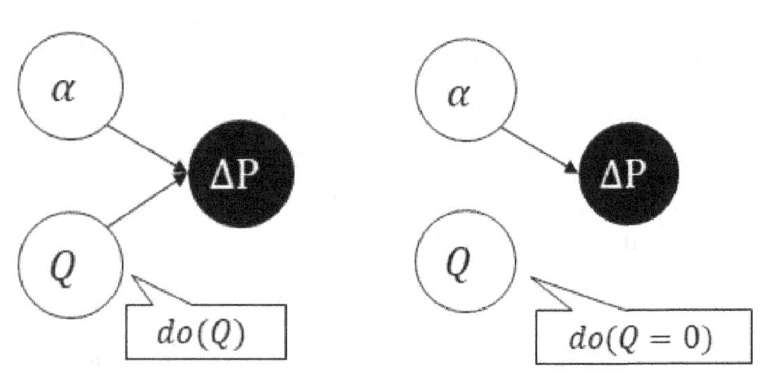

(a) Submitting random trades.

(b) Randomly not submitting trades.

FIGURE 6.3
Graph representations of the do-actions from Definition 6.3.7 on the causal structure $\mathcal{S}$.

Remark 6.3.8 (Causal regularization for Simpson's trading graph). *Section 5.3.2 from Chapter 5 introduces* causal regularization, *which uses* both *observational and interventional data. Section 6.3.4 applies causal regularization to Simpson's trading graph for two separate use cases.*

(a) When the trader does not observe historical α or cannot reproduce it. *For example, a broker evaluates trades of unknown origin. Causal regularization uses* random trades *to shrink price impact estimates accurately.*

(b) When the trader does not observe or trust Q. *For example, a distrustful alpha researcher submits trades through a broker but estimates their alpha signal without relying on their broker. Causal regularization uses* randomly unsubmitted trades *to shrink an alpha signal to consider price impact. This method* is model-free*: the researcher does not require an impact model or broker inputs to adjust their alpha signal for impact.*

6.3.4 A simulation example

This section illustrates standard A-B testing and causal regularization on synthetic trading data and follows Webster and Westray (2022)[234]. A simulation quantifies the bias-variance trade-off of price impact estimators in the presence of a causal bias. The study leverages Kolm and Westray's order simulator (2022)[133] to generate arbitrarily large observational and interventional data emulating real-life orders. Kolm and Westray simulate realistic orders using a Mean-Variance Optimization problem in the spirit of Markowitz (1952)[163]; the optimization problem trades off synthetic alpha signals against risk and practical portfolio and trading constraints to create daily orders for the Russell 3000.

The order simulator generates three types of data:

(O) 2.5 million orders *with* alpha.

The observational data (O) emulates data collected over ten years of trading.

(I) n orders *without* alpha.

Each interventional data (I) emulates a live trading experiment with a limited duration. For example, a team sets aside 3% of their flow over two months, leading to $n = 1250$ randomized orders. Indeed, traders launch A-B tests *after* they formulate experiments. Furthermore, opportunity cost, impatience, and competition between experiments limit the intervention's sample size and duration.

(V) Validation data (V) of 2.5 million orders *without* alpha.

(V) is unbiased validation data for the impact model and is only achievable in simulation. In practice, a live trading experiment of this magnitude is prohibitively expensive.

A statistician fits a family of ridge regression models

$$\lambda_\theta = \text{argmax}_\lambda \, ||\Delta P - \Delta I(\lambda, Q)||^2 + \theta||\lambda||^2$$

with meta-parameter θ in three ways.

a) The statistician uses *only observational data:* they run a least-squares regression on (O) with $\theta = 0$.

b) The statistician uses *only interventional data:* they run a least-squares regression on (I) with $\theta = 0$.

c) The statistician uses *causal regularization:* they run a ridge regression on (O) and tune the ridge penalty θ on (I).

Remark 6.3.9 (Precision and accuracy). *Using only observational data yields the most* precise *estimator. Assuming the Central Limit Theorem applies, the observational estimator's variance is* 200 *times smaller than the interventional estimator's. However, for the chosen simulation parameters, the observational estimator overstates λ by a third* regardless of the sample size. *This overestimate is due to prediction bias. Therefore, the interventional regression is more* accurate *than the observational regression. Causal regularization is precise* **and** *accurate, even with tiny interventions.*

The trading team repeats the estimation with independent samples and bootstraps three statistics.

(a) *The parameter's bias* measures the model's *inaccuracy.*

(b) *The parameter's t-stat* measures the model's *precision* using $\hat{\lambda}$'s bootstrapped distribution.

(c) *The model's validation* R^2 measures its performance. The team holds out (V) as independent, bias-free validation data and measures the model's validation R^2 with (V).

The bootstrap relies on simulation to produce sufficient interventional data and quantify the statistical estimators' performance. Therefore, Kolm and Westray's order simulator (2022)[133] emerges as an essential tool in experiment design: it quantifies the cost and benefit of proposed A-B tests *before* the team commits capital.

For instance, Table 6.1 summarizes the three methods' performance. The results are intuitive: first, observational data yields a precise but biased estimator. Second, interventional data yields a noisy but unbiased estimator. Finally, causal regularization yields a precise, unbiased estimator, even with tiny interventions. For example, causal regularization with $n = 250$ outperforms interventional data with $n = 1250$.

TABLE 6.1
Statistics across methods and intervention sizes when the trader does not observe historical alpha.

Intervention size	Fitting method	Bias	t-stat	$R^2(V)$
0	Observational data only	0.319	30.2	70bps
250	Interventional data only	−0.010	0.92	−10bps
250	Causal regularization	−0.010	3.32	70bps
1250	Interventional data only	0.05	2.40	63bps
1250	Causal regularization	0.01	4.12	72bps
12500	Interventional data only	−0.002	6.38	75bps
12500	Causal regularization	−0.005	6.66	75bps

Example 6.3.10 (When the trader does not observe Q). *Imagine an alpha researcher who does not collect trading data but relies on a third party, for example, a broker, to capture their trading data and estimate its price impact Consider the ridge loss function on* observational *data*

$$\beta_\eta = argmax_\beta \left|\left|\Delta P - \beta \cdot \alpha\right|\right|^2 + \eta\left|\left|\beta\right|\right|^2$$

with meta-parameter η. Here, α is a high-dimensional vector of predictive features, and $\beta \cdot \alpha$ is the final alpha prediction. With extensive observational *data, the dimension for β can reach millions of parameters, especially for advanced model architectures beyond linear regression. Because of price impact, the naive estimator* β_0 over-estimates *alpha. Waelbroeck et al. (2012)[231] remove this bias* under a given price impact model:

> *"the system may estimate corrected market prices that would have been observed had price impact not existed" ([231] p. 11).*

Using Waelbroeck's solution, the alpha researcher takes their broker's price impact model at face value and replaces ΔP with ΔS in their regression.

But what if the alpha researcher wants to calibrate their alpha without broker inputs? An alpha researcher using causal regularization can fit their signals without considering a third party's price impact model or data. Indeed, the researcher tunes η on randomly unsubmitted trades to adjust the alpha signal for impact in a model-free way. To quote Webster and Westray (2022)[234] on the necessity of causal regularization:

> *"The researcher cannot practically implement this approach without causal regularization: the opportunity cost of not submitting profitable trades is prohibitively high. Unfortunately, standard econometric techniques demand enormous interventional data. But causal regularization gets more for less and adjusts alpha for price impact in a model-free way with minimal opportunity costs" (p. 16).*

6.4 Synchronization Bias

Synchronization bias primarily affects the correct separation of internal and external impact. The section presents two applications.

a) *Crowding* and *leakage* model the correlation between the trader's orders and those from the rest of the market.

b) The *trade anonymity* assumption models the market's inability to distinguish the trader's impact from the market's.

6.4.1 Definitions

Bouchaud et al. (2018)[37] define synchronization bias.

> "**Synchronisation bias:** The impact of a metaorder can change according to whether or not other traders are seeking to execute similar metaorders at the same time" (p. 238).

To model synchronization bias, one includes M, the market's trades, into the causal graph. M extends Simpson's trading graph to model external trading, as per Definition 6.4.1.

Definition 6.4.1 (The external trading graph). *Define the causal structure $\mathcal{E}$:*

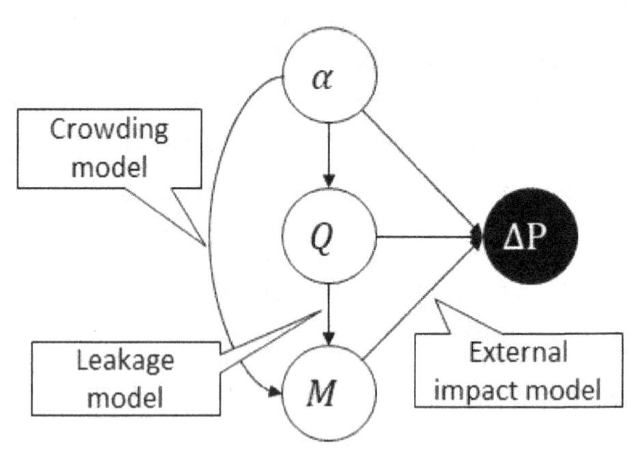

Name $\mathcal{E}$ the external trading graph.

In the external trading graph, both Q and M cause price impact on ΔP. Therefore, paths between Q and M bias the estimates of internal and external price impact.

Remark 6.4.2 (Crowding and leakage). *The causal structure $\mathcal{E}$ presents two paths linking Q to M.*

(a) The fork $Q \leftarrow \alpha \rightarrow M$ describes crowding.

Under this indirect *path, information common to both traders introduces a causal bias.*

(b) $Q \rightarrow M$ describes leakage.

Under this direct *path, other traders identify and react to the trades Q.*

The distinction between crowding and leakage is like that between alpha and impact. Crowding *trades in M occur* regardless *of Q: M reacts to α. On the other hand,* leakage *trades in M would* not *happen without Q: Q causes M. Bouchaud et al. (2004)[34] describe leakage when discussing live trading experiments.*

"However, following this procedure might induce 'copy-cat' trades" (p. 14).

Do-calculus formalizes the distinction between crowding and leakage through two counterfactuals:

(a) The crowding counterfactual

$$\mathbb{E}\left[\Delta M \mid do(Q=0), \alpha\right]$$

answers

How much would the market have traded if I had not traded?

(b) The leakage counterfactual

$$\mathbb{E}\left[\Delta M \mid do(Q)\right]$$

answers

How many market trades copy my trades?

Remark 6.4.3 (Causal assumptions). *The following statement summarizes all causal assumptions in the external trading graph.*

The trading system allows no feedback loops.

Ruling out feedback loops prevents the trader from using M inside the signal α. More realistically, impact-adjusted prices S avoid feedback loops between α and Q, and leakage-adjusted market trades prevent feedback loops between α and M.

Proposition 6.4.4 (Biases). *Under the causal structure $\mathcal{E}$, the following two biases hold.*

(a) Crowding extends prediction bias*:*

$$\mathbb{E}\left[\Delta P \mid do(Q), do(M)\right] \neq \mathbb{E}\left[\Delta P \mid Q, M\right].$$

(b) Leakage adds synchronization bias*:*

$$\mathbb{E}\left[\Delta P \mid do(M)\right] \neq \int \mathbb{E}\left[\Delta P \mid M, \alpha\right] p(\alpha) d\alpha.$$

Proof. (a) The paths emerging from α prevent identification without controlling for α.

(b) The fork $\Delta P \leftarrow Q \rightarrow M$ prevents identification without controlling for Q. □

6.4.2 Actions and counterfactuals

Traders use the external trading graph to state the following counterfactuals:

(a) *Impact with leakage counterfactual*

How much price impact are my trades causing?

(b) *Impact without leakage counterfactual*

How much price impact would my trades cause *if they did not leak?*

(c) *External impact counterfactual*

> How much price impact do the market's trades cause?

(d) *Anonymity assumption counterfactual*

> If the market had submitted the same trades as me, would they have caused the same price impact?

Definition 6.4.5 (Actions to measure). *Under the causal structure $\mathcal{E}$,*

(a) $p(\Delta P|\, do(Q))$ *defines the* impact with leakage counterfactual.

(b) $p(\Delta P|\, do(Q), do(M=0))$ *defines the* impact without leakage counterfactual.

(c) $p(\Delta P|\, do(M))$ *defines the* external impact counterfactual.

(d) $\forall q \in \mathbb{R}$, $p(\Delta P|\, do(M=q), do(Q=0)) = p(\Delta P|\, do(M=0), do(Q=q))$ *defines the* anonymity assumption counterfactual.

Counterfactuals (b), (c), and (d) are abstract: they do not correspond to a real-life action the trader can take. Indeed, the market flow M is not under the trader's control. However, the counterfactuals represent helpful thought experiments practitioners consider to improve their trading algorithms.[2]

Proposition 6.4.6 (Identifiability of actions). *The following identifiability equations hold under the causal structure $\mathcal{E}$:*

(a) Conditioning on the alpha but not external flow *identifies the* impact with leakage counterfactual:

$$p(\Delta P|\, do(Q)) = \int p(\Delta P|\, Q, \alpha)\, p(\alpha)\, d\alpha.$$

(b) Conditioning on alpha identifies the impact without leakage *and* anonymity assumption *counterfactuals:*

$$p(\Delta P|\, do(Q), do(M)) = \int p(\Delta P|\, Q, M, \alpha)\, p(\alpha)\, d\alpha.$$

(c) Conditioning on alpha and internal flow identifies the external impact counterfactual*:*

$$p(\Delta P|\, do(M)) = \int p(\Delta P|\, M, Q, \alpha)\, p(Q, \alpha)\, dQ d\alpha.$$

[2] A variant of the anonymity assumption can be tested on *non-anonymous* trading venues such as the Toronto Stock Exchange or the Foreign Exchange markets. The reader studie[s] Carmona and Leal (2021)[60] for an empirical study of non-anonymous trading.

Proof. (a) The conditioning variable $Z = \{\alpha\}$ satisfies the back-door condition from Theorem 5.2.34 in Chapter 5.

(b) The conditioning sequence $Z_1 = \{\alpha\}$, $Z_2 = \emptyset$ for the sequence $X_1 = Q$, $X_2 = M$ satisfies the sequential back-door condition from Theorem 5.2.36 in Chapter 5.

(c) The conditioning variables $Z = \{\alpha, Q\}$ satisfy the back-door condition from Theorem 5.2.34 in Chapter 5.

□

Remark 6.4.7 (Why estimating external impact is complex). *The reproducibility requirements to estimate external price impact are harsh. Indeed, a trader must condition on* α and Q *to identify external impact. The identifiability equation's complexity for external impact explains why the anonymity assumption, which states that external impact and internal impact models are the same, is helpful. It also increases the interventional data's value for estimating external price impact.*

Proposition 6.4.8 mirrors Proposition 6.3.5: leakage is to crowding as impact is to alpha.

Proposition 6.4.8 (Why estimating leakage is more complex than estimating crowding). *Assume given the causal structure* $\mathcal{E}$*.*

(a) The crowding counterfactual *has a naive identifiability equation:*

$$p\,(M|\,do(Q=0),\alpha) = p\,(M|\,Q=0,\alpha)\,.$$

(b) Conditioning on historical alpha identifies the leakage counterfactual*:*

$$p\,(M|\,do(Q)) = \int p\,(M|\,Q,\alpha)\,p(\alpha)d\alpha.$$

Proof. (a) follows from rule 2 of Theorem 5.2.26 in Chapter 5: M and Q are conditionally independent given α on $\mathcal{E}_{\underline{Q}}$.

b) The conditioning variable $Z = \{\alpha\}$ satisfies the back-door condition from Theorem 5.2.34 in Chapter 5.

□

Therefore, to estimate leakage, practitioners need reproducible signals or nterventional data. On the other hand, statisticians estimate crowding from bservational data. However, trading experiments may reduce colinearity and mprove crowding estimators.

6.4.3 Experiments

Remark 6.4.9 (Experiments). *The two real-life trading experiments that reduce* prediction bias *remain feasible.*

(a) Submit random trades, do(Q).

(b) Randomly not submit trades, do($Q = 0$).

The action do($M = 0$) is not achievable with a real-life trading experiment: M is not under the trader's direct control. Moreover, lacking interventions on M challenges a naive Bayesian identification of external impact.

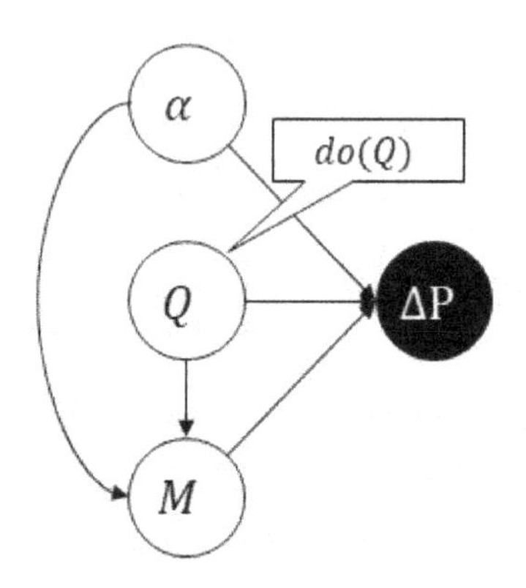

(a) Submitting random trades.

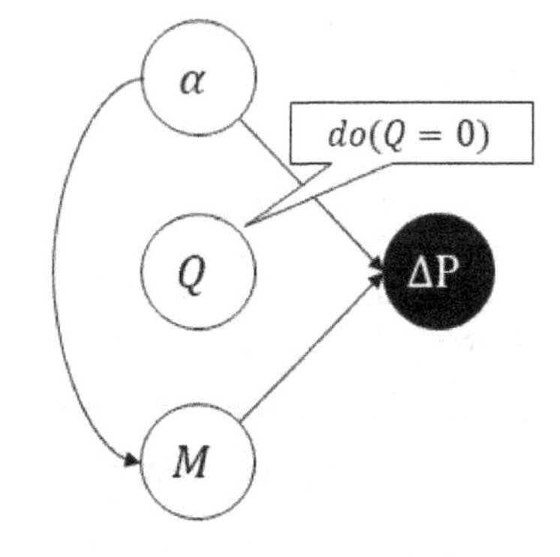

(b) Randomly not submitting trades.

FIGURE 6.4
Graph representations of the do-actions do(Q) and do($Q = 0$) from Remark 6.4.9 on the causal structure $\mathcal{E}$.

(a) Trades with *random* signs naively identify *leakage.*

(b) *Randomly* unsubmitted trades naively identify *crowding.*

Unfortunately, as per Remark 6.4.9, no intervention distinguishes internal and external price impact because the trader has no control over the market flow M. This hurdle makes the *anonymity assumption* crucial to disentangle one's impact from the market's. Bouchaud (2022)[31] empirically verified the anonymity assumption for CFM for short-term impact.

> "We have actually shown that the short term impact of CFM's trades is indistinguishable from the trades of the rest of the market" (p. 4).

Under the anonymity assumption, trades with *random* signs naively identify internal and external price impact.

6.5 Implementation Bias

Implementation bias affects the estimation of microstructure effects in price impact. The section presents an application: fitting price impact considering microstructure signals.

6.5.1 Definitions

Bouchaud et al. (2018)[37] define two implementation biases.

> "**Implementation bias 1:** [...] we have assumed that both the volume Q and execution horizon T are fixed before a metaorder's execution begins. In reality, however, some traders may adjust these values during execution, by conditioning on the price path" (p. 238).

> "**Implementation bias 2:** The impact path can become distorted for metaorders that are not executed at an approximately constant rate" (p. 239).

This section defines a general *implementation bias.*

> **Implementation bias:** Tactical deviations from the strategic trading trajectory introduce biases.

The minimal extension to Simpson's trading graph $\mathcal{S}$ capturing implementation bias adds two nodes. The low-latency alpha node α^{LL} leads to tactical deviations from the intended trading trajectory Q described by the realized trading node Q^r.

Definition 6.5.1 (The latency-aware graph)**.** *Define the causal structure $\mathcal{L}$:*

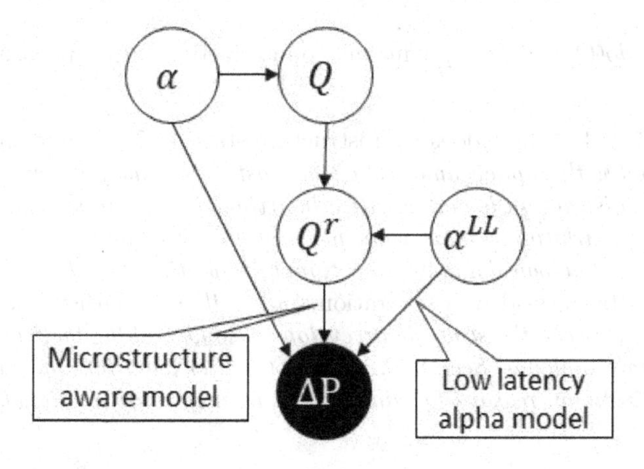

Name $\mathcal{L}$ the latency-aware graph.

Remark 6.5.2 (Causal assumptions). *The causal structure $\mathcal{L}$ rules out three links.*

- *The trader excludes $\alpha \to \alpha^{LL}$ and $Q \to \alpha^{LL}$ to simplify identifiability.*

  Concretely, both causal links lead to organizational challenges. *For example, the re-use of α in the tactical algorithm challenges attributing the alpha to the strategic and tactical algorithms. In addition, these links translate mathematically into complicated, non-intuitive identifiability equations.*

- *The trader excludes $\alpha^{LL} \to Q$ to reflect the strategic schedule's latency, as per Section 3.2.3 of Chapter 3.*

Unlike prediction and synchronization biases, a formal definition of *implementation bias* requires a careful definition of counterfactuals and actions.

6.5.2 Actions and counterfactuals

Traders distinguish two counterfactuals on the latency-aware graph:

(a) *Impact counterfactual with implicit microstructure*

> How much does the price move when I change my strategic schedule?

(b) *Impact counterfactual with explicit microstructure*

> How much price impact does an individual fill cause?

Definition 6.5.3 (Actions to measure). *Under the causal structure $\mathcal{L}$,*

(a) $p(\Delta P|\, do(Q))$ *defines the* impact counterfactual with implicit microstructure.

(b) $p(\Delta P|\, do(Q^r))$ *defines the* impact counterfactual with explicit microstructure.

Remark 6.5.4 (Where does microstructure start?). *The distinction between price impact with implicit and explicit microstructure may confuse the reader. Indeed, the distinction is mathematically artificial: one can shift features from α to α^{LL} and arbitrarily move elements from the strategic to the tactical level.*

From a technological and organizational perspective, the distinction is vital: it reflects a business-driven separation, *rather than a* mathematically-driven separation, *between the strategic execution schedule and the tactical deviations driven by micro-alphas. Section 3.2.3 from Chapter 3 outlines the technological and organizational reasons for singling out latency-sensitive signals.*

Practically, strategic decisions deal with the action $p(\Delta P|\, do(Q))$: they model microstructure effects implicitly. This approach covers most price impact applications from Chapters 3 and 4. On the other hand, tactical decisions deal with the action $p(\Delta P|\, do(Q^r))$ and model microstructure effects explicitly. For example, the OFI model of Cont, Cucuringu, and Zhang (2021)[73] in Section 4.2.1 of Chapter 4 includes all order book events.

Proposition 6.5.5 (Identifiability of actions). *The following identifiability equations hold under the causal structure $\mathcal{L}$:*

(a) $p(\Delta P|\, do(Q)) = \int p(\Delta P|\, Q, \alpha)\, p(\alpha)\, d\alpha.$

Controlling for α estimates the price impact model with implicit microstructure.

(b) $p(\Delta P|\, do(Q^r)) = \int p(\Delta P|\, Q^r, \alpha, \alpha^{LL})\, p(\alpha, \alpha^{LL})\, d\alpha d\alpha^{LL}.$

Controlling for α and α^{LL} estimates the price impact model with explicit microstructure.

Proof. (a) The conditioning variable $Z = \{\alpha\}$ satisfies the back-door condition from Theorem 5.2.34 in Chapter 5.

(b) The conditioning variables $Z = \{\alpha, \alpha^{LL}\}$ satisfy the back-door condition from Theorem 5.2.34 in Chapter 5.

□

Proposition 6.5.6 formalizes *implementation bias*: it affects the *price impact model with explicit microstructure* and extends prediction bias. Bershova and Rahklin (2013)[25] single out

> "[the] notorious bias of underestimating the market impact of large trades that were (1) initiated and executed because unusually high liquidity was available or (2) traded too slowly because of price limits" (p. 9).

Both are instances of implementation bias: the tactical deviations' profits reduce the measured price impact compared to the counterfactual, where the deviations did not occur. Farmer et al. (2008)[92] distinguish price impact across different tactical trading algorithms. When using limit orders, Farmer et al. warn of implementation bias due to adverse selection:

> "This can result in a biased view of impact, since for limit orders impact is felt largely by selective execution due to adverse information effects" (p. 47).

Proposition 6.5.6 (Implementation bias). *Under the causal structure $\mathcal{L}$, the* implementation bias

$$p(\Delta P|\, do(Q^r)) \neq \int p(\Delta P|\, Q^r, \alpha = x)\, p(\alpha = x)\, dx$$

holds.

Proof. The fork $Q^r \leftarrow \alpha^{LL} \rightarrow \Delta P$ prevents identification without conditioning on α^{LL}. □

Under the latency-aware graph $\mathcal{T}$, implementation bias does not affect the *price impact model with implicit microstructure*. The lack of bias follows from the clear separation between α and α^{LL}. To repeat the separation's *organizational* value: singling out microstructure signals allows traders to model them implicitly in their price impact model. Consequently, the strategic trading algorithm considers the *price impact model with implicit microstructure* and ignores implementation bias.

Remark 6.5.7 (Price impact with implicit microstructure). *Under the causal structure $\mathcal{L}$, there is no extension of the* prediction bias *for the* price impact model with implicit microstructure

$$p(\Delta P|\, do(Q)) = \int p(\Delta P|\, Q, \alpha)\, p(\alpha)\, d\alpha.$$

Furthermore, under the causal structure $\mathcal{L}$, there is no harm in including α^{LL} among the control variables in the above estimate.

Finally, consider the causal structure $\mathcal{L}'$, which Figure 6.5 illustrates, where one adds the link $Q \rightarrow \alpha^{LL}$ to $\mathcal{L}$. Under $\mathcal{L}'$, the negative result

$$p(\Delta P|\, do(Q)) \neq \int p\left(\Delta P|\, Q, \alpha, \alpha^{LL}\right) p\left(\alpha, \alpha^{LL}\right) d\alpha d\alpha^{LL}$$

holds. Therefore, adding the link $Q \rightarrow \alpha^{LL}$ introduces an additional causal bias and is not *recommended.*

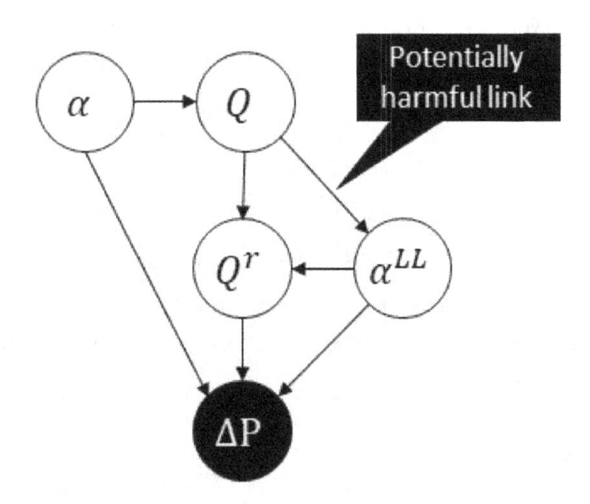

FIGURE 6.5
Graph representation of the causal structure $\mathcal{L}'$ obtained by adding the link $Q \rightarrow \alpha^{LL}$ to $\mathcal{L}$. See Remark 6.5.7.

6.5.3 Experiments and regularization

Definition 6.5.8 (Experiments). *Consider three experiments under the causal structure $\mathcal{L}$:*

(a) Submit random trades, $do(Q)$.

(b) Randomly trade on tactics only, $do(Q = 0)$.

The link $\alpha^{LL} \to Q^r$ allows tactical trading even in the absence of a strategic schedule. However, this experiment may not always be practical: business and technology considerations may constrain the trading team and prevent trading without a client order.

(c) Randomly do not execute tactical deviations, captured by the causal graph $\mathcal{L}''$ from Figure 6.6 (c).

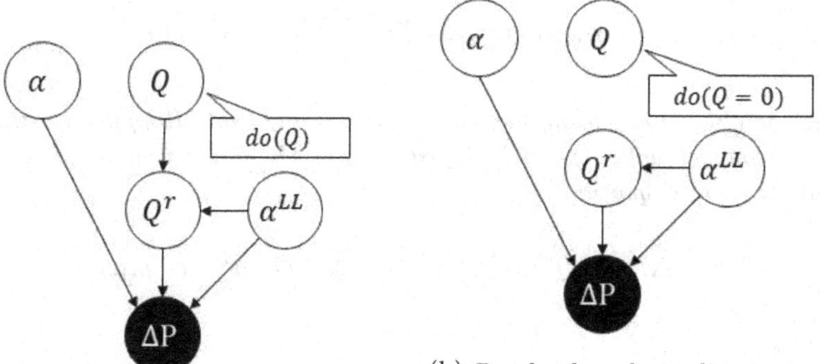

(a) Submitting random trades.

(b) Randomly only trading on tactics.

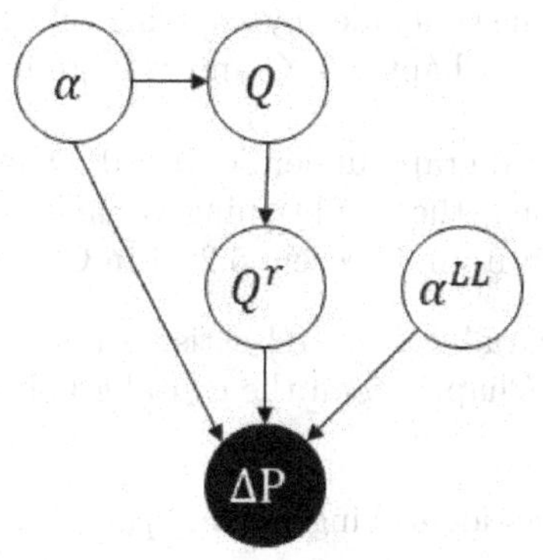

(c) Randomly not executing tactical deviations, $\mathcal{L}''$.

FIGURE 6.6
Graph representations of the three experiments from Definition 6.5.8.

This experiment tightly follows the strategic execution schedule, for instance, using a TWAP algorithm.

Because of live experiments' costs, it can be helpful to list multiple experiments that simplify a counterfactual's identification. However, while intersecting experiments further simplify identifiability, the corresponding interventional data may become tiny.

Proposition 6.5.9 (Bias-reduced data). *The following identifiability equations hold under the causal structure $\mathcal{L}$:*

(a) Random trades naively estimate price impact with implicit microstructure.

(b) Randomly *only trading on tactics and controlling for α^{LL} estimates price impact with explicit microstructure:*

$$p\left(\Delta P|\, do(Q^r), do(Q=0)\right) = \int p\left(\Delta P|\, Q^r, do(Q=0), \alpha^{LL}\right) p(\alpha^{LL}) d\alpha^{LL}.$$

(c) Randomly *not executing tactical deviations and controlling for α estimates price impact with explicit microstructure. On the causal graph $\mathcal{L}''$, the identification equation*

$$p\left(\Delta P|\, do(Q^r), \alpha^{LL}\right) = \int p\left(\Delta P|\, Q^r, \alpha^{LL}, \alpha\right) p(\alpha) d\alpha$$

holds.

Proof. (a) By Proposition 5.2.24 in Chapter 5, under do(Q), the graph is truncated to $\mathcal{L}_{\overline{Q}}$. One concludes by applying rule 2 of Theorem 5.2.26 and Proposition 5.2.24 in Chapter 5: Q and ΔP are independent under $\mathcal{L}_{\underline{\overline{Q}}}$.

(b) Consider the modified graph under do($Q=0$), where one removes all links from and to Q. Then, the conditioning variable $Z = \{\alpha^{LL}\}$ satisfies the back-door condition from Theorem 5.2.34 in Chapter 5.

(c) The conditioning variable $Z = \{\alpha\}$ satisfies the back-door condition from Theorem 5.2.34 in Chapter 5 on the causal graph $\mathcal{L}''$.

□

The two experiments identifying *price impact with explicit microstructure* present a trade-off:

(a) Trades *randomly* only trading on tactics naturally arise when making markets. In that case, the experiment is free. On the other hand, many team cannot trade solely on tactics for business or technology reasons: for instance, brokers can only trade their clients' orders.

(b) Trades *randomly* not executing tactical deviations are simple but costly to implement: trading teams can switch to a vanilla algorithm. However, microstructure signals contribute significantly to trading strategies.

Remark 6.5.10 (Causal regularization for price impact with explicit microstructure). *Propositions 6.5.5 and 6.5.9 show that identification formulas for price impact with explicit microstructure are challenging. They require control over both alphas through historical reproducibility or randomized experiments.*

Causal regularization improves the situation by leveraging observational and interventional data. A potent prior is available to the researcher if they trust their price impact model with implicit microstructure. *Assume that both statistical models have comparable functional forms, with parametrization* λ^i *and* λ^e *for the implicit and explicit models. Then a natural prior is that one samples* λ^i *and* λ^e *from* pairwise correlated *Gaussians. Given* λ^i, *the Bayesian posterior for* λ^e *centers around* λ^i *instead of zero. The Bayesian prior yields the regularization penalty*

$$\theta||\lambda^i - \lambda^e||^2.$$

This approach applies transfer learning *to causal regularization. The price impact model with explicit microstructure learns from the standard implicit model. Therefore, it focuses its learning on novel information and features compared to the standard impact model. The reader can study "A Concise Review of Transfer Learning" (p. 1) by Farahani et al. (2020)[91]:*

> *"In contrast to the traditional machine learning and data mining techniques, which assume that the training and testing data lie from the same feature space and distribution, transfer learning can handle situations where there is a discrepancy between domains and distributions. These characteristics give the model the potential to utilize the available related source data and extend the underlying knowledge" (p. 1).*

6.6 Issuer Bias

Issuer bias primarily affects the correct estimation of price impact for follow-up orders, which Section 3.3.4 of Chapter 3 studies. Finally, this section presents the causal regularization of follow-up impact based on a standard price impact model.

6.6.1 Definitions

Bouchaud et al. (2018)[37] define issuer bias.

> "**Issuer bias:** Another bias may occur if a trader submits several dependent metaorders successively" (p. 239).

Sections 3.2.4 and 3.3.4 from Chapter 3 study subsequent orders. The essential word in the definition of issuer bias is *dependent:* the *causal* link between two sequential orders causes issuer bias. The dependency can stem from two causal paths: either the first trade's *alpha* causes the second trade's alpha, or the first trade *directly* causes the second trade.

Definition 6.6.1 (The time-aware graph). *Define the causal structure $\mathcal{T}$:*

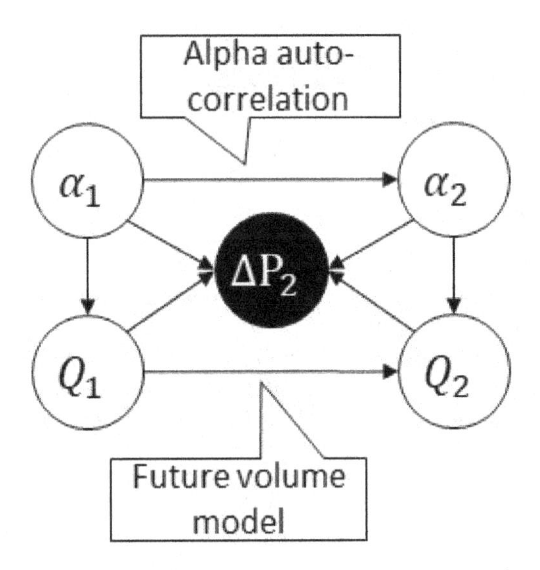

Name $\mathcal{T}$ the time-aware graph.

Remark 6.6.2 (Causal assumptions). *The causal structure $\mathcal{T}$ rules out two types of links*

(a) Traders exclude paths $\alpha_2 \to \alpha_1$ and $Q_2 \to Q_1$: they do not respect temporal causality.

(b) Traders exclude paths $\alpha_1 \to Q_2$ and $Q_1 \to \alpha_2$: they reflect a complicated trading infrastructure.

Example 6.6.3 (An example of each causal path). *The time-aware graph may seem abstract at first. For example, how does one distinguish between the indirect path $Q_1 \to \alpha_1 \to \alpha_2 \to Q_2$ and the direct path $Q_1 \to Q_2$?*[3]

---

[3]Bucci et al. (2018)[43] study follow-up orders and fit an impact model in that context. Bucci et al. link the effect they observe to order autocorrelation.

> "This plot illustrates how the autocorrelation of order flow can strongly bias the estimation of impact decay" (p. 10).

(a) *For instance, the path $\alpha_1 \to \alpha_2$ activates for mean-reversion signals, as features triggering a positive signal at time $t = 1$ mechanically trigger a negative signal at $t = 2$. This example motivates the terminology "alpha autocorrelation" for the path, even though it is* mathematically *a causal path.*

(b) *For instance, the path $Q_1 \to Q_2$ activates for a market maker constrained to end the day with no position: regardless of α_2, Q_2 trades opposite Q_1 to satisfy the end-of-day constraint.*

Proposition 6.6.4 (Issuer bias). *Under the causal structure $\mathcal{T}$, the* issuer bias

$$\mathbb{E}\left[\Delta P_2 | \, do(Q_2), do(Q_1 = 0)\right] \neq \mathbb{E}\left[\Delta P_2 | \, Q_2, Q_1 = 0\right]$$

holds.

Proof. The paths emerging from α_1, α_2 prevent identification without conditioning by α_1, α_2. □

6.6.2 Actions and counterfactuals

Under the time-aware graph, traders state three counterfactuals. The first is a variant of the price impact counterfactual. The latter two probe the system's internal behavior and help measure and communicate the trading infrastructure's complexity.

(a) *Follow-up impact counterfactual*

> How much price impact would the second order cause if I had not submitted the first order?

(b) *Alpha autocorrelation counterfactual*

> What part of the second alpha reflects added information?

(c) *Future volume counterfactual*

> How should I expect my second order to change if I change my first order?

The alpha autocorrelation counterfactual is abstract: it does not correspond to a real-life action, but is a thought experiment. Indeed, the underlying causes of future stock price moves are *not* under the algorithm's control, only how the trades react to the alpha signal's features.

Definition 6.6.5 (Actions to measure). *Under the causal structure $\mathcal{T}$,*

(a) $p(\Delta P_2 | \, do(Q_2), do(Q_1 = 0))$ defines the follow-up impact counterfactual.

(b) $p(\alpha_2 | \, do(\alpha_1 = 0))$ defines the alpha autocorrelation counterfactual.

(c) $p(Q_2 | \, do(Q_1))$ defines the future volume counterfactual.

Proposition 6.6.6 (Identifiability of actions). *The following identifiability equations hold under the causal structure $\mathcal{T}$:*

(a) Conditioning on both alphas identifies the follow-up impact counterfactual*:*

$$p\,(\Delta P_2|\, do(Q_2), do(Q_1 = 0)) = \int p\,(\Delta P_2|\, Q_2, Q_1 = 0, \alpha_1, \alpha_2)\, p(\alpha_1, \alpha_2) d\alpha_1 d\alpha_2.$$

(b) One naively identifies the alpha autocorrelation counterfactual:

$$p\,(\alpha_2|\, do(\alpha_1 = 0)) = p\,(\alpha_2|\, \alpha_1 = 0)\,.$$

(c) Conditioning on either alpha identifies the future volume counterfactual*:*

$$\begin{aligned} p\,(Q_2|\, do(Q_1)) &= \int p\,(Q_2|\, Q_1, \alpha_1)\, p(\alpha_1) d\alpha_1 \\ &= \int p\,(Q_2|\, Q_1, \alpha_2)\, p(\alpha_2) d\alpha_2 \\ &= \int p\,(Q_2|\, Q_1, \alpha_1, \alpha_2)\, p(\alpha_1, \alpha_2) d\alpha_1 d\alpha_2. \end{aligned}$$

Proof. (a) The conditioning sequence $Z_1 = \{\alpha_1, \alpha_2\}$, $Z_2 = \emptyset$ for the sequence $X_1 = Q_1$, $X_1 = Q_2$ satisfies the sequential back-door condition from Theorem 5.2.36 in Chapter 5.

(b) follows from Corollary 5.2.32 in Chapter 5.

(c) Either alpha blocks all indirect paths from Q_1 to Q_2. $Z = \{\alpha_1, \alpha_2\}$, $Z = \{\alpha_1\}$, and $Z = \{\alpha_2\}$ each satisfy the back-door condition from Theorem 5.2.34 in Chapter 5.

□

As in previous sections, price impact requires conditioning on historical alpha to be identifiable, while alpha is naively identifiable.

6.6.3 Experiments and regularization

The time-aware graph $\mathcal{T}$ has few noteworthy trading experiments: live trading experiments that isolate specific causal links are challenging to implement as the graph is heavily connected.

Definition 6.6.7 (Experiment). *Define submitting order sequences with random signs by the causal graph $\mathcal{T}'$ in Figure 6.7.*

Submitting order sequences with random signs differs from random trades: Q_1 and Q_2 retain their causal link, and the trader only randomizes the order sequence's initial sign. One does not independently randomize the two orders Q_1 and Q_2 to maintain the causal link $Q_1 \rightarrow Q_2$.

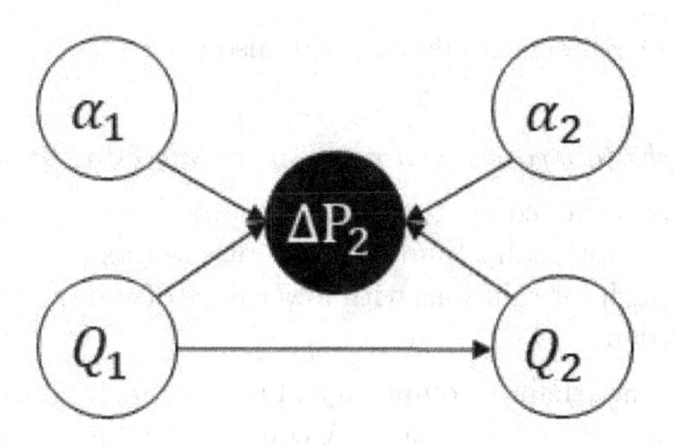

FIGURE 6.7
Graph representations of $\mathcal{T}'$ representing the experiment from Definition 6.6.7.

Proposition 6.6.8 (Bias-free data). *Under the causal structure $\mathcal{T}'$, order sequences with random signs naively identify the future volume counterfactual:*

$$p(Q_2|\, do(Q_1)) = p(Q_2|\, Q_1)\,.$$

Proof. Proposition 6.6.8 follows from rule 2 of Theorem 5.2.26 in Chapter 5: Q_1 and Q_2 are independent on $\mathcal{T}'_{\underline{Q}_1}$. □

Remark 6.6.9 (Regularization). *The same transfer learning method applies to the external trading graph and the latency-aware graph: one uses the price impact model under the straightforward causal structure as a prior for the more complex model.*

For the time-aware graph, if λ_t is the parameter for the impact model at time t, the prior on λ_2 centers around λ_1. The transfer learning prior yields the smoothing regularization penalty

$$\theta||\lambda_2 - \lambda_1||^2.$$

In the multi-period case, the penalty applies to the discretized derivative *of λ_t*

$$\theta \sum_t ||\lambda_{t+1} - \lambda_t||^2.$$

For the generalized OW model, the method shrinks the liquidity parameters' derivatives, leading to a spline smoothing model of the parameters. The reader studies Li (1986)[150] for detailed proof and application of regularization for smoothing.

6.7 Concluding Thoughts

The author hopes to have convinced the reader of the power and applicability of causal inference to trading: beyond providing technical fixes to biases in price impact models, causal inference can also drive change in trading organizations.

(a) *Causal graphs to measure and reduce a trading infrastructure's complexity*

Stakeholders have competing priorities for the trading infrastructure: traders favor optionality and manual interventions. On the other hand, developers push for solutions with fewer lines of code to reduce their maintenance burden.

Causal inference defines complexity in action units: which counterfactuals can the team measure or enact? A complex trading infrastructure allows fewer actions than a well-designed one. Therefore, quantitative researchers must voice design choices for the infrastructure and maximize their team's ability to measure and act on their trading strategies.

(b) *Causal graphs for experiment design in live trading*

The modern push for best execution puts A-B testing and live trading experiments in the spotlight. Unfortunately, live trading experiments are costly. Therefore, tools that formulate precise counterfactuals and design experiments to maximize their measurement are a competitive advantage in the field.

(c) *Causal graphs for simulated trading*

Simulation experiments complement live trading experiments: causal graphs implement simulations and improve fidelity to live systems.

6.8 Summary of Results

This section summarizes the four *trading biases* from Bouchaud et al. (2018)[37]. Let Q be a trade, α an alpha signal, and ΔP a contemporaneous return. Throughout the section, controlling for a variable implies either co-fitting with it or generating randomized interventional data from a live trading experiment.

This section's results hold for all statistical models consistent with the proposed causal graphs. In particular, the causal results are not reliant on traditional linear econometric models. Instead, practitioners can apply non-parametric fitting methods, such as machine learning, to estimate price impact or other causal expressions.

6.8.1 Prediction bias

Trades triggered by or considering an alpha signal exhibit bias when estimating their price impact.

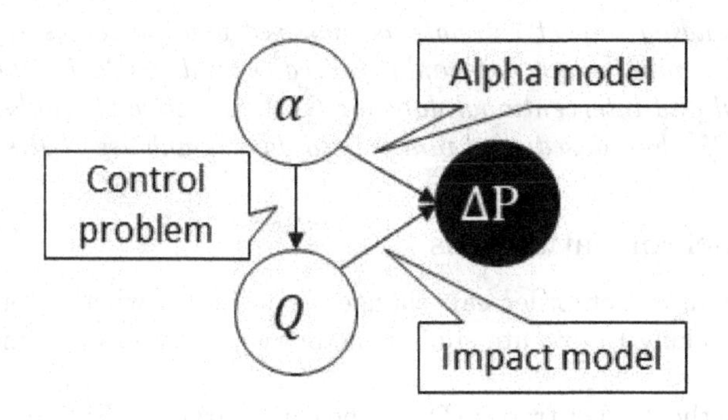

FIGURE 6.8
Causal graph for prediction bias.

The causal graph in Figure 6.8 captures prediction bias. To avoid prediction bias, one controls for α when estimating the price impact relationship $Q \to \Delta P$.

Example 6.8.1 (Impact-adjusted alpha research). *Consider a trader who can reproduce historical alphas but fits a novel model going forward. Let $\alpha_{historical}$ and α_{new} be the historical and new alpha signals. Let $I(\lambda, Q)$ be a price impact model parametrized by λ. Then, a two-step regression estimates the correct impact and novel alpha model.*

(a) The historical regression

$$\Delta P = \Delta I(\lambda, Q) + \beta^h \alpha_{historical} + \epsilon$$

yields the correct λ for impact.

(b) Given the impact model $I(\lambda, Q)$, one computes unperturbed *returns ΔS via $\Delta S = \Delta P - \Delta I(\lambda, Q)$. Finally, the regression*

$$\Delta S = \beta^n \alpha_{new} + \epsilon'$$

provides the correct β^n for the novel alpha model.

Example 6.8.2 (When alpha is unknown). *For example, a broker evaluates client trades with unknown alpha. The broker corrects for prediction bias in the data using causal regularization's three steps.*

(a) Train a regularized price impact model with meta-parameter θ on biased observational data.

(b) Submit randomized trades to collect unbiased interventional data.

(c) Calibrate the regularization meta-parameter θ on the randomized trades to accurately shrink *the price impact model.*

The corresponding impact estimate is unbiased and converges rapidly compared to using randomized data only. Indeed, causal regularization leverages observational and interventional data for their respective strengths: historical data is plentiful but biased, and randomized data is unbiased but scarce.

6.8.2 Synchronization bias

"The impact of a metaorder can change according to whether or not other traders are seeking to execute similar metaorders at the same time" (p. 238, [37]).

Let M be the market trades. Then, the causal graph in Figure 6.9 captures synchronization bias.

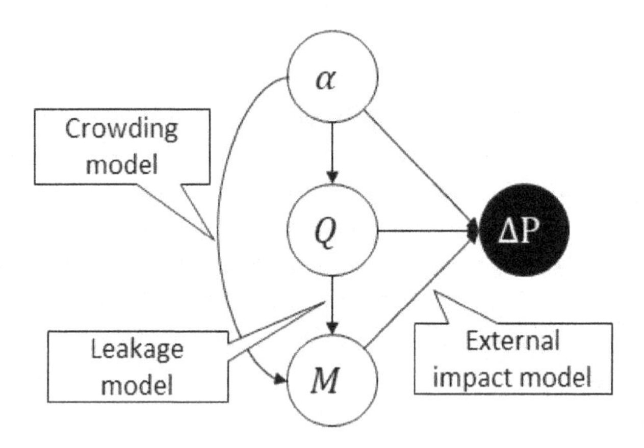

FIGURE 6.9
Causal graph for synchronization bias.

Synchronization bias has two sources:

(a) *Crowding,* $Q \leftarrow \alpha \rightarrow M$, describes the market trading in the same direction due to a common alpha signal.

(b) *Leakage,* $Q \rightarrow M$, describes copy-cat trades.

To avoid crowding bias, one controls for α. To prevent leakage bias, one controls for one's trades when estimating the external impact relationship $M \rightarrow \Delta P$.

6.8.3 Implementation bias

Tactical deviations from the strategic trading trajectory introduce biases.

Let Q^r be realized trades and α^{LL} be the low-latency alpha signals driving tactical deviations. Then, the causal graph in Figure 6.10 captures implementation bias. To avoid implementation bias, control for α, α^{LL} when estimating the price impact relationship $Q^r \rightarrow \Delta P$.

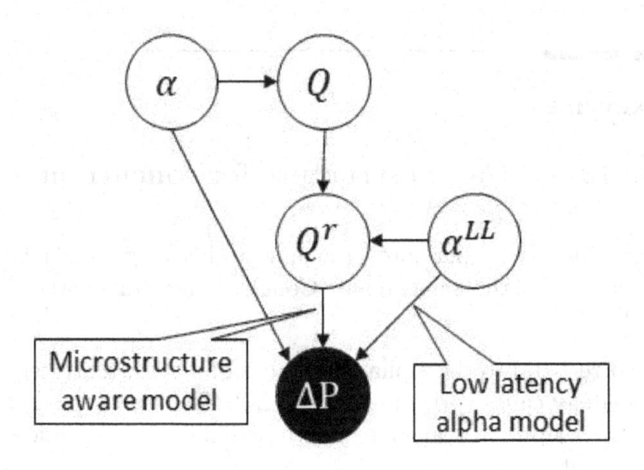

FIGURE 6.10
Causal graph for implementation bias.

6.8.4 Issuer bias

"Another bias may occur if a trader submits several dependent metaorders successively" (p. 239, [37]).

Split Q into Q^1, Q^2 and assume two alphas α^1, α^2 drive the successive trades Q_1, Q_2. ΔP_2 is the second order's contemporaneous return. Then, the causal graph in Figure 6.11 captures issuer bias. To avoid issuer bias, control for α^1, α^2 when estimating the second order's price impact $Q_2 \to \Delta P_2$.

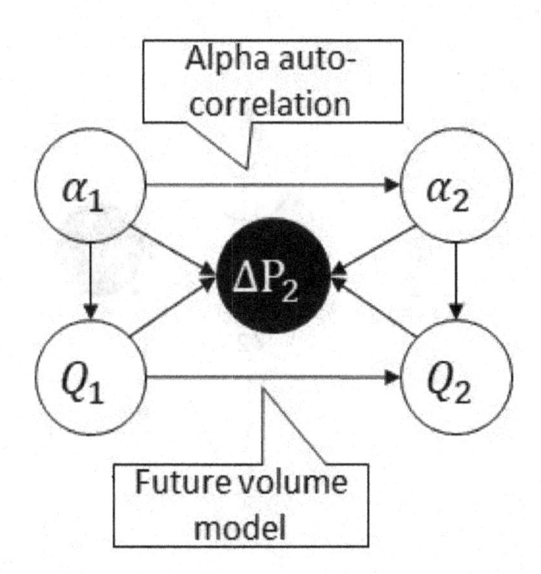

FIGURE 6.11
Causal graph for issuer bias.

6.9 Exercises

Exercise 19 Causal structure for concurrent strategies

This exercise delves deeper into crowding and leakage when both trading strategies belong to the same team. Consider the causal structure $\mathcal{C}$ from Figure 6.12.

(a) α_i, Q_i correspond to the alpha and trades of a trading strategy $i \in \{1, 2\}$. Each strategy causes price impact on ΔP. The strategies do not observe each other's alpha in real-time. The researcher can reproduce both α^i on historical data.

(b) α causes both α_i. Assume α is an unobserved confounder.

1. Prove that controlling for α_1 identifies the price impact of Q_1,

$$\mathbb{E}\left[\Delta P \middle| \operatorname{do}(Q_1)\right].$$

2. Assume Q_1 observes Q_2 before acting. Identify $\mathbb{E}\left[\Delta P \middle| \operatorname{do}(Q_1)\right]$.

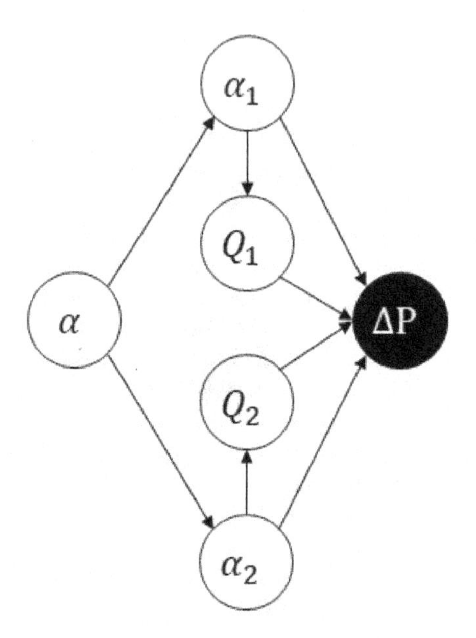

FIGURE 6.12
Causal graph $\mathcal{C}$ from exercise 19.

3. Assume Q_1 observes α_2 before acting. Identify $\mathbb{E}[\Delta P | \text{do}(Q_1)]$.
4. Assume Q_2 observes Q_1 before acting. Identify $\mathbb{E}[\Delta P | \text{do}(Q_1)]$. Interpret this counterfactual in terms of leakage.

Exercise 20 Causal structure for dark pools

This exercise presents a causal structure for trading in dark pools. For more general information on dark pools, the reader studies Lehalle and Laruelle (2018)[147], Section 1.4.1 "Mechanism of dark liquidity pools" (p. 102).

Consider the causal structure $\mathcal{D}$ from Figure 6.13. The nodes have the following interpretation.

(a) $Q, \alpha, \Delta P$ are like in Simpson's trading graph.

(b) An *unobserved* external flow M exhibits *crowding* with the trading process Q.

(c) D captures trades between M and Q that cross, making these trades observable.

This graph differs from the external trading graph in two ways: first, the external flow M interacting with Q over the dark pool is *unobserved.* Second, assume leakage is non-existent. The only source of confounding is crowding, represented by node α.

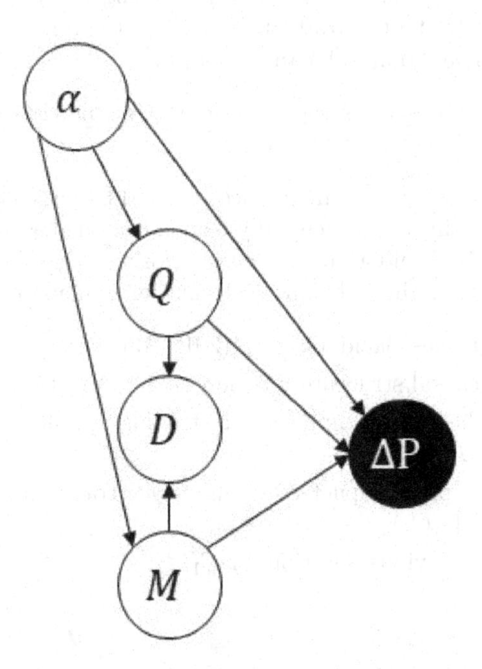

FIGURE 6.13
Causal graph $\mathcal{D}$ from exercise 20.

1. Show that Q and M are independent conditional on α.
2. Show that Q and M are dependent conditional on α, D. Explain the reason for this dependence.
3. Identify $\mathbb{E}\left[\Delta P \mid \text{do}(Q), D\right]$, the impact of Q conditional on observing a cross. How does it differ from the unconditional impact $\mathbb{E}\left[\Delta P \mid \text{do}(Q)\right]$?

Exercise 21 Causal structure for the closing auction

The closing auction is a daily auction where traders submit orders to buy or sell at the closing rather than the current price. Venues collect these orders throughout the day in parallel to public trading. The orders remain hidden, and the venues reveal the buy and sell imbalance at specific times of the day. The reader finds the details of the NYSE closing auction on their website [179] and in Section "1.3.4 The Closing Auction" (p. 24) of Velu, Hardy, and Nehren (2020)[227]. A simplified timeline follows:

(a) Traders submit orders to the closing auction before 2 pm, after which the venue reveals the outstanding closing imbalance. Therefore, let M_0 be the trade imbalance on the public trading tape at 2 pm and C_0 the auction imbalance at 2 pm.

(b) Further trading on the close and the public tape continues until 4 pm, after which the venue reveals the final closing imbalance and price P_T. Therefore, let M_1 be the trade imbalance on the public trading tape at 4 pm and C_1 the auction imbalance at 4 pm.

Bacidore (2020)[12] describes the price impact of the closing auction imbalances.

> "The market learns of any imbalances created by large closing auction orders at these times, and generally responds immediately. In fact, one empirical study found that the *entire* closing price impact actually occurs at the time the imbalance information is made public" (p. 53).

The reader also studies Bacidore et al. (2012)[13].

1. Suggest a causal structure for the variables M_0, C_0, M_1, Q_1, P_T.
2. Identify the price impact of trading publicly before 2 pm, $\mathbb{E}\left[P_T \mid \text{do}(M_0)\right]$.
3. Identify the price impact of submitting a trade on the close before 2 pm, $\mathbb{E}\left[P_T \mid \text{do}(C_0)\right]$.
4. Repeat the previous question for C_1.

7

Empirical Analysis of Price Impact Models

> *"If order flow is the dominant cause of price changes, "information" is chiefly about correctly anticipating the behavior of others, as Keynes envisioned long ago, and not about fundamental value."*
>
> – Bouchaud (2022)[31]

Price impact is concrete: straightforward to measure in real-time and essential to quantifying the actual cost of trading. The Committee on the Global Financial System (CGFS) published "Market-making and proprietary trading: industry trends, drivers and policy implications" (2014)[183] and defines liquidity in terms of price impact:

> "Market liquidity – the ability to rapidly execute large financial transactions with a limited price impact – is a key feature of financial market efficiency and functioning" (p. 4).

The CGFS uses price impact as a core liquidity measure.

> "Another popular liquidity measure considers the rise (fall) in price that typically occurs with a buyer-initiated (seller-initiated) trade. Such price impact measures are relevant to those executing large trades or a series of trades and, together with the bid-ask spread and depth measures, provide a fairly complete picture of market liquidity" (p. 52).

Studies measure price impact across

- Data, including
  - proprietary, order-level data (e.g., Almgren et al. (2005)[11], Bershova and Rhaklin (2013)[25], Bouchaud et al. (2015)[35], Tóth, Eisler, and Bouchaud (2017)[225], Frazzini, Israel, and Moskowitz (2018)[96])[1]
  - public, fill-level data (e.g., Bouchaud et al. (2004)[34], Cont, Kukanov, and Stoikov (2013)[74], Chen, Horst, and Tran (2019)[70])

[1]For a review and comparison of price impact estimations across proprietary data, the reader studies Zarinelli et al. (2015)[241]. For example, table one on page four provides "the approximate number of metaorders considered in previous studies, together with the corresponding trading institution where the orders originated." These include Citigroup, Morgan Stanley, CFM, and AllianceBernstein LP.

DOI: 10.1201/9781003316923-7

- Asset classes, including
  - US stocks (e.g., Bouchaud et al. (2016)[36], Cont, Cucuringu, and Zhang (2021)[73])
  - foreign stocks (e.g., Bouchaud et al. 2004 [34], Zhou (2012)[242])
  - fixed income (e.g., CGFS (2016)[184], Schneider and Lillo (2019)[206], Tomas, Matromatteo, and Benzaquen (2022)[224])
  - options (e.g., Said et al. (2021)[203], Kaeck, van Kervel, and Seeger (2021)[128], Tomas, Mastromatteo, and Benzaquen (2022)[223])
  - cryptocurrencies (e.g., Donier and Bonart (2015)[85])
- Timescales, including
  - tick timescale (e.g., Bouchaud et al. (2004)[34], Busseti and Lillo (2012)[47], Eisler, Bouchaud, and Kockelkoren (2018)[87])
  - hour to day timescale (e.g., Lillo and Farmer (2004)[151], Bouchaud, Kockelkoren, and Potters (2006)[33], Bershova and Rhaklin (2013)[25], Chen, Horst, and Tran (2019)[70])
- Model specifications

These papers find broad agreement on price impact's general shape and size. To quote Tóth, Eisler, and Bouchaud (2017)[225]:

> "The short-term price impact of trades is universal" (p. 1).

This chapter builds up the reader's empirical intuition on price impact models from the literature. A single methodology reproduces the models on shared data to enable comparison. However, the study presents three limitations.

(a) First, the reader should not construe the analysis as an attempt to find the *best overall price impact model.*

Indeed, price impact models behave differently depending on trading strategies and data. Therefore, each trading team can use the outlined methodology to fit and evaluate models adapted to their data, trading styles, and applications. The empirical results merely serve as a starting point based on the literature: each model suits different use cases, and the reader adapts the models to their data and strategies.

(b) The study only reproduces price impact models from the literature on publicly available trading data.

Some authors in the literature assess their impact models on their proprietary data. These may perform differently on the normalized methodology in this chapter. Furthermore, the data used throughout the chapter is *purely observational.* Chapters 5 and 6 cover causal biases in trading and their effect on price impact models.

(c) The list of price impact models is far from exhaustive.

The chapter reproduces a representative sample of single-stock price impact models from the literature. Section 7.5 discusses cross-impact and price impact for other asset classes.

A trading perspective complements the empirical study: the reader understands the models' statistical merits *and* their application to the problems covered in Chapters 3 and 4.

7.1 A Pedagogical Example

For traders, this chapter's essential takeaway is that, despite the plethora of price impact models, the literature finds a consensus on four empirical price impact facts. For teaching purposes, this section lists these empirical *sanity checks* for the reader to assess their own models and trading data.

(a) Depending on the prediction horizon, out-of-sample R^2 for price impact from the public trading tape models range upward of 10% when explaining contemporaneous returns.

(b) Price impact's magnitude for funds is around 30bps on liquid stocks. Generally, price impact is five to ten times larger than the bid-ask spread across all stocks.

(c) Price impact exhibits a pronounced time of day effect: trades cause the most price impact at the start of the day. This time of day effect aligns with similar time of day effects measured for volatility, market activity, and bid-ask spreads.

(d) Price impact is concave for sizable orders.

If the reader's analysis disagrees with the four broad statements, this may indicate a bias in their data or a non-standard fitting method for their model. Chapter 6 covers causal biases when fitting price impact models. Finally, this chapter delves into the implementation details of empirical price impact models in the literature.

7.2 Methodology

The Handbook reproduces the framework of Cont, Kukanov, and Stoikov (2013)[74]: to the author's best knowledge, Cont, Kukanov, and Stoikov provide the most reproducible details about their methodology. Furthermore, the reader finds additional descriptions in the follow-up paper by Cont, Cucuringu, and Zhang (2021)[73]. These details enable Section 7.4 to leverage a single, consistent, reproducible framework for assessing price impact models. Five steps structure the methodology.

(a) Pre-process the event-based data.

(b) Define the base features and binned data.

(c) Define the time kernel and compute the price impact model.

(d) Define the prediction horizon and training sample.

(e) Define the testing (in the case of regularization) and validation (for the out-of-sample evaluation) samples.

7.2.1 Pre-processing the event-based data

The analysis begins with the level 1 order book events from the LOBSTER database. LOBSTER [154] is "high-frequency, easy-to-use and latest limit order book data" for academic researchers. The analysis covers the S&P 500 constituents over the year 2019. Orderbook events fall into three categories, and *order flow imbalance* quantifies the events' directional effect, in line with Cont, Kukanov, and Stoikov (2013)[74].

(a) Trades

The contribution of trades to order flow imbalance is the signed fill quantity $q = a - b$, where a, b fill at the ask and bid.

(b) Limit order submissions

The contribution of limit order submissions to order flow imbalance is $b - a$, where a, b are submissions on the ask and bid.

(c) Limit order cancellations

The contribution of limit order cancellations to order flow imbalance is $a - b$, where a, b are cancellations on the ask and bid.

7.2.2 Definition of the base features and binned data

In line with Cont, Kukanov, and Stoikov (2013)[74], one bins the data using a bin-size $\delta t = 10s$. The order flow imbalance over an interval $[t, t + \delta t]$ is the sum of the OFI contributions. Denote the order flow imbalance by ofi_t. The trade imbalance is the sum of trade contributions only. Denote the trade imbalance by q_t.

One constructs four other variables from the events:

- The observed midprice P_t at the bin start and the return $r_t = \frac{P_{t+\delta t}}{P_t} -$ over the bin

- The amount of volume traded $|q_t|$ over the bin
- The average order book depth d_t over the bin

7.2.3 Definition of the time kernel and price impact computation

For simplicity and in line with the *generalized OW model,* the chapter studies an exponential time kernel under multiple clocks. Cont, Cucuringu, and Zhang (2021)[73] fit a fully non-parametric time kernel in Appendix E (p. 38) in the regular clock. Bouchaud et al. (2004)[34] fit a power-law time kernel in the trade clock. Chen, Horst, and Tran (2019)[70] fit an exponential kernel in the regular clock. Section A.3 of Appendix A implements an efficient algorithmic for an exponential kernel under an arbitrary clock.

One derives the impact state I_t and the instantaneous market activity v_t from the base features and the impact model. Under the chosen clock, the instantaneous liquidity v_t is the exponential moving sum of the volume traded $|q_t|$:

$$v_{t+\delta t} - v_t = -\beta v_t \delta t + |q_t|.$$

7.2.4 Definition of the prediction horizon and training samples

Price impact behaves differently across time horizons. For a time horizon h, define

- the horizon-specific return $\Delta_t^h P = \frac{P_{t+h}}{P_t} - 1$.
- the impact return $\Delta_t^h I = I_{t+h} - I_t$.

Generally, Δ^h is the variable's relevant increment operator over horizon h. The reader may notice a slight inconsistency: certain variables express as differences, while others express as returns. However, in practice, these inconsistencies are second-order.

A reader with an alpha signal α predicts the unperturbed price over horizon h by satisfying

$$\alpha_t^h = \mathbb{E}\left[\frac{S_{t+h}}{S_t} - 1 \middle| \mathcal{F}_t\right].$$

For example, see Sections 4.2.1 and 4.2.2 from Chapter 4. Finally, one runs the regression

$$\Delta^h P = \Delta^h I + \epsilon$$

over the model's free parameters to specify the price impact model.

Cont, Kukanov, and Stoikov (2013)[74] present their core results for $h = t = 10s$. Cont, Kukanov, and Stoikov (2013)[74] and Cont, Cucuringu, and Zhang (2021)[73] briefly mention timescales and prediction horizons ranging from fifty milliseconds to one hour (Figure twelve, p. 31, [74] and Figure seventeen, p. 40, [73]).

One runs the regression on

(a) a month of data to fit a monthly static model or

(b) thirty minutes of data, in line with the approach of Cont, Kukanov, and Stoikov (2013)[74], to fit a dynamic model.[2]

7.2.5 Definition of the testing and validation samples

Without regularization, the validation data is

(a) the following month's data, taking care not to overlap dates, or

(b) the following thirty minutes of data, taking care not to overlap time samples.

If one regularizes the regression, testing data calibrates the regularization meta-parameter. First, a statistician tests the model on the following monthly/thirty-minute sample. Then, another month/thirty-minute shift creates the validation data: it computes out-of-sample metrics, such as the regression's out-of-sample R^2.

7.3 Review of the Models in the Literature

Let σ, ADV estimate the stocks' daily volatility and traded volume. These variables normalize the price impact models, leading to comparable coefficients across stocks and time. The reader studies Section "4.1 Parameter scaling" (p. 18) of Almgren et al. (2005)[11].

7.3.1 The order flow imbalance (OFI) model

Cont, Kukanov, and Stoikov (2013)[74] first introduced Order Flow Imbalance (OFI). Section 4.2.1 of Chapter 4 introduces the model for alpha research. The original version did not present a time-decay kernel:

$$I_{t+h} - I_t = \lambda\sigma \sum_{s=t}^{t+h-\delta t} \frac{\text{ofi}_s}{\text{ADV}}.$$

[2] Calibrating a different model every thirty minutes is a simple method to obtain stochastic parameters. On the other hand, the approach creates time inconsistencies when evaluating price impact over longer timescales.

Cont, Cucuringu, and Zhang (2021)[73] extend the model with a non-parametric time-decay in a regular clock and a different normalization across stocks.

The OFI model is well-suited for short-term alpha modeling, covered in Section 4.2.1 of Chapter 4. By considering all order book events, OFI significantly outperforms price impact models based on trades only. Eisler, Bouchaud, and Kockelkoren (2018)[87] also fit event-based price impact models.

Event-based price impact models are less suited for optimal execution. The model does not answer trading counterfactuals: it lacks a causal structure of limit order book events. A trading algorithm cannot act on OFI's predictions without a causal structure articulating what-if scenarios. For example, because OFI includes both market and limit order events, one needs a joint model of the two events to assess their actions: do limit orders cause market orders, or is there a confounding variable causing both?

7.3.2 The original OW model

To the author's best knowledge, Chen, Horst, and Tran (2019)[70] first fitted the OW model. Section 2.3 of Chapter 2 introduces the model:

$$I_{t+\delta t} - I_t = -\beta I_t \delta t + \lambda\sigma \frac{q_t}{\text{ADV}}.$$

The original OW model is well-suited for establishing Nash equilibria, as covered in Section 4.4.2 from Chapter 4. The model's simplicity focuses the researcher on competition effects rather than the minutia of dynamic liquidity adjustments.

The model's simplicity prevents its use for two important regimes:

(a) when liquidity is highly variable, for example, when modeling time-of-day effects, such as in Cont, Kukanov, and Stoikov (2013)[74].

(b) when the order size is significant, and price impact exhibits concavity. Bouchaud, Farmer, and Lillo (2009)[32] review empirical studies on impact concavity.

7.3.3 The locally concave Bouchaud model

Bouchaud et al. (2004)[34] introduce and fit a propagator model with local concavity $g(x) \propto \log(x)$. Section 2.6.3 of Chapter 2 covers the model. Furthermore, Bouchaud, Farmer, and Lillo (2009)[32] review empirical studies of locally concave propagator models and find $g(x) \propto x^c$ with $c \in [0.2, 0.5]$. Finally, a standard locally concave model is the *square root* propagator where $c = 0.5$.

$$I_{t+\delta t} - I_t = -\beta I_t \delta t + \lambda\sigma \cdot \text{sign}(q_t)\sqrt{\frac{|q_t|}{\text{ADV}}}.$$

The locally concave propagator model by Bouchaud is well-suited for price simulation under new trading scenarios. Section 4.2.3 from Chapter 4 covers simulation and backtesting applications of price impact. The locally concave propagator model by Bouchaud outlines clear, fill-level counterfactuals. These what-if scenarios enable researchers to backtest trade fills' effects on prices.

Locally concave models are less suited for optimal execution. Control problems with local concavity exhibit price manipulation strategies and are intractable, as per Gatheral (2010)[104] and Curato, Gatheral, and Lillo (2017)[80]. Local concavity also does not capture order-level concavity: local concavity regards order speed and not size. Therefore, locally concave models may not fit sizable orders well.

7.3.4 The reduced-form model

To the author's best knowledge, Cont, Kukanov, and Stoikov (2013)[74] first assess time-varying and stochastic liquidity parameters. Section 2.6.3 of Chapter 2 introduces the reduced-form model and proves a link to the locally concave Bouchaud model from Section 7.3.3.

$$I_{t+\delta t} - I_t = -\beta I_t \delta t + \frac{\lambda \sigma q_t}{\sqrt{\mathrm{ADV} \cdot v_t}}$$

where

$$v_{t+\delta t} - v_t = -\beta v_t \delta t + |q_t|.$$

The reader studies Mertens et al. (2022)[166] for an alternative stochastic liquidity model. Indeed, their liquidity parameter is the "product of three components: 1) a daily price impact level, 2) a deterministic intraday pattern, and 3) a stochastic auto-regressive component" (p. 28).

Time-varying and stochastic liquidity parameters are essential in optimal execution. Section 3.2 from Chapter 3 details optimal execution applications. The reduced-form model is well-suited for stochastic control problems due to its tractability and ability to reproduce dynamic liquidity conditions.

The reduced-form model is less suited in two scenarios:

(a) For very brief timescales, the model does not directly model fill-level events.

(b) For long timescales, intraday liquidity variations are second-order, and the reduced-form model fails to capture order-level concavity.

7.3.5 The globally concave AFS model

Order-level concavity motivates the globally concave model of Alfonsi, Fruth and Schied (2010)[7]. Bouchaud, Farmer, and Lillo (2009)[32] review the empirical literature for order-level concavity. The general consensus finds a square root form for global concavity for sizable orders and a linear form for small orders. Section 2.6.4 of Chapter 2 introduces the AFS model.

$$I_t = \lambda\sigma \cdot \text{sign}(J_t)\sqrt{|J_t|}$$

where

$$J_{t+\delta t} - J_t = -\beta J_t \delta t + \frac{q_t}{\text{ADV}}.$$

The model is frequently fit on proprietary order data to capture sizable trading intentions. However, Section 7.4.4 presents a representative agent approach that generates realistic synthetic day orders from the public tape.

The globally concave AFS model is well-suited for studying price impact over larger timescales and for sizable orders. Section 4.3 from Chapter 4 introduces position inflation and other risk management applications that operate in that regime. It also provides tractable solutions for optimal execution problems focusing on sizable orders rather than intraday liquidity fluctuations. Exercises 6, 7, and 10 in Chapter 3 cover such problems.

The globally concave AFS model is less suited for applications focused on fill-level price impact. However, with effort, one can extend the AFS model to consider time-varying and stochastic liquidity.

7.4 Empirical Model Comparisons

Due to their wide range of applications outlined in Chapters 3 and 4, there are a dozen dimensions to consider when fitting and statistically evaluating price impact models. This section lists a subset of dimensions and reproduces empirical results from the literature using public trading data. In addition, readers with access to their own trading data can apply the methods outlined in Chapter 6 to enhance their price impact model. For instance, they can fit a model on their own trades or set up live trading experiments to improve on using public data alone.

This section delves into:

a) Timescales

   Different trading strategies and alpha signals operate on different timeframes. While price impact is relevant and measurable on all timescales, particular models perform differently depending on the timeframe of interest.

b) Time-of-day effects

   Liquidity is dynamic and exhibits notable time-of-day effects. Therefore, intraday trading and price impact are sensitive to these liquidity fluctuations.

(c) Clocks

Specific trading strategies slow down when the market slows down, for example, market-making. Other strategies rely on external catalysts operating on the regular clock. Therefore, price impact models should reflect a trading strategy's natural clock.

(d) Stock-specific effects

Price impact is a universal phenomenon across the stock market. This observation leads to a comparison between *universal* and *specialized* models. *Universal* models pool data across related stocks, for example, stocks with similar liquidity conditions. Specialized models fit a separate model per stock to capture the stock's idiosyncrasies.

(e) The magnitude of price impact

A statistician relates their price impact estimates to the applications from Chapters 3 and 4. They achieve this by measuring the magnitude of price impact, transaction costs, and position inflation.

7.4.1 Across timescales

Comparing impact models across timescales involves varying the horizon h. To cover a wide frequency range for trading strategies, consider the horizons $h \in \{1, 15, 60, 120\}$ in minutes. All statistical estimations in this section apply on a stock-by-stock basis using monthly training data, no regularization, and monthly validation data. One also refers to the validation data as the out-of-sample and the training data as the in-sample and repeats all statistical metrics in-sample and out-of-sample.

Notation 7.4.1 (Model t-stat). *Define the model's t-stat as*

$$\frac{mean\left(\hat{\lambda}\right)}{sd\left(\hat{\lambda}\right)}$$

where each $\hat{\lambda}$ is a realization of the fitted λ for each model on a given (month stock) pair. The t-stat appraises the model's robustness across the stock universe and period.

First, one calibrates the reversion parameter β. The OFI does not hav a reversion parameter, and this part of the analysis drops OFI. One considers half-lives of $\frac{\log 2}{\beta} \in \{1, 15, 60, 120\}$, in line with the prediction horizon $h \in \{1, 15, 60, 120\}$: for each model and β, a statistician estimates a separat λ. Figure 7.1 visualizes the half-lives' effect on the out-of-sample R^2 acros (month, stock) pairs:

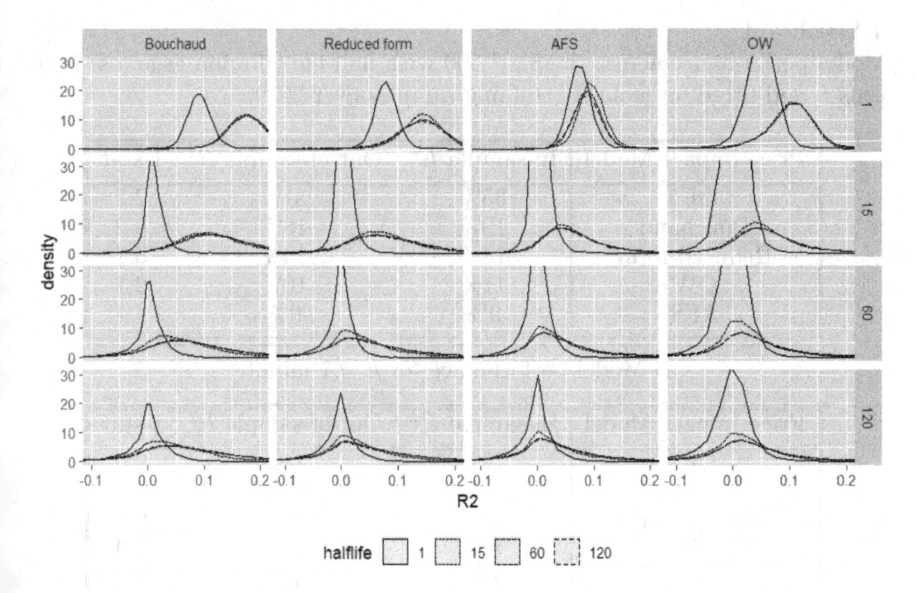

FIGURE 7.1
Distribution of R^2 across different choices of h (rows) and models (columns) for different betas (halflife).

(a) The shortest half-life $\frac{\log 2}{\beta} = 1$ performs poorly.

(b) The three other half-lives perform comparably, with a slight edge for $\frac{\log 2}{\beta} \in \{60, 120\}$ for longer prediction horizons.

For the following statistical estimations, one fixes the decay timescale to $\frac{\log 2}{\beta} = 60$ minutes and re-introduces the OFI model. Table 7.1 summarizes and compares the model performances for the horizons $h \in \{1, 15, 60, 120\}$. Figure 7.2 plots the time series of out-of-sample model performances.

Four broad observations follow.

a) The explanatory power of price impact models decreases with the horizon length h.

This decrease is in line with the broad study by Tomas, Mastromatteo, and Benzaquen (2022)[224]. In Figure eight (p. 15), the authors provide a sensitivity analysis of the impact models' out-of-sample R^2: the R^2 decreases from $\sim 30\%$ to $\sim 10\%$ as the timescale increases from one to sixty minutes.[3]

[3] Tomas, Mastromatteo, and Benzaquen (2022)[224] provide reproducible details of their analysis, for example, their filtering procedure. The filter is "removing the beginning and end of the trading period to focus on the intraday behavior of liquidity and volatility and circumvent intraday non-stationary issues" (p. 24). Section 7.4.2 illustrates these time-of-day effects on price impact models.

TABLE 7.1
Model performance across the S&P 500 with half-life 60 minutes across horizons h and fitted on monthly training samples over 2019.

Price Impact Model	In-sample R^2	Out-of-sample R^2	t-stat
OFI	35%	31%	2.2
Bouchaud	19%	18%	6.2
Reduced-form	15%	14%	3.7
OW	11%	10%	2.7
AFS	9%	9%	4.1

(a) Model performance for $h = 1$ minute.

Price Impact Model	In-sample R^2	Out-of-sample R^2	t-stat
OFI	29%	24%	2.0
Bouchaud	14%	13%	3.1
Reduced-form	11%	9%	2.1
OW	8%	6%	1.9
AFS	8%	7%	2.2

(b) Model performance for $h = 15$ minutes.

Price Impact Model	In-sample R^2	Out-of-sample R^2	t-stat
OFI	20%	15%	1.6
Bouchaud	10%	8%	2.0
Reduced-form	8%	5%	1.5
OW	7%	4%	1.3
AFS	7%	4%	1.4

(c) Model performance for $h = 60$ minutes.

Price Impact Model	In-sample R^2	Out-of-sample R^2	t-stat
OFI	17%	11%	1.4
Bouchaud	10%	6%	1.7
Reduced-form	8%	4%	1.3
OW	7%	3%	1.1
AFS	7%	3%	1.2

(d) Model performance for $h = 120$ minutes.

(b) OFI outperforms other models by an outstanding margin for in-sample an out-of-sample R^2. However, it underperforms the concave models along th t-stat dimension, indicating the model is more dynamic and less stabl over time and stocks. Sections 7.4.2 and 7.4.4 leverage this observation fo further improvements at the cost of additional model stability.

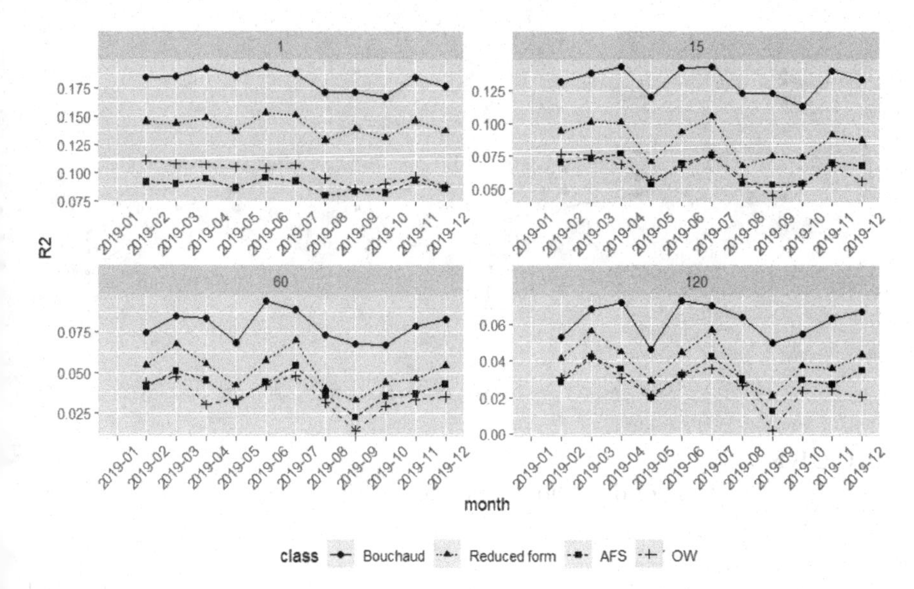

FIGURE 7.2
Average out-of-sample model R^2 across months and horizons h.

(c) The models with special local behavior (locally concave Bouchaud and dynamic reduced-form models) outperform the locally static models (AFS, OW), especially over briefer horizons.

(d) The concave models (Bouchaud, AFS) are more stable than their linear counterparts (Reduced-form, OW), as measured by their t-stat.

7.4.2 Across time of day

To the author's best knowledge, Cont, Kukanov, and Stoikov (2013)[74] first assessed price impact models by the time of day. Cont, Kukanov, and Stoikov (2013)[74] fit a separate model every thirty minutes of the day. The average model parameter $\hat{\lambda}$ as a function of the time of day visualizes the time-of-day effect, provided in Figure ten (p. 17, [74]). The model of Cont, Kukanov, and Stoikov (2013)[74] is non-parametric *and* dynamic: they estimate λ every thirty minutes. A static approach calibrates a single model over the month but measures the static model's *performance* across thirty-minute intervals. Another non-parametric method uses a month-long training sample but fits a different model per time-of-day.

This section illustrates the latter non-parametric method using 90-minute intervals for $h = 1$ and 150-minute intervals for $h = 15$ and compares it to

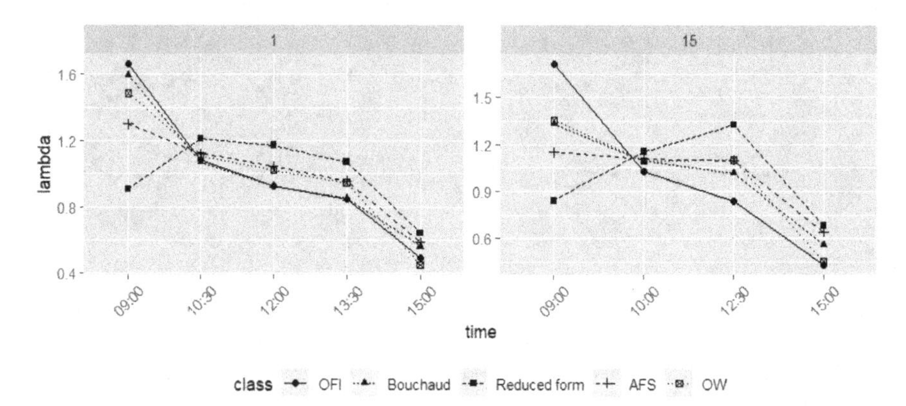

FIGURE 7.3
$\frac{\hat{\lambda}}{\text{mean}(\hat{\lambda})}$ across time of day under non-parametric fitting.

the static approach. Figure 7.3 shows the average time of day curve obtained by non-parametric fitting. Figure 7.4 plots the out-of-sample model R^2 as a function of time of day, comparing the static and non-parametric approaches. Finally, Table 7.2 summarizes performance under non-parametric fitting.

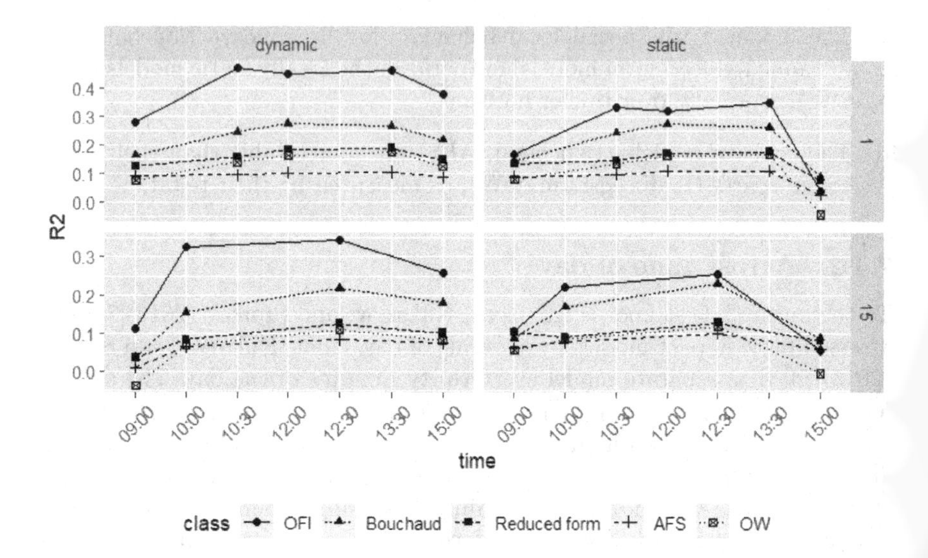

FIGURE 7.4
Average out-of-sample model R^2 across fitting methodologies and horizons a separate times of the day.

TABLE 7.2
Performance across the S&P 500 of *non-parametric* price impact models with a half-life of 60 minutes across horizons h fitted on monthly samples over 2019. One computes the t-stats by time of day. Therefore, t-stats only reflect (month, stock) variability.

Price Impact Model	In-sample R^2	Out-of-sample R^2	t-stat
OFI	49%	42%	1.6
Bouchaud	25%	24%	2.5
Reduced-form	18%	17%	2.5
OW	17%	14%	1.9
AFS	11%	10%	2.3

(a) Model performance for $h = 1$ minute.

Price Impact Model	In-sample R^2	Out-of-sample R^2	t-stat
OFI	40%	30%	1.2
Bouchaud	20%	17%	1.5
Reduced-form	13%	10%	1.5
OW	13%	8%	1.1
AFS	10%	7%	1.2

(b) Model performance for $h = 15$ minutes.

Three broad observations follow.

(a) The start and end of the day exhibit the most pronounced effects: performance improves most at the beginning and end of the day.

(b) OFI benefits the most from non-parametric fitting. Conversely, the reduced-form model benefits the least from non-parametric fitting. Indeed, the reduced-form model is the only model for which the non-parametric $\hat{\lambda}$ is smaller at the start of the day: its dynamic behavior parametrically adjusts for time-of-day effects.

c) The non-parametric models are less stable than their static counterparts across (month, stock) pairs for a given time-of-day.

7.4.3 Across clocks

To the author's best knowledge, Busseti and Lillo (2012)[47] first assessed price impact models under multiple clock specifications. Busseti and Lillo (2012)[47] observe that a time change is not the only way to reflect dynamic liquidity. As an alternative, the authors replace the volume imbalance q_t, v_n in their notation with the normalized volume imbalance $\frac{q_t}{v_t}$, v_n^{norm} in their notation.

> "The quantity v_n is the volume imbalance over the n-th time interval. This quantity may be very big in absolute value during time intervals of high market activity, for example near market opening and closing. Therefore, we consider a related variable, the *normalized* volume imbalance v_n^{norm}" (p. 10).

This normalization leads to a reduced-form model from Section 7.3.4 with $\lambda_t \propto v_t^{-1}$.

The calibration of β takes a different form under an alternative clock: to compare the decay parameters β across clocks, one normalizes time such that average half-lives match across clocks. For example, in the volume clock, one achieves an average half-life of one minute by setting

$$\frac{\log 2}{\beta} = \frac{ADV}{390}$$

as there are 390 minutes in a trading day for US stocks.

This section does not re-express horizon h in the new clock: h remains in the regular clock, and only the model specification changes. Busseti and Lillo (2012)[47] study the case where one expresses h in a trade clock and find a significant increase in R^2 for h in the trade clock. For instance, predicting the return over the following $h = 64$ trades increases R^2 by 50% compared to $h = 1$ minute, per table one on page 14 of Busseti and Lillo (2012)[47]. Finally, Table 7.3 summarizes the results: the volume clock models are on par with their regular clock counterparts when predicting regular clock horizons. Therefore, *applications* drive price impact models' estimation under different clocks, as the models are not intrinsically better for a given clock.

7.4.4 Across stocks

This section compares *universal* and *specialized* price impact models and raises the same question as Section 7.4.2:

> Should one fit a single model or a collection of models?

The former approach is stable, while the latter dynamically adjusts to data in this case, stock characteristics. Furthermore, multiple normalizations yiel universal models. For example, Cont, Kukanov, and Stoikov (2013)[74] use a stock's average order book depth d_t instead of its volatility and volume to normalize OFI.

To the author's best knowledge, Lillo, Farmer, and Mantegna (2003)[152 first fitted a universal price impact model. Zhou (2012)[242] contrast universa and specialized models and conclude in favor of using universal models a benchmarks:

> "universal price impact functions are unambiguous targets that any empirical model of order-driven markets must hit" (p. 13).

TABLE 7.3
Model performance across the S&P 500 under a volume clock with an *expected* half-life of 60 minutes across horizons h fitted on monthly training samples over 2019. The OFI model lacks a decay parameter, and the table omits OFI.

Price Impact Model	In-sample R^2	Out-of-sample R^2	t-stat
Bouchaud	18%	18%	6.1
Reduced-form	15%	14%	3.5
OW	11%	10%	2.7
AFS	9%	9%	4.0

(a) Model performance for $h = 1$ minute.

Price Impact Model	In-sample R^2	Out-of-sample R^2	t-stat
Bouchaud	14%	13%	3.1
Reduced-form	10%	9%	2.1
OW	9%	7%	2.0
AFS	7%	6%	2.2

(b) Model performance for $h = 15$ minutes.

Price Impact Model	In-sample R^2	Out-of-sample R^2	t-stat
Bouchaud	9%	7%	1.9
Reduced-form	8%	5%	1.4
OW	7%	4%	1.3
AFS	6%	4%	1.4

(c) Model performance for $h = 60$ minutes.

Price Impact Model	In-sample R^2	Out-of-sample R^2	t-stat
Bouchaud	8%	5%	1.6
Reduced-form	7%	3%	1.2
OW	7%	3%	1.1
AFS	6%	3%	1.2

(d) Model performance for $h = 120$ minutes.

Table 7.4 compares the in-sample and out-of-sample performance of universal and specialized models. Four broad observations follow.

a) By construction, specialized models perform better in-sample.

b) For OFI, in-sample performance translates out-of-sample.

c) For all other models, the universal version performs better *out-of-sample* than its stock-specific counterpart.

d) The performance gap increases with the horizon length h.

By construction, the universal model is more stable than the stock-specific one:

All else being equal, one prefers universal over specialized models.

Except for the OFI model, universal models also outperform stock-specific models out-of-sample. Therefore, they are excellent benchmarks.

Similarly to how Section 7.4.2 breaks down model performance by time of day, this section breaks down model performance across stocks. Indeed, such a breakdown answers an essential question:

Is price impact concave for sizable orders?

The question relates to order-level concavity and motivates the AFS model. Unfortunately, the public tape does not provide orders, only fills. Instead, for the public tape, one measures *large end-of-day imbalances*

$$\text{imb} = \frac{\sum_t q_t}{\sum_t |q_t|}$$

as a proxy for sizable orders. For each (stock, day) pair, one computes $\text{imb}(stock, day)$ and assesses models across imbalance levels. Assuming price impact is concave for marked order size, one expects concavity in $\text{imb}(stock, day)$. Figure 7.5 plots the model parameter λ across imbalances.

Remark 7.4.2 (Another use for universal price impact models). *Conditioning on the end-of-day imbalance is impossible for a specialized impact model: a single stock has insufficient data across imbalance levels to estimate concavity. So instead, a statistician pools together data and aims for a universal model by normalizing price impact across stocks. Then, they study finer, data-intensive properties, such as global concavity.*

Example 7.4.3 (Imbalance as an order size proxy). *The price impact coefficient for a (stock, day) pair with an imbalance of* 30% *proxies the price impact coefficient for a day order of* 30% *ADV.*

Two observations follow.

(a) For all models except AFS, λ decreases with the imbalance, indicating global concavity.

For instance, even the locally concave Bouchaud model *does not* accurately model the concave relationship between end-of-day imbalances and price impact. This observation empirically reinforces the distinction between *fill-level* and *order-level* concavity.

(b) The AFS model is *too* concave: λ increases as a function of imbalance.

One should not draw overly firm conclusions from the analysis: *prediction bias* likely affects significant end-of-day imbalances. Indeed, a statistician cannot identify this prediction bias without observing the underlying signal or running a live trading experiment. See Section 6.3 of Chapter 6 for details on prediction bias.

TABLE 7.4
Model performance across the S&P 500 with a half-life of 60 minutes across horizons h fitted on monthly training samples over 2019. Univ. stands for the universal and spec. for the stock-specific methodology.

Model	univ.	spec.
OFI	23%	35%
Bouchaud	18%	19%
Red. form	13%	15%
OW	10%	12%
AFS	9%	9%

(a) In-sample R^2 for $h = 1$ minute.

Model	univ.	spec.
OFI	22%	28%
Bouchaud	18%	17%
Red. form	14%	13%
OW	10%	9%
AFS	9%	8%

(b) Out-of-sample R^2 for $h = 1$ minute.

Model	univ.	spec.
OFI	17%	29%
Bouchaud	13%	15%
Red. form	9%	11%
OW	7%	10%
AFS	7%	8%

(c) In-sample R^2 for $h = 15$ minute.

Model	univ.	spec.
OFI	17%	23%
Bouchaud	14%	13%
Red. form	10%	9%
OW	8%	7%
AFS	7%	6%

(d) Out-of-sample R^2 for $h = 15$ minute.

Model	univ.	spec.
OFI	11%	21%
Bouchaud	8%	10%
Red. form	6%	9%
OW	5%	8%
AFS	5%	7%

(e) In-sample R^2 for $h = 60$ minute.

Model	univ.	spec.
OFI	11%	15%
Bouchaud	8%	7%
Red. form	6%	5%
OW	5%	4%
AFS	5%	4%

(f) Out-of-sample R^2 for $h = 60$ minute.

Model	univ.	spec.
OFI	9%	19%
Bouchaud	6%	9%
Red. form	8%	8%
OW	4%	8%
AFS	4%	7%

(g) In-sample R^2 for $h = 120$ minute.

Model	univ.	spec.
OFI	9%	12%
Bouchaud	7%	5%
Red. form	5%	4%
OW	5%	3%
AFS	4%	3%

(h) Out-of-sample R^2 for $h = 120$ minute.

.4.5 The magnitude of price impact

"We estimate that mutual and pension funds trade around $90 billion each day[...]. That adds to around $23 trillion over a year.

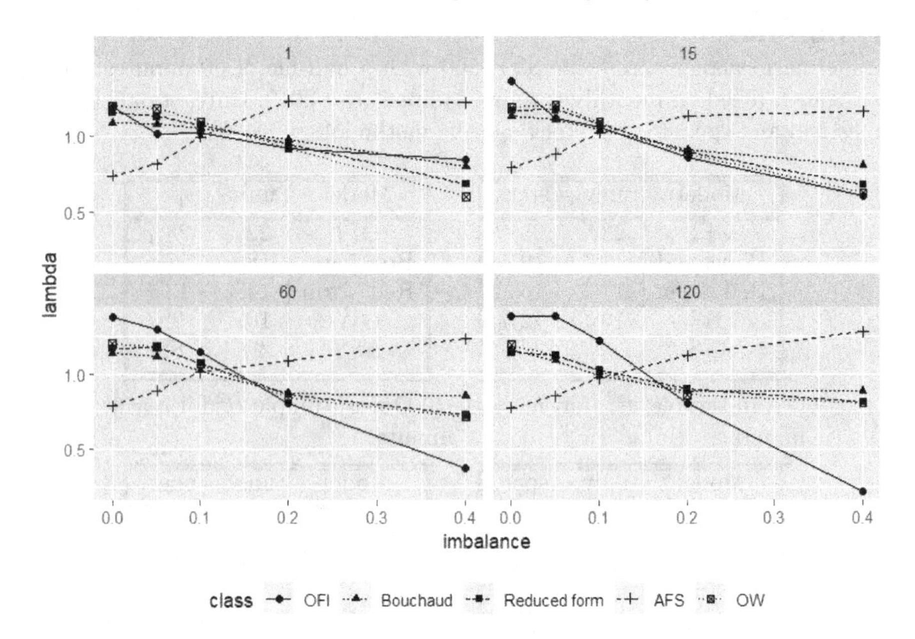

FIGURE 7.5
Normalized model λ across end-of-day imbalances.

Even though average trade costs are reported at just 0.31%, that adds to around \$70 billion each year in trading costs." ("The 2022 Intern's Guide to Trading", Nasdaq (2022)[159])

One complements a model's R^2 with three metrics measuring price impact's magnitude:

(a) Express the end-of-day impact state as a proportion of the stock's daily volatility.

Figure 7.6 visualizes the distribution of end-of-day impact states as a proportion of daily volatility, and Table 7.5 displays summary statistics.

(b) Compute the transaction costs of a set of trades.

As an example trade set, Remark 7.4.4 constructs the representative agent's trades on the public tape. A representative agent is a proxy that hides massive variations across market participants, for example, market makers and mutual funds.

TABLE 7.5
Average magnitude of the end-of-day impact state.

Price Impact Model	$\mathbb{E}[\lvert I_T \rvert]$	$\mathbb{E}\left[\frac{\lvert I_T \rvert}{\sigma}\right]$
Bouchaud	43bps	34%
Reduced-form	25bps	20%
OW	28bps	23%
AFS	23bps	19%

(c) Compute a portfolio's position inflation.

Remark 7.4.5 constructs a daily representative agent's position from the public tape. Daily position inflation significantly underestimates the longer-term build-up of position inflation for substantial portfolios.

A multi-day price impact model, for example, the power-law time kernel of Bouchaud et al. (2004)[34], estimates price impact's long-term effects. This section leverages universal price impact models with exponential half-lives of 60 minutes.

Remark 7.4.4 (Transaction costs of the representative agent). *One can estimate transaction costs*

$$TC = \int_0^T I_t dQ_t + \frac{1}{2}[I, Q]_T$$

using the market process M as a proxy for Q. This method uses the public tape as a representative agent. *This approach's advantage is that it does not rely on proprietary trading data. The disadvantage is that the* representative agent *hides a vast heterogeneity across market participants. For example, one can rescale the transaction costs using two denominators: the end-of-day imbalance emulates a day order, and the total trade notional emulates a market-maker. Because these two market participants do not trade alike, the two numbers differ tellingly. Table 7.6 estimates a representative agent's trading costs.*

Remark 7.4.5 (Position inflation of the daily representative agent). *One can estimate the trading footprint*

$$F_T = I_T Q_T$$

using the market process M as a proxy for Q. The advantages and disadvantages listed in Remark 7.4.4 still apply.

Furthermore, another limitation of the representative agent is that the study estimates position inflation based only on a single day's impact and accumulated trades. From a risk-management perspective, actual portfolios' positions and price impact reflect materially longer timescales. The reader studies Caccioli, Bouchaud, and Farmer (2012)[53], Cont and Schaanning (2017)[75],

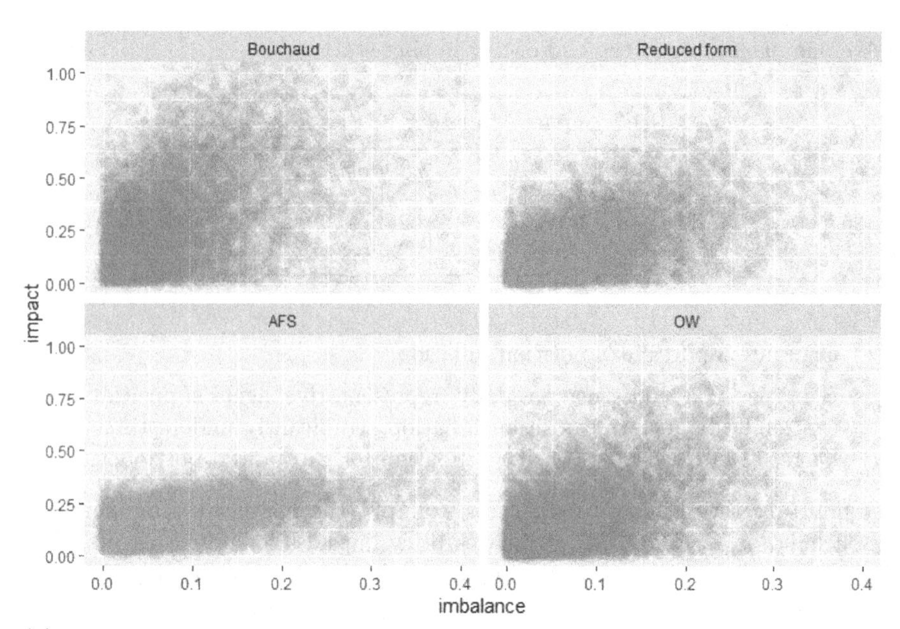

(a) Scatterplot with normalized end-of-day trade imbalances on the x-axis and impact states on the y-axis.

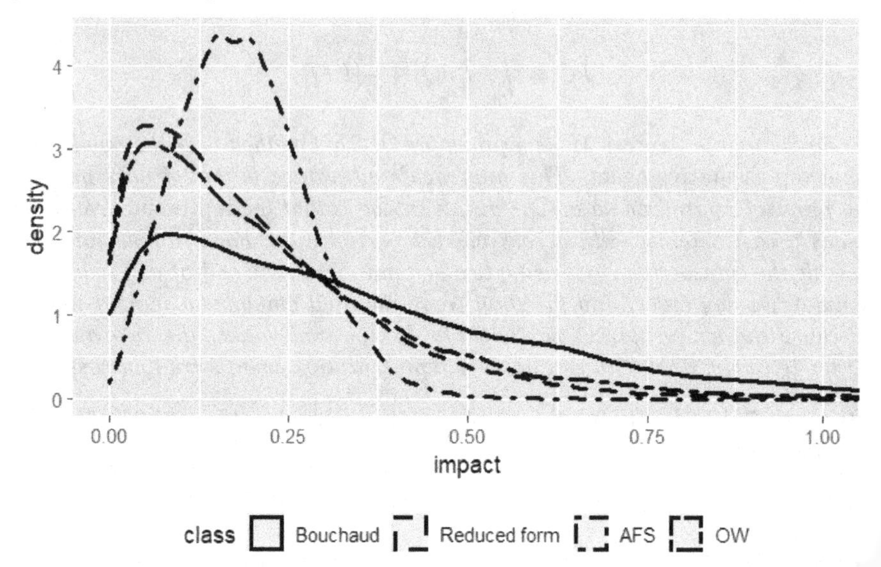

(b) Density function of normalized end-of-day impact states.

FIGURE 7.6
Distribution of end-of-day impact states across (stock, day) pairs. One com-putes the impact state daily as a proportion of the twenty-day stock volatilit

TABLE 7.6
Average magnitude of price impact based on a representative agent. $\sum TC$ are the total impact costs on the S&P 500 in 2019. $\frac{TC}{imb}$ are the transaction costs per unit of imbalance, a proxy for day orders. $\frac{TC}{ntl}$ are the transaction costs per unit of trade notional, a proxy for market-making. F_T is the daily *representative agent's* trading footprint.

Price Impact Model	$\sum TC$	$\frac{TC}{imb}$	$\frac{TC}{ntl}$	F_T
Bouchaud	1.7bil	87bps	13bps	57bps
Reduced-form	1.2bil	58bps	8bps	30bps
OW	1.1bil	55bps	8bps	34bps
AFS	1.7bil	86bps	13bps	23bps

and Roncalli et al. (2021)[197] for examples. Therefore, the daily representative agent's trading footprint underestimates actual trading footprints. Finally, Table 7.6 estimates a daily representative agent's position inflation.

Four broad observations follow.

(a) The average price impact state ranges from 20bps to 43bps, representing 20-30% of the daily stock volatility, in line with the associated models' R^2.

The Bouchaud model estimates price impact states with a heavy-tail distribution. On the other hand, the AFS model estimates a narrow range of impact states and strongly aligns impact with end-of-day imbalances.

(b) Computed impact costs total over one billion dollars. However, this total cost is a severe underestimate.

- It only includes the S&P 500, and considering less liquid stocks yields a multiplier of four.
- It does not include the bid-ask spread, adding another $\approx$ 5bps.
- It does not include multi-day impact; a conservative estimate by Bershova and Rahklin (2013)[25] provides a multiplier of two.
- Lastly, LOBSTER data only includes trades on Nasdaq, leading to another multiplier to consider trades on competing trading venues. The reader consults the CBOE website [89] to estimate the share of trades on NASDAQ against other venues. For April 2022, Nasdaq represented $118 billion of $579 billion in the total consolidated volume for the US equities market. This volume estimate yields an additional multiplier of five.

When considering the multipliers, the numbers align with Mackintosh (2022)[159]. Mackintosh estimates 30bps of trading costs for large-cap stocks, 50bps for mid-cap stocks, and 75bps for small-cap stocks (see chart four, [159]), yielding a yearly trading cost estimate of $70 billion for mutual and pension funds.

(c) When rescaled to emulate a day order, price impact cost estimates span from 55bps to 87bps. When rescaled to emulate market making, price impact cost estimates span from 8bps to 13bps. Concave models exhibit higher transaction costs due to them penalizing small trades.

(d) The daily trading footprint, an estimate of the representative agent's P&L at risk, span from 23bps to 57bps. This footprint is a marked underestimate, as actual price impact and positions build up over extended periods and do not reset each trading day.

7.5 Cross-Impact

Cross-impact is a novel, exciting, and challenging research area in quantitative trading. Two insights motivate cross-impact:

(a) From an *action* perspective, an increasing number of investment strategies involve the correlated trading of multiple assets, for example, an ETF, a bond basket, or an options portfolio.

   TheStreet.com (2021)[216] defines ETFs:

   > "What Is an ETF (Exchange-Traded Fund) in Simple Terms?
   >
   > An exchange-traded fund, or ETF, is a collection of securities that can be bought and sold in shares on a stock exchange just like an individual stock" (p. 1).

   Blackrock's iShares (2022)[120] motivates factor ETFs with their transparency and liquidity.

   > "Factors are the persistent and well-documented asset characteristics that have historically driven investment risk and return.
   >
   > Factors are not new - they have been present in portfolios for decades. But exchange traded funds (ETFs) helped revolutionize how investors access these historically rewarded strategies by capturing the power of factors (sometimes called "smart beta") in a transparent and cost-effective way" (p. 1).

   For instance, Blackrock publishes a market insights series, "Flow and Tell" linking ETF trading flows with market and investment trends. The January 2022 issue by Chaudhuri (2022)[68] highlights:

   > "Total ETF volumes averaged over $250 billion trading on exchange during January - well above the 2021 average of $141 billion. In all, ETFs accounted for 33% of all equity trading during the month" (p. 1).

The November 2021 issue by Chaudhuri (2021)[69] leverages ETF trading patterns to identify sector rotations and provide a glimpse into the theses behind investors' sector preferences.

> "Investors have expressed more granular views via sectors ETFs" (p. 1).

(b) From an empirical perspective, flow and return correlations link across stocks:

> Assets that trade together move together.

To the author's best knowledge, Wang, Schäfer, and Guhr (2015)[232] first introduced and empirically measured cross-impact using high-frequency data.

> "We shed light on the price impact from trades in different stocks by discussing the efficiency of the financial market and by analyzing how the stocks respond to the whole market and to different economic sectors. We thereby present a first complete view of the response in the market as a whole" (p. 1).

The two observations are *mutually reinforcing:* because of price impact, the more correlated the trading of assets, the more correlated the asset returns. But conversely, the more assets move together, the more ETFs and factor strategies *bundle* them together, creating cheaper supply for correlated trading and increasing demand. Da and Schive (2017)[81] argue this feedback loop theory:

> "Using a large panel of 549 US equity ETFs and 4,887 stocks from July 2006 to December 2013, we show that ETFs contribute to equity return comovement. An ETF-level analysis reveals that the higher turnover an ETF has, the more its component stocks move together" (p. 2).

Madhavan and Morillo (2018)[160] make the same empirical observation but propose a causal explanation *without* feedback: macroeconomic trends are a common cause of stock correlations *and* the demand for factor trading.

> "[T]his research indicates that the rise in cross-stock correlations is due to the macro environment, not ETF growth" (p. 1).

For traders, the first observation leads to natural generalizations of price impact applications to multiple assets. For example, these include:

(a) Optimal execution, covered in Section 3.2 of Chapter 3.

Portfolio teams trading a basket of assets need to consider the basket's overall transaction costs. To the author's best knowledge, Schied and Schöneborn (2008)[204] first solved a control problem with cross-impact. Exercise 24 provides an example based on the *EigenLiquidity model* of Mastromatteo et al. (2017)[165].

(b) TCA, covered in Section 3.3 of Chapter 3.

TCA reports for orders with a significant factor component require experiments establishing factor transaction costs.

(c) Risk management, covered in Section 4.3 of Chapter 4.

The concepts of position inflation and liquidation costs apply to factors. Moreover, asset correlations increase during liquidity crises, further reinforcing cross-impact in stress periods.

(d) Combining trading costs, covered in Section 4.4 of Chapter 4.

Portfolio teams may not always overlap when trading stocks but are likely to overlap when trading factors: competition effects propagate across factors and correlated stocks.

The second observation shows that one can empirically test cross-impact. This section focuses on the latter aspect of cross-impact.[4] Three parts review the empirical cross-impact literature.

(a) Cross-impact exhibits a *potential causal bias.*

(b) Cross-impact for factor trading focuses on modeling a *large number of cross-impact terms.*

(c) Cross-impact for pairs trading focuses on modeling a *small number of cross-impact terms.* This section also extends price impact outside of the equities market.

7.5.1 Causal bias for cross-impact

Most papers do not directly tackle the causal structure of cross-impact and focus on observational results or applications under an implied causal model. However, two articles outline a causal interpretation for cross-impact:

(a) Bouchaud et al. (2016)[36] define and study cross-impact. In Section 4. "Direct and cross impact" (p. 12), Bouchaud et al. implicitly propose causal structure and explicitly estimate the causal path contributions t the observed cross-impact.

(b) Capponi and Cont (2020)[56] explicitly "reexamine this empirical evidenc from a causal standpoint" (p. 1). Capponi and Cont propose a causa structure and identify cross-impact under this causal structure.

[4]Four papers cover cross-impact's mathematical implications for trading. All four articl focus on optimal execution and ruling out price manipulation strategies: Mastromatt et al. (2017)[165], Schneider and Lillo (2019)[206], Tomas, Mastromatteo, and Benzaqu (2022)[224], and Rosenbaum and Tomas (2022)[223]. Exercises 24 and 26 outline theoretic results for cross-impact.

The primary question Capponi and Cont (2020)[56] solve is the acceptance or rejection of pair-wise cross-impact links. The analysis begins with the causal structure from Definition 7.5.1, found in Figure one (p. 3, [56]).

Definition 7.5.1 (The Capponi and Cont causal structure for cross-impact). *Define the causal structure $\mathcal{C}$:*

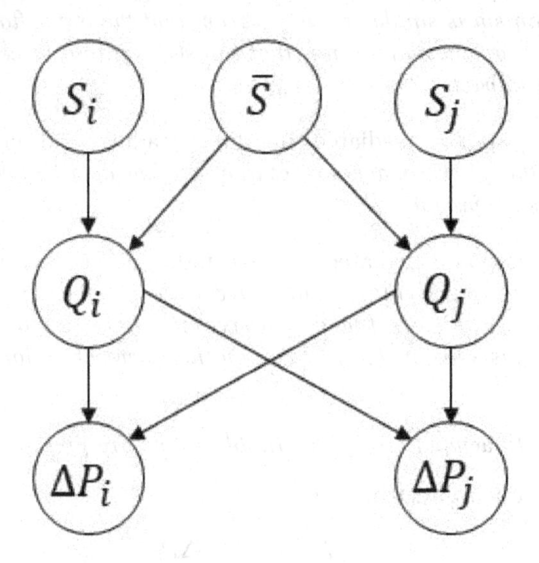

for two stocks i and j with

- *returns* $\Delta P_i, \Delta P_j$,
- *public trading tapes* Q_i, Q_j,
- *underlying trading strategies* S_i, S_j, *and basket strategy* $\bar{S}$.

Section 7.5.2 defines an alternative causal structure, $\mathcal{E}$ based on the *EigenLiquidity model* of Mastromatteo et al. (2017)[165] that reconciles a broader set of empirical results on cross-impact in equities than $\mathcal{C}$.

Remark 7.5.2 (Causal assumptions). *On page 13, Bouchaud et al. (2016)[36] provide names for causal paths but do not explicitly define a causal structure.*

> *"Below each term, we have indicated its relative contribution to the average self/cross-response/covariance. The interpretation of each term in the response is as follows.*
>
> *(a1)* Self-response via direct impact. *This is the classic term considered in most works on impact: trading in product i impacts the price of i itself.*

(a2) Self-response mediated by cross-trading and cross-impact. *This term is induced by the order flow on all the stocks k that are correlated to i. This causes market makers to include this extra information in their price for i.*

(b1) Cross-response mediated by cross-trading and direct-impact. *The mechanism is similar to (a1), except that the order flow on j now induces an imbalance on i, that translates into a price change via direct impact.*

(b2) Cross-response mediated by direct-trading and cross-impact. *Here the market makers react to order flow on j by updating their quotes on product i.*

(b3) Cross-response mediated by cross-trading and cross-impact. *Trading in a stock j that is correlated with a large number of other stocks k. The market maker observes the order flow of all of those, and adjusts his quote of i based on this aggregate information" (p. 13).*

Most paths Bouchaud et al. (2016)[36] outline are present in $\mathcal{C}$.

(a1) Self-response via direct impact

$$S_i \to Q_i \to \Delta P_i$$

(a2) Self-response mediated by cross-trading and cross-impact

$$Q_i \leftarrow \bar{S} \to Q_j \to \Delta P_i$$

(b1) Cross-response mediated by cross-trading and direct-impact

$$Q_i \leftarrow \bar{S} \to Q_j \to \Delta P_j$$

(b2) Cross-response mediated by direct-trading and cross-impact

$$Q_i \to \Delta P_j$$

(b3) Cross-response mediated by cross-trading and cross-impact

The distinction between (b3) and (b1) is moot for the two-asset example. I the case of a large basket, it corresponds to the distinction between pai *trading, covered in Section 7.5.3, and factor trading, covered in Sectio 7.5.2.*

The alternative causal model $\mathcal{E}$, introduced in Definition 7.5.8 and bas on the EigenLiquidity model *of Mastromatteo et al. (2017)[165], provid a causal pathway (b3). Therefore, one motivates $\mathcal{E}$ from the empirical observed value of (b3) in Remark 7.5.10.*

The introduction of causal structure $\mathcal{C}$ allows the reader to define cross-impact using do-calculus and recast cross-impact estimation as an *identifiability* problem.[5] Exercise 22 illustrates the identifiability problem by working through Capponi and Cont's example (2020)[56] in Section 2.2 "are cross-impact coefficients identifiable?" (p. 6).

The back-door criterion from Theorem 5.2.34 in Chapter 5 solves the identifiability problem under the causal structure $\mathcal{C}$. First, a trader defines the counterfactual of interest.

How much does the price of stock j move when I trade on stock i?

This counterfactual yields a clear do-action.

Definition 7.5.3 (Definition of pair-wise cross-impact). *Define the pair-wise cross-impact counterfactual by*

$$\mathbb{E}\left[\Delta P_j \,|\, do(Q_i)\right].$$

Proposition 7.5.4 (Identifiability of pair-wise cross-impact under $\mathcal{C}$). *Pair-wise cross-impact exhibits a causal bias under the causal structure $\mathcal{C}$:*

$$\mathbb{E}\left[\Delta P_j \,|\, do(Q_i)\right] \neq \mathbb{E}\left[\Delta P_j \,|\, Q_i\right].$$

Conditioning on $\bar{S}$ identifies it:

$$\mathbb{E}\left[\Delta P_j \,|\, do(Q_i)\right] = \int \mathbb{E}\left[\Delta P_j \,|\, Q_i, \bar{S} = s\right] p\left(\bar{S} = s\right) ds.$$

Proof. $\bar{S}$ satisfies the back-door criterion from Theorem 5.2.34 in Chapter 5. The fork $Q^i \leftarrow \bar{S} \rightarrow Q^j$ prevents the naive identification of pair-wise cross-impact. □

To quote Capponi and Cont (2020)[56]:

"cross-impact coefficients may not be identified solely from the covariance of returns with order flow" (p. 9).

Based on the causal structure $\mathcal{C}$ and Proposition 7.5.4, the identification of pair-wise cross-impact requires conditioning by the basket strategy $\bar{S}$. This identification strategy is the vital insight of Capponi and Cont (2020)[56]. Capponi and Cont implement the pair-wise cross-impact identification formula by creating a public proxy for the otherwise unobservable $\bar{S}$.

[5]Under the causal structure $\mathcal{C}$, there is no *prediction bias*: the links $S_i \rightarrow \Delta P_i$, $S_j \rightarrow \Delta P_j$, and $\Delta P_i \leftarrow \bar{S} \rightarrow \Delta P_j$ are missing. However, a statistician can extend the causal ructure to include prediction bias. For example, the trading strategies $S_i, S_j, \bar{S}$ are alpha gnals predicting their respective returns. Exercise 23 proposes a causal structure that nsiders cross-impact *and* prediction bias.

Remark 7.5.5 (Link between identifiability and regression). *The identifiability problem does not arise if one fits the cross-impact model as a multi-dimensional regression model. For example, Bouchaud et al. (2016)[36], Cont, Cucuringu, and Zhang (2021)[73], and Tomas, Mastromatteo, and Benzaquen (2021)[223] estimate cross-impact with a multidimensional regression. Cross-impact exhibits causal bias if one estimates the pair-wise cross-impact terms $Q_i \to \Delta P_j$ from independent one-dimensional regressions. Direct pair-wise estimation does not adjust for the confounding between Q_i and Q_j, while co-fitting does.*[6] *Exercise 22 details causal statements, regression methods, and the pitfalls of naively estimating pair-wise cross-impact. Unfortunately, co-fitting cross-impact is computationally more burdensome than estimating cross-impact pair-wise.*

Capponi and Cont derive a "common order flow factor" (p. 18) using principal component analysis of the order flow imbalance across stocks. The shared order flow imbalance proxies $\bar{S}$ and removes the causal bias in pair-wise cross-impact.

> "failing to account for the presence of common factors in order flow leads to an identification problem and makes it difficult to draw conclusions regarding the presence of cross-impact as a phenomenon distinct from correlation in order flow.
>
> We account for the commonality in order flow by performing a principal component analysis (PCA) on the correlation matrices of returns and OFIs. We adjust the correlation in order flow using the first principal component of OFI, then test for the statistical significance of any residual cross-impact effect" (p. 13).

Interestingly, the primary *trading motivation* for cross-impact applications, the widespread use of basket and factor strategies, is also the primary *sourc* *of causal bias* in its estimation. With this methodological point cleared up this section presents empirical cross-impact results from the literature.

7.5.2 Price impact for factor trading

Three papers estimate cross-impact considering equity factors:

(a) Bouchaud et al. (2016)[36] focus on establishing factor impact's explana- tory power for the observed cross-impact matrix.

(b) As per Section 7.5.1, Capponi and Cont (2020)[56] focus on proving tha pair-wise cross-impact is negligible when considering factors.

[6]To the author's best knowledge and understanding, this seems to be the case in t original work by Wang, Schäfer, and Guhr (2015)[232], which was the first to introduce a estimate cross-impact. The authors assess the pair-wise correlations $\mathbb{E}[\text{sign}(\Delta Q_i)\,\Delta P_j]$ a $\mathbb{E}[\text{sign}(\Delta Q_i)\,\text{sign}(\Delta Q_j)]$ directly but, again, to the author's best knowledge, do not see to estimate a joint model.

(c) Tomas, Mastromatteo, and Benzaquen (2022)[224] provide the broadest range of functional forms for cross-impact models. Furthermore, Tomas, Mastromatteo, and Benzaquen link each functional form to appealing mathematical properties, for example, the absence of price manipulation strategies.

While the three papers' causal interpretations may differ, the empirical estimations agree. The three papers observe factor impact dominating pairwise cross-impact.

> "[T]he dominating mechanism is (b3), implying that most of the cross-response is mediated by *delocalized modes* (such as the market mode, or large sectors). This is one of the central messages of this paper" (p. 13, [36]).

> "[W]e have shown that returns and order flows are driven by highly correlated common factors, and that this commonality in order flow is the dominant contribution to the covariance between order flow imbalance and return across assets" (p. 21, [56]).

> "Precisely because of the importance of the market mode, the scores reported in Table 5 strikingly show that cross-impact models can explain market-wide moves up to twice as well as direct models" (p. 21, [224]).

Remark 7.5.6 (Simplifying the functional form of cross-impact). *The empirical studies suggest two techniques to simplify cross-impact's functional form.*

(a) First, only estimate factor impact. For example, if $N \approx 1000$ is the number of stocks and $M \approx 10$ is the number of factors, this yields a dimension reduction of at least 100.

(b) Second, factor cross-impact into a static cross-impact matrix of size N^2 and one-dimensional time kernel of size T. This method avoids a separate time kernel for each cross-impact term and reduces the model dimension from $T \cdot N^2$ to $T + N^2$.

Bouchaud et al. (2016)[36] compare a non-parametric and a factorized form for equities cross-impact.

> *"**Fully non-parametric** The most general propagator model specifies the $N^2(T+1)$ parameters defining Eq. (6). This corresponds to the absence of any prior about the structure of G^{ij}_τ.*
>
> ***Factorized** A simpler model is obtained under the assumption $G^{ij}_\tau = G^i j \phi_\tau$, where ϕ_τ given by Eq. (9). The dimensionality of the model is then reduced to $2N^2 + T$.*

[...]
the good in-sample performance of the fully non-parametric model does not generalize out-of-sample. The scores displayed by the lower dimensional models are roughly the same in and out-of-sample, thus validating the practical use of the factorized and homogeneous propagator models" (p. 15).

The homogenous propagator model simplifies the factorized model along the market factor. G_τ^{ij} is the empirical cross-impact matrix from stock i to stock j with lag (time decay) τ.

The last simplification is specific to Capponi and Cont (2020)[56] and relates to the causal bias inherent to cross-impact, covered in Section 7.5.1.

Remark 7.5.7 (Causal interpretation of pair-wise cross-impact). *In summary, Capponi and Cont's analysis (2020)[56] states:*

Q_i and ΔP_j are independent conditional on $\bar{S}$.

This observation rules out direct pair-wise links such as $Q_i \to \Delta P_j$. Under the causal structure $\mathcal{C}$, only self-impact exists, and order flow correlation drives observed cross-impact. In conclusion, the causal graph $\mathcal{C}$ simplifies to $\mathcal{C}_0$, illustrated in Figure 7.7.

"Once this common factor in order flow is accounted for, introducing cross-impact terms provides little or no additional explanatory power" (p. 22, [56]).

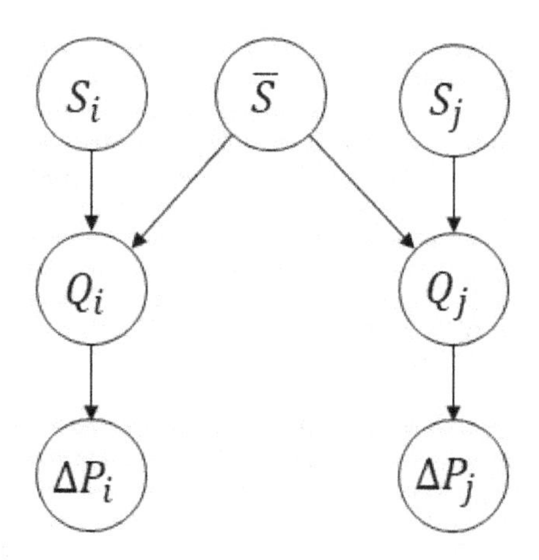

FIGURE 7.7
Simplified causal structure $\mathcal{C}_0$ based on the study by Capponi and Co[nt] (2020)[56] and presented in slide ten of [72].

7.5.2.1 Causal graph for the EigenLiquidity model

This section proposes an alternative causal structure $\mathcal{E}$ where cross-impact travels along factors, in line with Bouchaud et al. (2016)[36], Mastromatteo et al. (2017)[165], and Tomas, Mastromatteo, and Benzaquen (2022)[224].

Definition 7.5.8 (Alternative causal model for cross-impact). *Define the causal structure $\mathcal{E}$:*

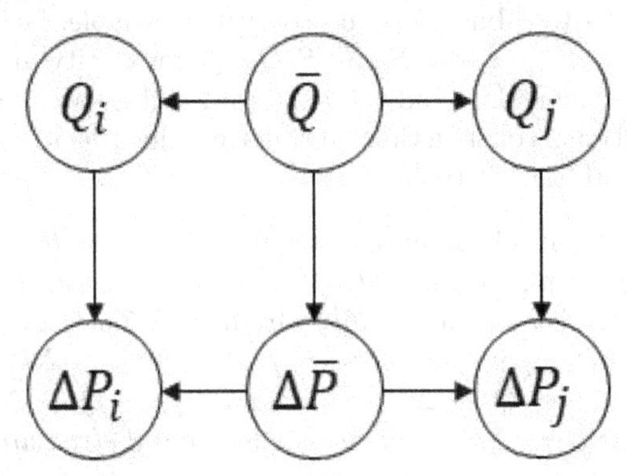

for two stocks i, j driven by a shared factor with

- *returns $\Delta P_i, \Delta P_j$ and factor return $\Delta \bar{P}$,*
- *public trading tapes Q_i, Q_j, and factor flow $\bar{Q}$.*

Under the causal structure $\mathcal{E}$, Capponi and Cont (2020)[56] do not empirically contradict Bouchaud et al. (2016)[36] and Tomas, Mastromatteo, and Benzaquen (2022)[224] and even reinforce their analysis. Proposition 7.5.9 proves that the conditional independence premise of Capponi and Cont (2020)[56] is consistent with the causal structure $\mathcal{E}$.

Proposition 7.5.9 (No pair-wise cross-impact). *Under any probability measure consistent with the causal structure $\mathcal{E}$, Q_i and ΔP_j are independent conditional on $\bar{Q}$.*

Proof. The node $\bar{Q}$ d-separates Q_i and ΔP_j under the causal structure $\mathcal{E}$. Theorem 5.2.15 from Chapter 5 concludes. □

Unlike under causal graph $\mathcal{C}$, under $\mathcal{E}$ the conditional independence of Q_i and ΔP_j given $\bar{Q}$ does not rule out all cross-impact but only pair-wise cross-impact.

Remark 7.5.10 (Causal contribution analysis). *Bouchaud et al. (2016)[36] break down cross-impact into the causal path contributions, named in Remark 7.5.2.*

) pair-wise cross-impact, term (b2) from Remark 7.5.2, contributes only 7% (p. 13, [36]) to the observed empirical cross-impact matrix.

(b) *pair-wise flow correlation, term (b1) from Remark 7.5.2, contributes* 17% *(p. 13, [36]).*

(c) *factor impact, term (b3) from Remark 7.5.2, contributes* 76% *(p. 13, [36]).*

The observation that pair-wise cross-impact contributes little to observed cross-impact is consistent with Capponi and Cont (2020)[56] and Proposition 7.5.9.

$\mathcal{E}$ closely relates to $\mathcal{C}$ but adds an observable variable: factor returns. The graph also re-factors the nodes $S_i, S_j, \bar{S}, Q_i, Q_j$ for clarity and is directly actionable for factor investing. Indeed, exercise 24 illustrates an application of $\mathcal{E}$ for factor portfolio construction considering the *EigenLiquidity model* of Mastromatteo et al. (2017)[165].

Remark 7.5.11 (Causal assumptions under $\mathcal{E}$). *Both the causal inference analysis of Capponi and Cont (2020)[56] and the causal path contribution analysis of Bouchaud et al. (2016)[36] from Remark 7.5.10 motivate the causal structure $\mathcal{E}$. $\mathcal{E}$ removes pair-wise cross-impact links $Q_i \to \Delta P_j$ and $Q_j \to \Delta P_i$ in favor of paths $Q_i \leftarrow \bar{Q} \to \Delta\bar{P} \to \Delta P_j$ and $Q_j \leftarrow \bar{Q} \to \Delta\bar{P} \to \Delta P_i$. The observable factor return $\Delta\bar{P}$ distinguishes the causal structure $\mathcal{E}$ from $\mathcal{C}$.*

7.5.2.2 Distinction between do(Q_i, Q_j) and do($\bar{Q}$)

Readers new to do-calculus may be confused as to the distinction between the two actions

(a) do(Q_i, Q_j)

(b) do($\bar{Q}$)

and which one to use when estimating price impact via counterfactual analysis.

Example 7.5.12. *For instance, consider a trade of size q on stock i. Assume the trade is* not *factor-neutral, yielding the decomposition*

$$q = q_i + \bar{q},$$

where $\bar{q}, q_i$ are the trade's factor and factor-neutral components.

To estimate this trade's impact counterfactual, should one use the do action

$$do(Q_i = q)$$

or

$$do(Q_i = q_i, \bar{Q} = \bar{q})?$$

This distinction may seem a philosophical question. However, it is no[t] causal inference formalizes such distinctions. Recall from Definition 5.2.1[…] and Proposition 5.2.24 from Chapter 5 that the do-action do($Q_i \doteq q$) is *shorthand for the truncated causal graph* $\mathcal{E}_{\overline{Q_i}}$. Under the truncated caus[al] graph $\mathcal{E}_{\overline{Q_i}}$, Q_i and $\bar{Q}$ are independent.

Therefore, $\text{do}(Q_i = q)$ directly contradicts the causal structure. A similar analysis shows that $\text{do}(Q_i = q_i, \bar{Q} = \bar{q})$ is admissible.[7]

Concretely, to compute an order's price impact using the EigenLiquidity model, one must

(a) decompose the order in its factor and factor-neutral components,

(b) establish their impact separately, and

(c) add the factor and factor-neutral impact for a given stock to obtain its price impact.

For instance, consider a one-period toy price impact model

$$I^i(Q^i) = \lambda^i Q^i; \quad \bar{I}(\bar{Q}) = \bar{\lambda}\bar{Q}; \quad \Delta P^i = I^i + \rho^i \bar{I} + \epsilon.$$

An order with factor decomposition $q = q^i + \bar{q}$ causes price impact

$$\mathbb{E}\left[\Delta P^i \middle| \text{do}(Q_i = q_i, \bar{Q} = \bar{q})\right] = \lambda^i q^i + \rho^i \bar{\lambda} \bar{q}$$

and not

$$\mathbb{E}\left[\Delta P^i \middle| \text{do}(Q_i = q)\right] = \lambda^i q.$$

In conclusion, applying the stock-component impact model to the entire order size is inconsistent with the EigenLiquidity model: the causal structure mandates a decomposition.

7.5.2.3 Counterfactuals under $\mathcal{E}$

The counterfactual associated with cross-impact changes slightly. One can state the counterfactual at two levels:

a) *the stock level.*

How much does the price of stock j move when I trade the factor?

b) *the factor level.*

How much does a factor move when I trade it?

Definition 7.5.13 (Definition of cross-impact). *Under the causal structure $\mathcal{E}$, define*

a) the stock-level cross-impact counterfactual by

$$\mathbb{E}\left[\Delta P_j \middle| do(\bar{Q})\right].$$

[7]To be precise, Definition 5.2.19 defines $do(X = x)$ for constant interventions x. The definition extends to a random variable x consistent with the truncated causal structure.

(b) the factor-level cross-impact counterfactual by

$$\mathbb{E}\left[\Delta\bar{P}\middle|\, do(\bar{Q})\right].$$

Proposition 7.5.14 (Identifiability of cross-impact under $\mathcal{E}$)**.** *Under the causal structure $\mathcal{E}$, one naively identifies stock-level and factor-level cross-impact.*

Proof. $\bar{Q}$ has no parents. Therefore, the naive identification condition from Corollary 5.2.32 of Chapter 5 applies. □

Remark 7.5.15 (New causal interpretation of past empirical studies)**.** *Under the causal structure $\mathcal{E}$, Proposition 7.5.14 proves that the empirical cross-impact matrices of Bouchaud et al. (2016)[36], Capponi and Cont (2020)[56], and Mastromatteo et al. (2017)[165] reconcile.*

> *The proposed causal interpretation is that cross-impact solely travels through factors but is distinct from direct price impact.*

Statisticians naively identify factor cross-impact $\bar{Q} \to \Delta\bar{P}$, exclude pair-wise *cross-impact links, and estimate cross-impact paths $Q_i \leftarrow \bar{Q}... \to \Delta P_j$.*

The causal structure $\mathcal{E}$ lends itself well to the applications from Mastromatteo et al. (2017)[165]. Mastromatteo et al. propose the *EigenLiquidity* model (ELM) for factor and portfolio trading.

> "To quantify the slippage incurred by the strategy, we introduce the EigenLiquidity model (ELM). This model is directly related to statistical risk factors that have been used for portfolio risk management for several decades. Based on a principal component analysis of the correlation matrix, which provides a practical method to quantify the different kinds of market risk (long the market, sectorial, etc) one can trade while staying within a prescribed budget of transaction cost" (p. 1).

7.5.2.4 Cross-impact for risk-management and factor research

The causal structure $\mathcal{E}$ has two practical applications to risk management and factor research:

(a) factor level impact $\bar{Q} \to \Delta\bar{P}$ directly models factor trading costs.

Mastromatteo et al. (2017)[165] estimate factor transaction costs in Figure three on page 5. g^a is the price impact parameter for factor a and π^a the corresponding factor-mimicking portfolio.

> "Each eigenvalue g^a is interpreted as the cost of trading a dollar of risk in the portfolio π^a" (p. 5).

This factor trading cost model leads to factor-level optimal execution and portfolio construction without simulating individual stock trades and their costs.

(b) Under the causal structure $\mathcal{E}$, trades confound the factor model $\Delta\bar{P} \to \Delta P_j$.

This observation implies that covariance matrices estimated from factor models need to control for the factor flow $\bar{Q}$ to avoid confounding through the back-door path $\Delta\bar{P} \leftarrow \bar{Q} \to Q_j \to \Delta P_j$. Proposition 7.5.16 proves this identification result.

Proposition 7.5.16 (Factor model estimation in the presence of cross-impact.). *Under the causal structure $\mathcal{E}$, controlling for factor flow identifies a factor model:*

$$\mathbb{E}\left[\Delta P_j |\, do(\Delta\bar{P})\right] = \int \mathbb{E}\left[\Delta P_j |\, \Delta\bar{P}, \bar{Q}\right] p\left(\bar{Q} = q\right) dq.$$

One does not naively identify factors' causal effect on individual stock prices,

$$\mathbb{E}\left[\Delta P_j |\, do(\Delta\bar{P})\right] \neq \mathbb{E}\left[\Delta P_j |\, \Delta\bar{P}\right].$$

Proof. The result follows from the back-door $\Delta\bar{P} \leftarrow \bar{Q} \to Q_j \to \Delta P_j$ and Theorem 5.2.34 from Chapter 5. □

Considering (b), when estimating risk, the causal structure $\mathcal{E}$ distinguishes between two types of stock correlations:

(a) $\Delta P_i \leftarrow \Delta\bar{P} \to \Delta P_j$ model fundamental correlations driven by the causal links between factors and stocks.

(b) $\Delta P_i \leftarrow Q_i \leftarrow \bar{Q} \to Q_j \to \Delta P_j$ and $\bar{Q} \to \Delta\bar{P}$ model *flow-based* correlations driven by endogenous market mechanisms.

Furthermore, Proposition 7.5.16 proves that observed risk matrices *blend both correlation types*, and one can only distinguish between fundamental and flow-based correlations through causal inference. And while, during regular times, fundamental and flow-based correlations are alike, the two diverge during liquidity crises. Indeed, fundamental stock correlations remain stable during liquidity crises, as they reflect companies' underlying economics, for instance, their supply chains.[8] Conversely, flow-based correlations drastically increase during liquidity crises, are endogenous to the market, and do not reflect the companies' fundamentals. Cont and Wagalath (2016)[78] model impact-driven, endogenous increases in observed stock correlations:

[8] Conversely, crises not driven by liquidity but real-world fundamentals drastically change companies' underlying economics. For example, the 2020 Covid crisis massively disrupted supply chains and swiftly reshaped fundamental stock correlations.

"We show that feedback effects can lead to significant excess realized correlation between asset returns and modify the principal component structure of the (realized) correlation matrix of returns. Our study naturally links, in a quantitative manner, the properties of the realized correlation matrix — correlation between assets, eigenvectors and eigenvalues — to the sizes and trading volumes of large institutional investors" (p. 1).

7.5.3 Price impact for pairs trading

This section focuses on the opposite scenario from Section 7.5.2. Instead of studying price impact across many assets, the emphasis is pair-wise price impact on a small number of assets. In equities, Remark 7.5.7 shows pair-wise cross-impact is second-order compared to factor impact. This observation raises the threshold for introducing pair-wise cross-impact into trading. There are two cases where the relative importance of factor impact and pair-wise cross-impact inverts.

- Asset classes outside equities may present higher pair-wise correlations and less factor-driven trading.
- Specific trading strategies, for example, pairs trading, focus on the pair-wise behavior of assets by trading in a *factor-neutral* way.

Five papers empirically explore these scenarios:

(a) Schneider and Lillo (2019)[206] "estimate cross-impact among sovereign bonds" (p. 1).

(b) Said et al. (2021)[203] empirically study "market impact in the options market" (p. 1).

(c) Tomas, Mastromatteo, and Benzaquen (2022)[224] study three markets in particular, "a universe of three instruments: two liquid NYMEX Crude Oil future contracts and the corresponding Calendar Spread contract" (p. 14).

(d) Rosenbaum and Tomas (2021)[199] quantify cross-impact "for two maturities of E-Mini SP500 futures traded on the CME" (p. 13).

(e) Cont, Cucuringu, and Zhang (2021)[73] provide an example of a *sparse* cross-impact model on US equities, which is well-suited for factor-neutral pairs-trading.

The articles also derive implications for pairs-trading strategies on the respective assets. This section focuses on summarizing the studies' *empiric* emphasizing *asset-specific* methodologies.

7.5.3.1 Bonds in Schneider and Lillo (2019)[206]

Schneider and Lillo study bond cross-impact. Schneider and Lillo motivate the focus on bonds with their higher pair-wise correlation over equities.

> "We choose to estimate cross-impact between bonds instead of equities since we expect the strength of cross-impact among sovereign bonds of the same issuing country, especially of similar maturity, to be bigger than the one between e.g. stocks or indices. Sovereign bonds of one country typically have a similar underlying risk and their prices are implicitly connected via the yield curve, a link that we deem stronger than e.g. a common factor between the stocks of a same sector" (p. 12).

An interesting methodological point is the model's clock: Schneider and Lillo leverage a "combined market order time" (p. 18) that aggregates trade events across bonds. This approach is numerically intractable when asset numbers are vast, such as in Section 7.5.2, but is well-suited for small asset numbers. Schneider and Lillo contrast this method with the regular clock.

> "Previous approaches avoided potential pitfalls by estimating the propagator in calendar time and binning trades. The estimation is sensitive to the bin width" (p. 17).

Schneider and Lillo fit a *locally concave* model, in line with the Bouchaud model from Section 7.3.3: the concave function applies element-wise to every market fill across all assets. The non-linearity introduces a scaling factor based on the time discretization bin, which Section 2.6.3 of Chapter 2 shows to be $\sqrt{\Delta t^N}$.

Schneider and Lillo find a self-impact coefficient of 6bps per asset traded, where each asset normalizes to one hundred euros face value. The cross-impact coefficient is a fifth of self-impact across assets. The authors also provide a brief causal analysis of the empirically observed cross-impact.

> "Having established the evidence for cross-impact, we investigate its possible origin: Is this due to correlated trades across assets (e.g. a strategy trading several bonds simultaneously) or is it mostly due to quote revision following a trade, leading to changes of the mid-price of a bond in the absence of trades?" (p. 15).

)ne recasts the question as an identifiability problem, like Capponi and Cont 2020)[56]. Based on their causal analysis, Schneider and Lillo show that both ausal paths $Q_i \leftarrow \bar{S} \rightarrow Q_j \rightarrow \Delta P_j$ and $\Delta Q_i \rightarrow \Delta P_j$ contribute to cross-npact, and the latter is five to ten times smaller than the former.

.5.3.2 Options in Said et al. (2021)[203]

aid et al. leverage a proprietary set of "orders executed by the BNP Paribas ading desk for the 2-year period from June 2016 through June 2018" (p. 4). his data enables the authors to define an "option metaorder" (p. 3): a proxy r a strategy trading across multiple related options.

> "An **option metaorder** with respect to an implied volatility parameter θ is a series of orders sequentially executed during the same day and having those same attributes:
>
> - *agent* i.e. a participant on the market (an algorithm, a trader...);
> - *underlying product id* i.e. the underlying financial instrument;
> - *direction* regarding the sign of $\mathcal{S}^\theta := s \times Q \times \frac{\partial \mathcal{O}}{\partial \theta}$ where s is the sign of the trade, Q and $\mathcal{O}$ the quantity and the price of the option traded;
>
> [...] As a matter of fact, trading an option with a given strike K and maturity T also affects those with nearby strikes and maturities, so that trades on options with different strikes and maturity can very well belong to the same metaorder" (p. 3).

Said et al. divide their data into two sets of option metaorders with distinct trading theses. The authors note that "the option metaorders presented here last a few dozen seconds on average", bolstering the case that a single thesis spanning multiple assets drives the order, in line with this section's theme.

(a) *At the money forward volatility* option metaorders trade the implied volatility curve's level.

(b) *At the money forward skew* option metaorders trade the implied volatility curve's skew.[9]

Said et al.'s approach is like Mastromatteo et al.'s EigenLiquidity model (2017)[165]. Exercise 25 makes the parallel explicit by proposing a causal structure for options price impact. Instead of predicting individual options' impact, for example, by estimating the link $\Delta Q_i \rightarrow \Delta P_j$, the authors focus on the two option factors and their underlying strategies. They assess the price impact $\Delta\bar{Q} \rightarrow \Delta\bar{P}$ using a standard model on the two factors. The paper yields an options variant of the *EigenLiquidity model*, where the EigenLiquidity factors are the implied volatility curve's level and skew.

Said et al. fit a concave price impact model at the order level with $h(x) \propto x^{0.56}$ for the level and $h(x) \propto x^{0.53}$ for the skew of the implied volatility curve. The authors conclude with a parallel to equities price impact.

> "The results presented in this section are certainly the most important of the article. They confirm the consistency of the *Square-Root Law* already observed in the equity market" (p. 19).

[9] Gatheral (2006)[103] covers the "dynamics of the volatility surface" (p. 101) in Chapter eight of his book "The volatility surface, a practitioner's guide". Gatheral states that the skew and level factors are independent.

> "Empirical studies of the dynamics of the volatility skew show that $\frac{\partial}{\partial k}\sigma(k,t)$ is approximately independent of volatility level over time" (p. 101).

7.5.3.3 Commodity futures in Tomas, Mastromatteo, and Benzaquen (2022)[224]

The pre-processing step covers a methodological point for commodities futures.

> "**Pre-processing: accounting for non-stationarity** Overall, the front month contract CRUDE0 is by far the most liquid, followed by the subsequent month contract CRUDE1 and the calendar spread CRUDE1_0. However, there are strong, seasonal dependencies which are shown in Figure 3. For example, the subsequent month contract becomes more liquid as one approaches the maturity of the front month contract. Global estimators of Σ, Ω and R would thus be biased by this varying liquidity ω (σ also appears to follow a non-stationary pattern, but is not shown here). Thus we used local (daily) estimators of price volatility σ_t and liquidity ω_t, and built local covariance estimators Σ_t and Ω_t by assuming stationarity of the correlations $\rho = \text{diag}(\sigma_t)^{-1}\Sigma_t\text{diag}(\sigma_t)^{-1}$ and $\rho_\Omega = \text{diag}(\omega_t)^{-1}\Omega_t\text{diag}(\omega_t)^{-1}$" (p. 23).

In this chapter's terminology, Tomas, Mastromatteo, and Benzaquen fit a price impact model with a liquidity parameter that is dynamic over time but has a fixed correlation structure across instruments. Indeed, due to the contract expiry, multi-day effects drive this dynamic liquidity rather than the time-of-day effects of Section 7.4.2.

The paper provides two essential empirical takeaways:

(a) First, individual contracts miss a portion of the total price impact.

> "Variants of direct models account for 33% and 40% of the variance of market wide moves. Cross-impact models slightly improve on direct models (scoring around 46%). This is somewhat surprising: despite the concentration of liquidity in the front month contract and the large correlation between the front and subsequent month contracts, accounting for the off-diagonal elements of Σ and Ω matters" (p. 10).

(b) Second, a one-dimensional model combining all three contracts is the most parsimonious price impact model for this market:

> "this three-dimensional system roughly behaves like a one-dimensional system" (p. 11).

7.5.3.4 Equities futures in Rosenbaum and Tomas (2021)[199]

The analysis of Rosenbaum and Tomas (2021)[199] follows up on the bond and indexes cross-impact analysis of Tomas, Mastromatteo, and Benzaquen (2022)[224]. The same methodology as in Section 7.5.3.3 deals with the non-stationarity of the instruments' liquidity. The empirical novelty in Rosenbaum and Tomas (2021)[199] is estimating a *separate time kernel* for each cross-flow and cross-impact term. The different time kernels lift the *factorized functional form* assumption from Remark 7.5.6 made in the context of equities cross-impact. The resulting cross-impact model is significantly more sophisticated.

> "The order flow auto-covariances $\Omega_{11}(\tau)$ and $\Omega_{22}(\tau)$ are slowly decaying in τ, although $\Omega_{11}(\tau)$ exhibits faster decay than $\Omega_{22}(\tau)$. Furthermore, we observe that $\Omega_{12}(\tau) \approx \Omega_{21}(\tau)$ so that there are no lead-lag effects in the order flows. This shows that Ω cannot be easily factorized" (p. 13).

The non-factorized form can lead to price-manipulation strategies, unlike the factorized form. Rosenbaum and Tomas (2021)[199] provide additional conditions to rule out price manipulation under this more complex model and fit the cross-impact function within this class.

Exercise 26 provides a second family of non-factorized cross-impact models without price manipulation based on Fruth, Schöneborn, and Urusov (2013, 2019)[97, 98].

7.5.3.5 Sparse equities cross-impact in Cont, Cucuringu, and Zhang (2021)[73]

Section 4.2.1 of Chapter 4 and Section 7.3.1 introduce the OFI model. Cont, Cucuringu, and Zhang (2021)[73] explore pair-wise cross-impact with OFI.

> "Under the sparsity assumption of cross-impact coefficients, we use LASSO to describe such a structure and compare the performances with the price-impact model, which only utilizes a stock's own OFIs" (p. 30).

The LASSO regularization penalty focuses the model on the most consequential pair-wise cross-impact terms rather than the entire cross-impact matrix.

Cont, Cucuringu, and Zhang highlight that their results are sensitive to the fitting methodology. For example,

(a) cross-impact improves model performance, both in and out-of-sample compared to the regular OFI model.

(b) cross-impact improvements vanish when considering deeper limit order book levels.

(c) a dynamic cross impact model along the market factor, fitted on thirty minute samples, performs poorly in the next thirty-minute out-of-sample

Finally, the average number of cross-impact terms for a given stock is ten.

7.6 Summary of Results

This section summarizes empirical price impact results based on the public trading tape. Let δt be a time bin, q_t the volume traded in a bin, σ a volatility estimator for the stock price, ADV an estimator of the stock's daily volume, and v_t a market activity measure (e.g., a predicted volume curve or a moving average of $|q_t|$).

7.6.1 Discrete formulas for price impact models

7.6.1.1 The original OW model

$$I_{t+\delta t} - I_t = -\beta I_t \delta t + \lambda\sigma \frac{q_t}{\text{ADV}}.$$

7.6.1.2 The locally concave Bouchaud model

$$I_{t+\delta t} - I_t = -\beta I_t \delta t + \lambda\sigma \cdot \text{sign}(q_t)\sqrt{\frac{|q_t|}{\text{ADV}}}.$$

7.6.1.3 The reduced-form model

$$I_{t+\delta t} - I_t = -\beta I_t \delta t + \frac{\lambda\sigma q_t}{\sqrt{\text{ADV} \cdot v_t}}.$$

7.6.1.4 The globally concave AFS model

$$I_t = \lambda\sigma \cdot \text{sign}(J_t)\sqrt{|J_t|}$$

where

$$J_{t+\delta t} - J_t = -\beta J_t \delta t + \frac{q_t}{\text{ADV}}.$$

7.6.2 Summary table

Table 7.7 summarizes the four models' essential statistics.

TABLE 7.7
Average magnitude of the end-of-day impact state, out-of-sample R^2 for one-minute predictions.

Price Impact Model	$\mathbb{E}\left[\lvert I_T\rvert\right]$	$\mathbb{E}\left[\frac{\lvert I_T\rvert}{\sigma}\right]$	R^2
Bouchaud	43bps	34%	18%
Reduced-form	25bps	20%	14%
OW	28bps	23%	10%
AFS	23bps	19%	9%

7.6.3 The EigenLiquidity model

The EigenLiquidity model is the simplest model capturing cross-impact's core economic idea:

Assets that trade together move together.

Consider stocks i, j with returns $\Delta P^i, \Delta P^j$ linked by a common factor $\Delta \bar{P}$. Let Q^i, Q^j be the stock-specific trades, and $\bar{Q}$ the corresponding factor trade. The EigenLiquidity model follows the causal graph in Figure 7.8 to describe how trading Q_i affects returns ΔP_j. All cross-impact travels along

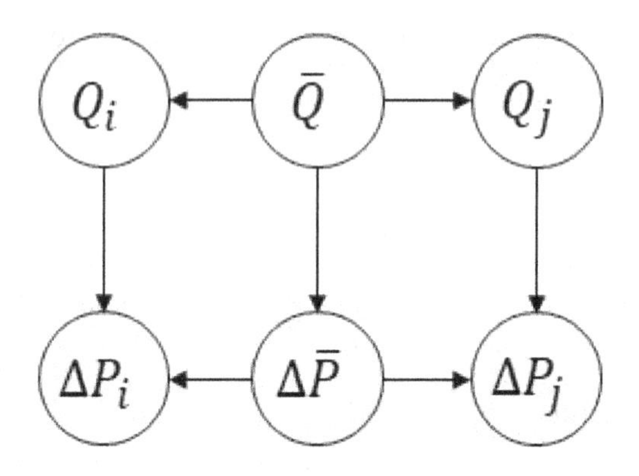

FIGURE 7.8
Causal graph for the EigenLiquidity model.

factors: conditional on observing factor impact, the price impact of stocks i, j are independent.

7.7 Exercises

Exercise 22 Identifiability under the causal model $\mathcal{C}$

This exercise replicates a cross-impact identifiability counterexample fror Capponi and Cont (2020)[56] under the causal structure $\mathcal{C}$ from Definitio 7.5.1. The corresponding Section 2.2 "Are cross-impact coefficients identif able" (p. 6) in Capponi and Cont (2020)[56] spans pages six to nine. Promo the causal structure $\mathcal{C}$ to a causal model by adding the following probabili model:

(a) The functional form for the price impact links $Q_i \to \Delta P_i$, $Q_j \to \Delta I$ $Q_i \to \Delta P_j$, and $Q_j \to \Delta P_j$ is

$$\Delta P = \Lambda \Delta I(Q) + \epsilon$$

where, for a given data point, $\Delta P, \Delta I(Q), \epsilon$ are vectors of dimension $n = 2$.

(b) The actual cross-impact matrix Λ is

$$\Lambda = \begin{pmatrix} \lambda_{11} & \lambda_{12} \\ \lambda_{21} & \lambda_{22} \end{pmatrix}.$$

(c) The distributions of $\Delta I(Q), \epsilon$ are independent Gaussians with mean zero and covariance matrices

$$L = \begin{pmatrix} l_1^2 & \rho l_1 l_2 \\ \rho l_1 l_2 & l_2^2 \end{pmatrix}; \quad N = \begin{pmatrix} \sigma_1^2 & 0 \\ 0 & \sigma_2^2 \end{pmatrix}.$$

Now consider the following two identification equations for $\mathbb{E}\left[\Delta P_2 | \operatorname{do}(Q_1)\right]$:

$$\hat{\lambda}_{21} \Delta I(Q_1) = \mathbb{E}\left[\Delta P_2 | Q_1\right]$$

and

$$\hat{\lambda}'_{21} \Delta I(Q_1) = \int \mathbb{E}\left[\Delta P_2 | Q_1, Q_2 = q\right] p\left(Q_2 = q\right) dq.$$

1. Compute the naive cross-impact term $\hat{\lambda}_{21}$ and the corrected cross-impact term $\hat{\lambda}'_{21}$ as a function of the actual cross-impact and flow covariance matrices Λ, L.
2. Consider the two scenarios

$$\Lambda^0 = \begin{pmatrix} 0.6 & 0.3 \\ 0.3 & 0.6 \end{pmatrix}; \quad L^0 = \begin{pmatrix} 1 & 0 \\ 0 & 1 \end{pmatrix}$$

and

$$\Lambda^1 = \begin{pmatrix} 0.6 & 0 \\ 0 & 0.6 \end{pmatrix}; \quad L^1 = \begin{pmatrix} 1 & 0.5 \\ 0.5 & 1 \end{pmatrix}.$$

Compute the corresponding naive cross-impact terms $\hat{\lambda}^0_{21}, \hat{\lambda}^1_{21}$.

3. What are the path contributions for each scenario?
4. Provide a purely statistical interpretation to explain the difference between $\hat{\lambda}_{21}$ and $\hat{\lambda}'_{21}$ by distinguishing one-dimensional and two-dimensional regression models.

Exercise 23 Causal model for cross-impact and prediction bias

Definition 7.7.1 (Cross-impact with prediction bias). *Define the causal structure $\mathcal{E}'$*

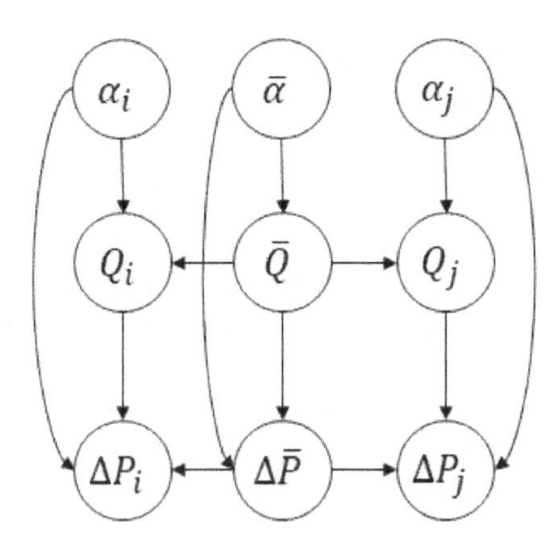

for two stocks i, j driven by a shared factor with

- *returns $\Delta P_i, \Delta P_j$ and factor returns $\Delta \bar{P}$,*
- *public trading tapes Q_i, Q_j, and factor flow $\bar{Q}$,*
- *stock-specific alpha signals α_i, α_j, and factor signal $\bar{\alpha}$.*

1. Provide an identification formula for the cross-impact counterfactual

$$\mathbb{E}\left[\Delta P_2 | \operatorname{do}(\bar{Q})\right].$$

2. Assume the portfolio only has stock-specific alphas. Risk considerations drive the factor flow $\bar{Q}$, and $\bar{Q}$ has no alpha. What is the identification formula for cross-impact?

Exercise 24 Factor trading

This exercise implements an insight by Mastromatteo et al. (2017)[165] about the EigenLiquidity model. From a causal perspective, it operates under graph $\mathcal{E}$ from Definition 7.5.8.

> "This construction implies that the cost $\mathcal{C}$ can be calculated by first projecting the strategy $q(t)$ on the portfolios π^a via Eq. (12) and then by taking the sum of an impact cost per mode g^a' with a weighting factor $||\tilde{q}^a||^2$" (p. 3).

Assume given n stocks and a single market factor. For $i \leq n$, project each stock's trading process Q^i into a factor-neutral $\mathring{Q}^i$ and factor component $\tilde{Q}^i$

$$Q^i = \mathring{Q}^i + \tilde{Q}^i.$$

The factor trading process is

$$\bar{Q} = \sum_{i \leq n} \tilde{Q}^i$$

where the $\tilde{Q}^i$ correspond to the projection $\tilde{q}^a$ in Mastromatteo et al. (2017)[165].

The factor price impact is

$$d\bar{I}_t = -\beta \bar{I}_t dt + \bar{\lambda} d\bar{Q}_t.$$

The factor-neutral price impact is

$$dI_t^i = -\beta I_t^i dt + \lambda d\mathring{Q}_t^i.$$

Therefore, the total transaction costs over a trading period $[0, T]$ are

$$\sum_{i \leq n} \int_0^T I_t^i d\mathring{Q}_t^i + \frac{1}{2}[I^i, \mathring{Q}^i]_T + \int_0^T \bar{I}_t d\bar{Q}_t + \frac{1}{2}[\bar{I}, \bar{Q}]_T.$$

If the trading strategy trades only the factor, the corresponding transaction costs are

$$\int_0^T \bar{I}_t d\bar{Q}_t + \frac{1}{2}[\bar{I}, \bar{Q}]_T.$$

1. Solve the optimal execution problem for a pure factor trade of size $\bar{Q}_T$.

2. Solve the optimal execution problem for a basket trade

$$\forall i \leq n, \quad Q_T^i = \mathring{Q}_T^i + \tilde{Q}_T^i.$$

3. Solve the optimal execution problem for a basket trade with factor-neutral alpha signals $\left(\alpha^i\right)_{i \leq n}$ and factor alpha signal $\bar{\alpha}$. Assume all alphas are deterministic and the trader is risk neutral.

Exercise 25 Causal model for options impact

This exercise complements the analysis of Said et al. (2021)[203] from Section 7.5.3.2 with a causal structure.

1. Propose a causal structure for options price-impact that considers the level and skew factors.
2. Extend the above causal structure to include prediction bias.
3. Provide an identification formula for cross-impact under both proposed causal structures.

Exercise 26 OW model with cross-impact

This exercise extends the proof of Fruth, Schöneborn, and Urusov (2013, 2019)[97, 98] to cross-impact. This proof provides an alternative method to build cross-impact models without price manipulation and derives optimal execution strategies for portfolios of correlated assets.

Consider the case of n assets. Let B, Λ be two n-dimensional matrices and define the OW model with cross-impact as the solution to the n-dimensional SDE

$$dI_t = -B \cdot I_t dt + \Lambda \cdot dQ_t$$

with an initial condition I_0.

1. Under what conditions is the map from Q to I invertible?
2. Consider a risk-neutral trader. Map their objective function in impact space. Under what condition is there no price manipulation strategy?
3. Now consider the case where B, Λ are stochastic processes, with $\Lambda = e^{\Gamma}$ for a differentiable, matrix-valued process Γ. Furthermore, assume Γ, Γ' commute. Map the objective function in impact space.
4. Under what conditions on the stochastic processes B, Γ is there no price manipulation strategy?

Part IV

Appendix

A

Using Kdb+ for Trading Models

> *"The kdb+ database and its underlying programming language, q, are the standard tools that financial institutions use for handling high-frequency data."*
>
> – Novotny et al. (2019)[178]

A.1 A Gentle Introduction to Kdb+

This appendix covers the database kdb+ and q, the programming language used within kdb+. The material is a succinct introduction rather than a textbook. However, two books cover kdb+ and q for quantitative research:

- *"Machine Learning and Big Data with kdb+/q"* by Novotny et al. (2019)[178] and
- *"Fun Q"* by Psaris (2020)[192].

This introduction covers the *why, how, and what* of using kdb+ and q for quantitative trading models:

a) Why do financial institutions build their trading infrastructure using kdb+?

b) How can quantitative researchers and traders use kdb+ for their daily work?

c) What models and applications exemplify the power of kdb+?

eaders familiar with kdb+ and q may skip to Sections A.3 and A.4: using db+ for price impact models.

.1.1 What is kdb+ and why does it matter to quants?

x systems develop kdb+, and dozens of financial institutions deal with trad- g and high-frequency data using kdb+. Novotny et al. (2019)[178] recount kdb+'s history in their preface: "Kx Systems was under an exclusive agreement with UBS. It expired in 1996", leading to kdb+'s widespread use at banks and hedge funds. Efinancialcareers.com (2020, 2021)[49, 50] state,

> "if you want to be assured of a job in finance, it will benefit you to learn the coding languages K and Q. K and Q underpin the Kdb+ database system which is used increasingly by banks, hedge funds and high frequency trading houses"

> "Developers proficient in both Q and kdb+, the database system that goes with it, tend to be both hard to find and in constant demand globally."[1]

Kx systems describe kdb+ on their web page "Developing with kdb+ and the q language" [135]:

- "a high-performance cross-platform historical-time series columnar database
- an in-memory compute engine
- a real-time streaming processor
- an expressive query and programming language called q"

The first and last bullet points describe data access and transformation with kdb+ and q. Kdb+ focuses on time series, high performance, and expressive queries. In practice, users *query* kdb+ databases:

```
select time, stock, price, ret from returnTable where date =
   2021.12.10

select avg slippage by date from orderTable where date >
   2020.12.31

select time, stock, bid, ask, return from quoteTable lj
   returnTable
```

The Kx website [139] name such intuitive coding patterns "qsql" queries, as th query "closely resembles conventional SQL". The similarity to SQL guarantee this syntax is expressive and easy to learn: qsql provides newcomers with practical data manipulation grammar like R's dplyr and python's pandas.

> "dplyr is a grammar of data manipulation, providing a consistent set of verbs that help you solve the most common data manipulation challenges" (Dplyr website [115]).

[1]A search for the keyword "kdb" on the job-search website LinkedIn [153] displays t demand for kdb+ developers and users as of 2021. The search yields over 350 results acr financial institutions such as Millennium, Morgan Stanley, Bank of America, JPMorg Chase, Deutsche Bank, Citi, UBS, and Barclays.

> "Since many potential pandas users have some familiarity with SQL, this page is meant to provide some examples of how various SQL operations would be performed using pandas" ("Comparison with SQL", pandas website [185]).

Behind the scenes, kdb+ massively outperforms SQL, dplyr, and pandas for trading applications.

Remark A.1.1 (The competitive edge of kdb+). *Kdb+'s extreme performance stems from two principles.*

(a) Q natively *optimizes all data structures and functions for vector and time series manipulation.*

In comparison, native R and Python are not competitive. Both high-level languages rely on external packages to optimize vector operations in a lower-level language such as C. The latter observation makes specific *Python and R packages competitive. Reliance on non-native solutions shifts code from R or Python to the lower-level language. Consequently, quantitative researchers and traders require a vast support team of developers to analyze their trading data in R or python. Conversely, researchers and traders using q nearly exclusively use qsql and can inspect their functions' inner workings without studying a separate, lower-level language.*

(b) Kdb+ strongly incentivizes data locality.

Data locality is an essential principle for dealing with large datasets:

> *Models should run computations* where one stores the data.

Velu, Hardy, and Nehren (2020)[227] discuss microservices as part of modern research technology.

> *"An approach that considers leveraging a micro-services middleware is growing in popularity in recent years" (p. 409).*

The interested reader studies KX's webinar (2023)[142] on "kdb Microservices for Ultra-High Velocities on a Postage Stamp Footprint". The benefits of data locality and microservices follow from not transferring data from storage to computation: it is already there. Indeed, the more massive the data, the higher its upfront transfer cost is. This data marshaling cost is prohibitive for the iterative coding style of researchers and traders. Furthermore, the transfer cost prevents researchers and traders from implementing data-heavy machine learning models. Standard packages do not scale to trading applications and require developers to build bespoke solutions for high-frequency data.

elu, Hardy, and Nehren (2020)[227] note: "KDB continues to maintain its minance in large scale trading operations but its reign is no longer as certain it used to be" (p. 402) as other solutions start incorporating principles such as native time series manipulation and data locality into their core features. Kx is reacting by including support for other languages.

> *"KX built a reputation for working with big data back before "big data" was a thing. [...] But the q language has limited adoption of Kdb+ database. [...] By enabling data analysts and data scientists to use SQL and Python, respectively, to access data stored in the KX environment, KX has significantly widened the pool of developers who will be exposed to KX" (Woodie (2021)[238]).*

The second and third bullet points describe using kdb+ for real-time *streaming* applications, for instance, live trading systems or analytics. Quantitative researchers and traders do not build such applications independently but rely on developers for such production-grade code. Nevertheless, the two points are crucial to quantitative researchers and traders: models and strategies aim to move from research to production. When production and research share their technology, which kdb+ strongly incentivizes, the team reaps the following benefits:

- Sharing the same infrastructure simplifies the transition and feedback between research and production.

  A shared coding language *accelerates the deployment* of signals and models to production and *reduces costly errors* in live trading.

  In his presentation *Real Time Trading Signals*, Almgren (2018)[9] outlines the leading role kdb+ plays at Quantitative Brokers for implementing signals.

  > "Signal generator (Kdb+)
  > The signal generator receives market data, performs computations to predict prices, and feeds the results to the algorithmic engine to improve trade execution" (p. 9).

  Almgren explains that using the same high-performance infrastructure, i this case, kdb+, to generate signals in research and production is essentia to Quantitative Brokers' trading architecture. Similarly, Velu, Hardy, an Nehren (2020)[227] point out:

  > "[A] frequent complication is that many of these processes are usually built within the research environment that is often different from the production installation setup" (p. 404).

  Infrastructures that significantly differ in production and research, for e ample, using a research language unsuited for live trading, lead to expe sive delays and errors in a model's deployment.

  The live trading performance of alpha models decreases over time and exceedingly sensitive to errors. Therefore, the speed and accuracy of mod implementation is a crucial competitive advantage in algorithmic tradi

As a result, the Financial Industry Regulatory Authority (FINRA) issued multiple regulatory notices on algorithmic trading [94]. For instance, Regulatory Notice 15-09 (2015)[95] emphasizes testing during a strategy's development and deployment.

> "As an initial matter, firms must have appropriate policies and procedures in place to review and test any trading algorithms they use, including development, deployment and post-implementation monitoring of algorithmic strategies."

Algorithmic trading relies on models: therefore, quantitative researchers and traders must involve themselves in testing. For example,

> "firms should consider: [...]
> - establishing a quality assurance process such that testing is performed independently of code development.
> - implementing and periodically evaluating test controls to confirm their adequacy and reliability.
> - implementing data integrity, accuracy and workflow validation testing processes."

Section 1.1.4 of Chapter 1 introduces four core modeling principles: the emphasis on *testable* and *robust* models matches FINRA's focus on testing trading algorithms.

- Production data seamlessly move to historical research databases.

  Kdb+ guarantees that data used in research is of high fidelity and *accurately reflects live trading.* The Kx website [140] illustrates an implementation under

  https://code.kx.com/q/learn/startingkdb/tick/

  > "[A] standard setup might consist of
  > - a tickerplant to capture and log incoming data
  > - a historical database (HDB) to access all the data prior to the current day
  > - a real-time database (RDB) to store the current day's data in memory and write it to the HDB at the end of day."

  High-fidelity trading data powers simulations and A-B tests, defined in Section 1.2.8 of Chapter 1. Section 4.2.3 of Chapter 4 covers simulations. Section 5.3.1 of Chapter 5 covers A-B tests.

- Conversely, real-time analytics access historical data, models, and simulations to provide enhanced context on live trading activity and performance.

  Real-time analytics increase trading models' *audience.* Never underestimate a real-time Graphical User Interface (GUI) for stakeholders to *visualize a trading strategy or model:*

> "Though I did ultimately improve the model, the traders benefited most from the friendly user interface I programmed into it. This simple ergonomic change had a far greater impact on their business" (p. 8, Derman (2004)[84]).

A.1.2 First steps in kdb+

This section sets the reader up with kdb+ to follow the appendix's examples. The Kx website [141] provides a general introduction on the web page

https://code.kx.com/q/learn/startingkdb/.

First, the reader downloads the 32-Bit Personal Edition of kdb+ from the Kx website [137] under

https://kx.com/developers/download-licenses/.

By default, q installs in folders:

```
// unix
~/q/l32

// mac
~/q/m32

// windows
c:\q\w32
```

Then, running the executable in a shell or command prompt yields a terminal-like interface, waiting for the reader's q-command. See Figure A.1.

```
c:\q\w32>q.exe
KDB+ 3.6 2019.04.02 Copyright (C) 1993-2019 Kx Systems
w32/ 16()core 4095MB webst desktop-jfb94cd 10.0.0.213 NONEXPIRE

Welcome to kdb+ 32bit edition
For support please see http://groups.google.com/d/forum/personal-kdbplus
Tutorials can be found at http://code.kx.com
To exit, type \\
To remove this startup msg, edit q.q
q)
```

FIGURE A.1
Q's command line.

Finally, "Hello World!" tests the installation.

```
show "Hello World!"
```

Running code from the q-terminal is challenging for larger scripts or pr grams. Four Integrated Development Environments (IDEs) solve this issue

- KX Developer [136], developed by KX itself.
- Qpad [240], developed by Zhakarov O.
- JupyterQ [138], highlighted on the KX website.
- A kdb+/q plugin on a general-purpose IDE such as Visual Studio [168].

A native approach without IDE saves the script in a .q file, then loads the code directly into the terminal. For example, the reader saves the following test.q file in the working directory.

```
show "Hello World!"
```

Listing A.1: test.q

The script runs from the q-terminal, as per Figure A.2.

```
q)\l test.q
"Hello World!"
q)
```

FIGURE A.2
Running a q-script.

When the user encounters an error, q automatically enters debug mode. The reader determines that q entered debug mode by paying attention to the q-terminal:

```
\terminal when not in debug mode
q)

\terminal when in debug mode
q))
```

The backslash command exits debug mode, and double backslash exits q altogether. See Figure A.3.

```
q)f:{1+`}
q)f[1]
'type
  [1]  f:{1+`}
          ^
q))\
q)\\

c:\q\w32>
```

FIGURE A.3
Example q error.

A.1.3 Basic operations in Q

While quantitative researchers and traders work in qsql, a little *raw* q is unavoidable. This section describes basic operations in q with remarks, highlighting q's idiosyncrasies that surprise newcomers.

A.1.3.1 Q does not follow the traditional order of operations

For performance reasons, q evaluates statements *from right to left, regardless of the standard order of operations.* This property leads to counter-intuitive results, for instance, see Figure A.4.

```
q)1+5*1
6
q)5*1+1
10
```

FIGURE A.4
Counter-intuitive q example.

While q-developers learn to read expressions from right to left, this convention increases newcomers' barriers to entry. Therefore, paying attention to the order of operations, refactoring code, and adding parentheses to avoid misleading computations are good practices.

A.1.3.2 Assignments and other basic operators

Q has an extensive list of simple operators, ranging from addition to more sophisticated operations, for example, the search operator *find.* Unfortunately, operators are *heavily overloaded:* they behave differently depending on the argument type. Kx's reference page

https://code.kx.com/q/ref/

lists base operators and their overloaded meanings. For instance, colons assig variables, and the equal sign tests for equality, per Figure A.5.

FIGURE A.5
Variable assignement example.

A.1.3.3 Atoms, lists, and dictionaries

Q firmly separates data types. Furthermore, unlike in R, q functions do n *upgrade* atoms into single element lists and are particular about data types

(a) Atoms represent an indivisible data unit of a single type. The reader fin atom types under

https://code.kx.com/q/basics/datatypes/.

One changes an object's type using the $ operator and checks an object's type using the function "type", which returns the type's integer representation. See Figure A.6.

```
q)type 5
-7h
q)`float$5
5f
q)type `float$5
-9h
```

FIGURE A.6
Type example.

(b) A list is a collection of objects. One uses lists to apply a transformation or aggregation to collected objects. There are three list subtypes:

- *Vectors* are lists of atoms with a single type.

  Q vectorizes essential functions and operations: when performed on vectors, these functions significantly outperform traditional loops. See Figure A.7.

```
q)1 2 3 4
1 2 3 4
q)1 + til 4
1 2 3 4
```

FIGURE A.7
Vector example.

- *Mixed lists* are lists of varying types.

  One uses mixed lists as function arguments and avoids their use for massive datasets.
- *Nested lists* are lists of lists or lists of dictionaries.

  Matrices are an example of nested lists: a list of vectors with uniform length and type implements a matrix in q. See Figure A.8. Nested lists are efficient if the reader guarantees properties about the nested lists.

```
q)(1+ til 4; 2 + til 4; 2* til 4)
1 2 3 4
2 3 4 5
0 2 4 6
```

IGURE A.8
atrix example.

Dictionaries take in two lists: keys and values. Q assumes keys are unique, and the dictionary defines a map from keys to values. See Figure A.9.

```
q)(`a`b`c)!(1 1 3)
a| 1
b| 1
c| 3
q)dic: (`a`b`c)!(1 1 3)
q)dic[`b]
1
```

FIGURE A.9
Dictionary example.

The reader manually promotes an atom to a single element list using the function "enlist". One commonly uses enlist to shape data for use as function arguments.

A.1.3.4 Strings and symbols

Strings are a common data type in programming languages. In q, strings are atomic character lists; therefore, a string list is nested. While the string representation in q allows for more flexibility, an alternative data type performs faster for a frequent but reduced operation set.

Symbols are an enumeration type: q internally encodes them with integers but displays symbols like strings with a backtick prefix. Symbols perform better with assignments and searches, but modifying strings is straightforward. One converts strings to symbols with the '$ operator and symbols to strings with the "string" function, per Figure A.10.

```
q)a: "Hello World!"
q)a
"Hello World!"
q)b:`$a
q)b
`Hello World!
q)string b
"Hello World!"
```

FIGURE A.10
String example.

A.1.3.5 Functions and loops

Q implements loops by wrapping an operation in a function and iterating th function over a list. This pattern is like R's lapply and python's map.

```
/define a function that displays its arg and returns arg+1
f:{[arg]
  show arg;
  :arg+1; / colon acts as the return operator in this context
  };

/test on an atom
f[0]
```

```
/loop over a list
f each 0 1 2

/common iteration pattern
f each til 3
```

A common pattern in q is *projection*, where one projects a function with multiple arguments down to a single argument by *freezing* other arguments.

```
\the following function adds two numbers together
g:{[a;b]
  :a+b;
  };

\f is a projection of g, with the first argument fixed to 1
f: g[1;];

\we can now loop over the second argument
f each til 3

\one can loop g without defining the projection f
g[1; ] each til 3
```

A.1.3.6 Tables are flipped dictionaries of lists

Kdb+ tables behave like sql tables and R and pandas dataframes. To define a table, one uses the syntax:

```
tbl: ([]stock: 'AAPL'GOOG; time: 09:31:01.000 09:35:05.000; price
    : 101 312);
```

Internally, q defines tables as flipped dictionaries. Flip is the kdb+ term for nathematics' transpose operator. The dictionary's keys are the table's column names. The dictionary's values are vectors representing the table's columns. 'or q to recognize a dictionary as a table, the dictionary must contain vectors f identical length: the length corresponds to the table's number of rows. This bservation implements a table programmatically:

```
bl: flip ('stock'time'price)!('AAPL'GOOG;  09:31:01.000
   09:35:05.000; 101 312);
```

.1.4 Setting up a small database

db+ classifies databases as Historical DataBases (HDBs) or Real-time ataBases (RDBs): this section focuses on the former. For convenience, the pendix provides a function to load sample trading data into a small kdb+ tabase. The reader can skip this section if they have access to a kdb+ tabase in their trading infrastructure.

The code below is neither interesting nor relevant to a quantitative rearcher. Download the data from LOBSTER data samples [155]. The level one samples from https://lobsterdata.com/info/DataSamples.php suffice. LOBSTER [154] is "high-frequency, easy-to-use and latest limit order book data" for academic researchers.

```
\the following function takes in the name of a stock, loads the
    corresponding csvs, and formats them into a table
readCsv:{[stockName]
  \msgFile is the file capturing all the messages on the public
    tape
  msgFile: `$ (string stockName), "_2012-06-21
    _34200000_57600000_message_1.csv";
  \the function 0: reads a csv file. See https://code.kx.com/q/
    ref/file-text/#load-csv
  msg:  0:[("fjjjjj"; ","); msgFile];
  \we create the table as a dictionary of vectors
  msg: (`time`eventType`orderId`size`price`direction)!msg;

  \lobFile is the file capturing the best bid and ask at each
    message time
  lobFile: `$ (string stockName), "_2012-06-21
    _34200000_57600000_orderbook_1.csv";
  lob:  0:[("jjjj"; ","); lobFile];
  lob:  (`ask`askVolume`bid`bidVolume)!lob;

  \we join the two dictionaries together, then flip to get our
    joined table
  tbl: flip msg, lob;
  \some minor reformatting of the columns to have the right type
    and units
  tbl: update stock: stockName, time: `time$time*1000, price:
    price%10000, ask: ask%10000, bid: bid%10000 from tbl;
  :tbl;
  };

\we run the function on each stock, then merge all the results
    with raze. See https://code.kx.com/q/ref/raze/
tbl: raze readCsv each `AAPL`AMZN`GOOG`INTC`MSFT;

\finally, we store the table as a splayed database. See https://
    code.kx.com/q/kb/splayed-tables/
tbl: .Q.en[`:/q/tbl/] tbl;
`:/q/tbl/ set tbl;
```

Advanced kdb+ trading infrastructures employ deeper features to boost an hdb's performance. Examples include:

- *Partitioning:* the reader finds the Kx documentation on the web page

  https://code.kx.com/q/kb/partition/.

  Partitions store data sections separately. For instance, a frequent partition is the date partition, which improves queries that filter or group by date.

- *Splaying:* the reader finds the Kx documentation on the web page

  https://code.kx.com/q/kb/splayed-tables/.

A splayed table stores columns separately. When a user queries a table q only loads the required columns into memory.

- *Compression:* the reader finds the Kx documentation on the web page

  https://code.kx.com/q/kb/file-compression/.

  Compression reduces an hdb's footprint on disk. However, compression introduces a trade-off: either write or read time increases.

- *Attributes:* the reader finds the Kx documentation on the web page

  https://code.kx.com/q/ref/set-attribute/.

  Attributes guarantee data properties. Examples include the sorted and unique attributes guaranteeing sorted and unique column values. By ensuring data properties, attributes significantly speed up functions and operations.

Quantitative researchers and traders assume their trading infrastructure is well set up and do not need to consider these properties, except for partitions. Remark A.1.2 explains partitions' practical use for researchers and traders.

Remark A.1.2 (Query date-partitioned tables with care)**.** *Data often exceeds a machine's RAM for date-partitioned tables; therefore, one filters tables with a date range clause. Furthermore, this "where date within" clause should be the query's* first *where clause to minimize query run time.*

A.2 A Cheat-Sheet for Quantitative Trading

A.2.1 Data wrangling in kdb+

A quantitative researcher's iteration loop breaks down into three high-level steps:

(a) Normalize data.

(b) Define a model.

(c) Run analytics.

Ideally, step (a) is a one-time effort: most research iterations focus on testing model specifications in (b) and evaluating their merits in (c). Furthermore, manipulating large tables is a core skill for all three steps. Because dealing with massive data is challenging, practitioners refer to the skill colloquially as *data wrangling*.

A.2.1.1 Qsql queries

Queries manipulate data. The reader finds the Kx documentation on the web page

https://code.kx.com/q/basics/qsql/.

Three building blocks structure query syntax and evaluate in order:

(a) The where clause.

The where clause filters the table rows to satisfy the where clause. Where clauses chain with commas, interpreted as "and" and evaluated left to right. Q evaluates the left-most where clause first: it drives the filter's run time.

(b) The by clause.

After filtering the table, q groups data using the by clause. Then, q computes all future operations, for example, averages or cumulative sums, separately by groups. If the reader specifies no by clause, q considers the data as one group.

(c) The action statement.

Action is one of "select", "exec", "update", and "delete", followed by functions one wishes to apply to table columns. For quantitative research, "select" and "update" are frequent action statements. Select keeps only selected rows and columns, deleting the rest. Update updates selected rows and columns, keeping the remaining data untouched.

A significant pattern is the behavior of "select" and "update" with by clauses.

- A select statement *with a by clause* should only use aggregation functions that aggregate the group into one data point: with a by clause, select returns *a single entry* per group. Therefore, *select reduces the table rows* when grouping. See Figure A.11.

```
q)select avg size by stock from tbl where time < 10:00:00
stock| size
-----| --------
AAPL | 87.62302
AMZN | 86.2926
GOOG | 97.90174
INTC | 372.7082
MSFT | 408.5571
```

FIGURE A.11
Select statement example.

- An update statement guarantees the table row count remains unchanged *even with a by clause:* an update can accept functions that do not aggregate data. However, update statements also accept aggregation functions: it simply repeats results across a group's rows, guaranteeing the table row count remains unchanged. See Figure A.12.

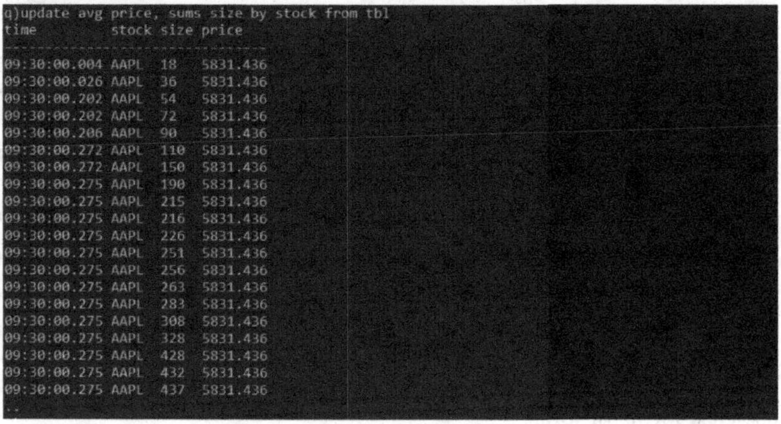

```
q)update avg price, sums size by stock from tbl
time         stock size price
------------------------------
09:30:00.004 AAPL  18   5831.436
09:30:00.026 AAPL  36   5831.436
09:30:00.202 AAPL  54   5831.436
09:30:00.202 AAPL  72   5831.436
09:30:00.206 AAPL  90   5831.436
09:30:00.272 AAPL  110  5831.436
09:30:00.272 AAPL  150  5831.436
09:30:00.275 AAPL  190  5831.436
09:30:00.275 AAPL  215  5831.436
09:30:00.275 AAPL  216  5831.436
09:30:00.275 AAPL  226  5831.436
09:30:00.275 AAPL  251  5831.436
09:30:00.275 AAPL  256  5831.436
09:30:00.275 AAPL  263  5831.436
09:30:00.275 AAPL  283  5831.436
09:30:00.275 AAPL  308  5831.436
09:30:00.275 AAPL  328  5831.436
09:30:00.275 AAPL  428  5831.436
09:30:00.275 AAPL  432  5831.436
09:30:00.275 AAPL  437  5831.436
..
```

FIGURE A.12
Update statement example.

A.2.1.2 Joins

Joining two tables is an essential data analysis operation. Indeed, quantitative researchers focus on four joins, ranging from the straightforward to the advanced:

a) Comma join (,).

The comma operator appends two conforming data structures. For tables, if tbl1 and tbl2 share columns and data types,

```
tbl1, tbl2
```

appends tbl2 rows after tbl1 rows.

The following pattern joins the columns of two tables. Tbl1 and tbl2 must have the same number of rows.

```
flip (flip tbl1), (flip tbl2)
```

) Left join (lj): the reader finds the Kx documentation on the web page

https://code.kx.com/q/ref/lj/.

Left join merges two tables based on shared columns, called keys. Rows from the two tables with matching keys merge into one row. The syntax for joining tables tbl1 and tbl2 based on keys col1, col2 is:

```
tbl1 lj `col1`col2 xkey tbl2
```

As of join (aj): the reader finds the Kx documentation on the web page

https://code.kx.com/q/ref/aj/.

Unfortunately, time series data is frequently *asynchronous.* For example, let tbl1 represent trades on AAPL and tbl2 trades on GOOG. Trades on AAPL and GOOG are unlikely to happen exactly simultaneously. Therefore, using time as a key variable leads to few matches when using precise timestamps. Few matches may tempt users to round down timestamps to "get more matches". However, modifying timestamps is a *bad practice:* it severely misrepresents the data and leads to considerable look-ahead bias.

Instead, a model requires the value of tbl2 as-of the latest time on tbl1. This approach assumes that each value in tbl1 *persists in time* until the next row and makes the closest previous timestamp valid for joining. As-of joins are ubiquitous in trading, and q's implementation boasts performance unmatched by other high-level languages. Section A.4 provides an example use case for as-of joins.

The most potent property of q's as-of join is that users can perform them *with tbl2 remaining on disk.* The frequent use-case is for trades and quotes: a given day's S&P 500 trades are about a million rows, tiny by kdb+ standards, and easily fit in memory. The corresponding quotes are *multiple* orders of magnitude larger. Therefore, loading quotes in memory incurs a considerable upfront cost or is impossible in many cases. But practically, users do not care for every quote. Depending on the use case, the last prevailing quote before, after, or with a given offset from each trade time is sufficient. With on-disk aj, this data enrichment runs *without loading the quote table in memory.* Other solutions not leveraging data locality incur a severe performance penalty, as they load and transfer the massive quotes data before joining. Non-local solutions promptly discard most quotes after their transfer, highlighting inefficient memory use.

(d) Window join (wj): the reader finds the Kx documentation on the web page

https://code.kx.com/q/ref/wj/.

Window joins generalize as-of joins. However, the reader should use window joins as a last resort: map-reduce and as-of joins solve most use case more efficiently. For example, Novotny et al. (2019)[178] cover q's ma reduce patterns in Chapter 8 "Parallelisation" (p. 151): q's data loca ity principle effortlessly enables map-reduce methodology. Nevertheles traders frequently use window joins, and the reader should learn to rea window-join statements.

A.2.1.3 Generalizing qsql

Qsql queries are intuitive and expressive: queries excel at shaping and nc malizing data with little code. However, a limitation of qsql queries is t case where the reader only has programmatic access to column names. F example, one aims to generalize the query

```
select avg ret1, avg ret2, avg ret3 from tbl
```

The pattern is clear, but qsql does not provide grammar to generalize the query to arbitrary "retx". Q provides a solution: *functional qsql*. The reader finds the Kx documentation on the web pages

https://code.kx.com/q/basics/funsql/

https://code.kx.com/q/wp/parse-trees/.

Functional qsql generalizes queries to broad use cases: the topic is vast. Therefore, this section outlines a method for generalizing qsql queries. The reader adapts the solution to arbitrary use cases. The process has three steps:

(a) Write down an example qsql query to generalize:

```
select avg ret1, avg ret2, avg ret3 from tbl
```

(b) Use the function "parse" to translate the query into raw q:

```
parse "select avg ret1, avg ret2, avg ret3 from tbl"

?[tbl; (); 0b; (`ret1`ret2`ret3)!((avg; `ret1); (avg; `ret2);
    (avg; `ret3))]
```

(c) Pattern-match the example to generalize the query:

```
colNames: `${[x] "ret", string x} each til n;
formulas: {[x] (avg; x)} each colNames;
?[tbl; (); 0b; colNames!formulas]
```

A.2.2 Long or wide format?

As an alternative solution to the previous example, one can store data in a different shape. Instead of storing returns in separate columns, the reader holds the same information with only two columns: return name and value. Practitioners refer to this shape as *long format* data: the format increases the table length. They refer to the original shape as *wide format* data: the format increases the number of columns.

For a long format table, a regular qsql query replaces the functional qsql query: furthermore, this solution automatically scales to any number of return names.

```
select avg retValue by retName from longTbl
```

Conversely, a wide format lends itself to matrix operations. The following pattern turns a wide format table into a matrix:[2]

```
value exec ret1, ret2, ret3 from widetbl
```

[2]The reader studies the web page
https://code.kx.com/q/ml/toolkit/fresh/
further machine learning and matrix patterns in q.

Matrix operations, such as the "mmu" implementation of matrix multiplication, are potent and essential to implementing machine learning algorithms. The reader finds the Kx documentation on the web page

https://code.kx.com/q/ref/mmu/.

Furthermore, q's vector operations work on vectors of matrices. For example,

```
ema[0.5; matVec]
```

computes the element-wise exponential moving average of a vector of matrices. Finally, vector operations applied to a single matrix perform the operation column-wise.

Pivoting tables between wide and long formats incurs a one-time cost proportional to their size: the data structures are different. The Kx website discusses pivoting at

https://code.kx.com/q/kb/pivoting-tables/.

Switching between a wide table and a matrix is instantaneous: the data formats are internally consistent.

A.2.3 Vectorized operations and parallelism in kdb+

As discussed previously, two main performance drivers in kdb+ are its vector-centric approach and its focus on data locality. Other languages, such as python, use parallelism to speed up computations. Furthermore, q also leverages parallelism, including large-scale distribution: one may even argue that q makes it easier than other languages. This statement stems from three patterns. All three require awareness of how the infrastructure team sets up the underlying hardware. While the hardware setup is out of scope, the patterns enabled by a full setup are part of quantitative researchers' toolbox.

(a) *Implicit native parallelism.*

Kx introduced native parallelism in 2020 as per its website [134]:

> "Kdb+ 4.0 adds an additional level of multithreading via primitives. It is fully transparent to the user, requiring no code change by the user to exploit it."

When enabling sub-threads, primitive functions implicitly and automat cally parallelize computations across available sub-threads. For examp the statement "select f[x] by stock from tbl" automatically parallelizes t function f across stocks without any input from the user.

(b) *Explicit parallelism across sub-threads.*

The reader finds the Kx documentation on the web page

https://code.kx.com/q/ref/peach/.

Peach explicitly controls parallelism: peach parallelizes a loop across sub-threads using the standard looping pattern each.

(c) *Distribution across sub-processes.*

The reader finds the Kx documentation on the web page

https://code.kx.com/q/basics/peach/.

Peach also distributes loops across sub-processes: this pattern requires a one-time setup by the trading infrastructure team. The advantage is that sub-processes do not need to live on a single machine. Therefore, using peach, the user automatically distributes a loop across multiple devices and the data locally present on them.

A.3 An Efficient Implementation of the Generalized OW Model

This section computes price impact states from trade data using the generalized OW model introduced in Chapter 2. Assume given a set of parameters for a generalized OW model and a table capturing all trading fills.

A.3.1 Key mathematical idea

The naive computation of a propagator model with a long fill history is prohibitively expensive. Indeed, for each fill, one computes price impact using a vast history, leading to quadratic computation costs. The OW model is a special case: one computes it using a simple recursive formula, leading to linear costs. For equidistant times $t_i = i\Delta t$, define the OW model with decay parameter β and push λ as

$$I_{i+1} = \rho \cdot I_i + \lambda \Delta_{i+1} Q$$

where $\rho = (1 - \beta\Delta t)$. The recursive formula generalizes to arbitrary event times t_i, if the event times are a super-set of the fills:

$$I_{t_{i+1}} = \rho^{t_{i+1}-t_i} \cdot I_{t_i} + \lambda \Delta_{i+1} Q. \tag{A.1}$$

For a generalized OW model, β and λ are not constants. Therefore, one generalizes the recursion formula A.1

$$I_{t_{i+1}} = \frac{\rho_{t_{i+1}}}{\rho_{t_i}} I_{t_i} + \lambda_t \Delta_{i+1} Q \tag{A.2}$$

where one defines ρ_t as

$$\rho_t = e^{-\int_0^t \beta_s ds}.$$

More practically, a quantitative researcher specifies ρ_t and retrieves β_t using

$$\beta_t = \Delta_t \log \rho.$$

A.3.2 Key algorithmic idea

Python and R have efficient exponential kernel implementations for equidistant event times via their exponential moving average (ema) functions. Q also boasts an ema function that natively optimizes the exponential kernel computation.

However, these pre-built functions do not generalize to arbitrary event times or non-constant parameters. This lacking functionality leads quantitative researchers to force a grid on their data or make other trade-offs. For example, while a grid structure allows high-performance pre-built ema functions, a grid reduces accuracy when events do not fall exactly onto the grid.

Q has a pattern that allows the native implementation of general recursive formulas. Novotny et al. (2019)[178] highlight this potent pattern in Section 3.2.5.1 "EMA: The Exponential Moving Average" (p. 69). The reader finds the Kx documentation on the web page

https://code.kx.com/q/ref/accumulators/.

If x is a scalar, y is either a function or a vector, and z is a vector, then the pattern `{x y\z}` implements a recursive formula:

- If y_t is a vector, the formula reads

$$x_{t+1} = y_{t+1} \cdot x_t + z_{t+1}.$$

- If y is a function only using operations mapped to the c-level (e.g., a[illegible] mathematical functions), the formula reads

$$x_{t+1} = y\left(x_t, z_{t+1}\right).$$

The initial condition for both formulas is $x_0 = x$.

To implement A.2, one sets

$$y_{t_{i+1}} = \frac{\rho_{t_{i+1}}}{\rho_{t_i}}$$

and

$$z_{t_{i+1}} = \lambda_{t_i} \Delta_{t_{i+1}} Q.$$

The recursion implements the generalized OW model for arbitrary event tim

A.3.3 Computing impact states

The two ideas combine to provide a straightforward, high-performance algorithm for computing the price impact caused by a set of fills. Assume given a fill table "trdTbl" with columns:

- 'time

  Time is the timestamp at which the fill event occurred.

- 'volume

  Volume is the signed volume filled.

- 'rho

  Section A.3.1 defines rho. Rho captures the generalized OW model's impact decay.

- 'lambda

  Lambda is the fill's push factor.

Then the following code computes the generalized OW model's impact state:

```
update impact: {[x;y;z] x y\z}[0; rho%prev rho; lambda*volume]
   from tradeTbl
```

If the reader computes impact states for timestamps other than fill times, for example, on a regular time grid, there are two options:

a) Add synthetic fill rows with *zero volume* to tradeTbl.

```
/newTbl is a table with the same columns as tradeTbl, but zero
    volume.
tbl: 'time xasc tradeTbl, newTbl;
\the table must remain  sorted by time for the recursion to
    make sense
update impact: {[x;y;z] x y\z}[0; rho%prev rho; lambda*volume]
    from tbl
```

b) A higher performance solution when newTbl is significantly larger than tradeTbl, or if the user previously computed tradeTbl, is to use an as-of join, then, decay the corresponding impact values.

```
impactTbl: update impact: {[x;y;z] x y\z}[0; rho%prev rho;
    lambda*volume] from tradeTbl;
impactTbl: select time, impact, oldRho: rho from impactTbl;
tbl: aj['time; newTbl; impactTbl];
tbl: update impact: impact*rho%oldRho from tbl;
```

This solution relies on the as-of time guaranteeing no fills occur since the as-of join time. Without fills in the interval $(s, t]$, the impact state decays according to the formula:

$$I_t = \frac{\rho_t}{\rho_s} I_s.$$

A.4 An Efficient Implementation of TCA

TCA takes orders and enriches them with returns and impact states at times of interest. The reader finds the business case for TCA in Chapter 1, Section 1.3.1. The times of interest include

- the start time and end time of the order,
- each individual fill time, and
- fixed offsets of the above times, for example, one second after the start time.

From a code perspective, the main task is to compute returns and impact states at arbitrary time intervals. For simplicity, assume returns are log-returns, making them additive. The user input is a table of interval start and end times, named "intervalTbl" and the trade and market data. Because the user will average returns and impact columns by order groups, the enrichment must preserve grouping variables.

The output is an enriched table with returns over the intervals and the impact state at the interval start and end times.

```
\example schema for intervalTbl
intervalTbl: ([] startTime: 09:31:00 09 31:00; endTime: 09:35:01
    09:40; weight: 1 1);

\expected schema for the outputTbl
outputTbl: ([] startTime: 09:31:00 09 31:00; endTime: 09:35:01
    09:40; weight: 1 1; logRet: 0.0015 0.0075; impactStart:
    -0.0005 -0.0005; impactEnd: 0.0010 0.0015);
```

A.4.1 Key algorithmic idea

Given the input table "intervalTbl", the task is a join with a table of return and impact states "stockTbl".

```
\minimal schema for stockTbl
intervalTbl: ([] time: 09:30:01 09:30:02; logRet: 0.0001 -0.0002;
    impactState: 0.0 0.00005);
```

The naive solution uses a window join. This section leverages as-of joi instead: while this approach requires additional lines of code, the flexibili and performance are worth the effort. Indeed,

- A window join traverses *every entry of stockTbl between the interval st and end times.*
- Two as-of joins directly locate the entries of stockTbl at the start and e times.

Therefore, computing returns or impact states via a window-join scales poorly with stockTbl's size. Quantitative researchers employ approximations to reduce the entries to traverse. Furthermore, these return aggregations include redundant computations.

If, instead of computing interval returns from stockTbl's individual entries, one has access to cumulative log-returns "startSumRet" and "endSumRet" at each interval's start and end times, the return computation becomes instantaneous.

```
logRet: endSumRet - startSumRet
```

A.4.2 Computing TCA returns

The previous observation points to a three-step procedure. The first step incurs a one-time cost: cumulative returns can be stored in the database and re-used for multiple TCA reports. Therefore, steps (b) and (c) drive performance.

```
\ Step (a): compute cumulative returns from stockTbl. This
    computation is independent from intervalTbl: it is a one-time
    cost.
stockTbl: update SumRet: sums logRet from stockTbl;

\ Step (b): Apply two as-of joins to enrich cumulative returns
    and impact states against the interval start and end times
tmpTbl: select startTime: time; startSumRet: SumRet from stockTbl
    ;
intervalTbl: aj[`startTime; intervalTbl; tmpTbl];

tmpTbl: select endTime: time; endSumRet: SumRet from stockTbl;
intervalTbl: aj[`endTime; intervalTbl; tmpTbl];

 Step (c) Finalize log return computation using the start and
    end states
 utputTbl: update logRet: endSumRet-startSumRet from intervalTbl;
```

The reader can join stockTbl from disk to improve performance further. 'he reader finds the Kx documentation on the web page

https://code.kx.com/q/ref/aj/

nder Section "Performance". The on-disk solution does not load stockTbl in emory, leading to massive performance gains. Practically, one injects tmpTbl rectly in the as-of join function for an on-disk solution.

```
 Step (a): compute cumulative returns from stockTbl. This
   computation is independent from intervalTbl: it is a one-time
   cost.
ockTbl: update SumRet: sums logRet from stockTbl;

Step (b): Apply two as-of joins to enrich cumulative returns
   and impact states against the interval start and end times
tervalTbl: aj[`startTime; intervalTbl; select startTime: time;
   startSumRet: SumRet from stockTbl];
```

```
intervalTbl: aj[`endTime; intervalTbl; select endTime: time;
    endSumRet: SumRet from stockTbl];

\ Step (c) Finalize log return computation using the start and
    end states
outputTbl: update logRet: endSumRet-startSumRet from intervalTbl;
```

B

Functional Convergence Theorems for Microstructure

B.1 Or How to Deal with Local Non-Linearities in Microstructure

Two results of Jacod and Protter (2012)[122] apply directly to the reduced form modeling of market microstructure.

(a) The *functional law of large numbers* estimates non-linear instantaneous trading costs. Section B.2 covers the result, and Section 2.4.2 of Chapter 2 applies it to the self-financing portfolio equation.

b) The *functional central limit theorem* computes the limit of locally non-linear processes, such as price impact. Section B.3 covers the result, and Section 2.6.3 of Chapter 2 applies it to Bouchaud's locally concave propagator model.

These results derive reduced-form models from microstructure assump-ions. Reduced-form models are significantly more tractable than their discrete ounterparts. By deriving reduced-form models from the limits of microstruc-ure models, one mathematically maps discrete assumptions to continuous-ime models. The maps remove the *guesswork* when using continuous-time quations for microstructure phenomena.

The following two notations simplify the results' formulations.

otation B.1.1 (Time discretization). *A fixed integer N discretizes time. et*

$$t_n^N = \frac{n}{N}T = n\Delta t^N$$

d

$$X_n^N = X_{t_n^N}$$

a time interval $[0, T]$ and a stochastic process X. Similarly, define

$$\Delta_n X^N = X_n^N - X_{n-1}^N.$$

I: 10.1201/9781003316923-B

Notation B.1.2 (Gaussian kernel). *Mathematicians frequently encounter a random function F_t and random process σ_t and express Gaussian integrals*

$$\int_{-\infty}^{\infty} F_t(x)\phi_{\sigma_t}(x)dx$$

where ϕ_σ is the density function for a Gaussian with mean zero and variance σ^2.

One extends the probability space to include an independent Gaussian variable Z with mean zero and variance one. To trim down the notation for Gaussian kernels, define $\tilde{E}$ as the expectation operator that only *averages over Z:*

$$\tilde{\mathbb{E}}\left[F_t(\sigma_t Z)\right] = \int_{-\infty}^{\infty} F_t(x)\phi_{\sigma_t}(x)dx.$$

B.2 Functional Law of Large Numbers

Theorem 7.2.2(b) on page 217 of Jacod and Protter (2012)[122] covers the functional law of large numbers. This section simplifies the theorem for the use cases outlined in Chapter 2.

Theorem B.2.1 (Theorem 7.2.2(b), p. 217 [122]). *Assume given an Itô process X with the dynamics*

$$dX_t = b_t dt + \sigma_t dW_t$$

where b_t is locally bounded, and σ_t is càd-làg and uniformly bounded away from zero.

Let F be a measurable function on $(\Omega \times \mathbb{R}_+ \times \mathbb{R}, \mathcal{F} \times \mathcal{B}(\mathbb{R}_+) \times \mathcal{B}(\mathbb{R}))$ such that

(a) for almost all ω, the function $(t, x) \mapsto F(\omega, t, x)$ is Lebesgue-almost everywhere continuous on $\mathbb{R}_+ \times \mathbb{R}$ (7.2.1, p. 217, [122]).

(b) the function F satisfies the uniform growth condition

$$\forall(\omega, t, x) \in (\Omega \times \mathbb{R}_+ \times \mathbb{R}), \quad |F(\omega, t, x)| \leq C\left(1 + x^2\right)$$

for a constant C (7.2.2 condition (b), p. 217, [122]).

Then, the discrete process

$$\Delta t^N \sum_{n=1}^{\lceil Nt \rceil} F_{t_n^N}\left(\frac{\Delta_n X^N}{\sqrt{\Delta t^N}}\right)$$

converges u.c.p. to the continuous-time process

$$\int_0^t \tilde{E}\left[F_s(\sigma_s Z)\right] ds.$$

Proof. F satisfies assumption 7.2.1 and condition (b) from Theorem 7.2.2, on page 217 of Jacod and Protter (2012)[122]. Hence, Theorem B.2.1 directly applies Jacod and Protter's theorem. □

B.3 Functional Central Limit Theorem

Theorem 10.3.2(a) on page 285 of Jacod and Protter (2012)[122] covers the functional central limit theorem. This section simplifies the theorem for the use cases outlined in Chapter 2.

Assumption B.3.1 (Assumptions on the process). *Assume given a process X with the dynamics (Assumption (K) with the jumps removed, p. 284, [122]):*

$$dX_t = b_t dt + \sigma_t dW_t$$

where b_t is locally bounded, and σ_t is càd-làg (Assumption (H), p. 273, [122]). Furthermore, assume the volatility process is a semimartingale

$$d\sigma_t = \tilde{b}_t dt + \tilde{\sigma}_t dW_t + dM_t$$

for a local martingale M independent *of W, with quadratic variation $[M,M]_t = \int_0^t a_s ds$. Assume $a, \tilde{b}, \tilde{\sigma}$ are progressively measurable and locally bounded.*

Assumption B.3.2 (Assumptions on the local non-linearity). *Define $\bar{F} : [0,T] \times \mathbb{R} \times \mathbb{R} \to \mathbb{R}$ as a general function of time, current states, and future state changes. Assume the function $\bar{F}$ satisfies the following conditions for a function g with at most polynomial growth and a constant $\gamma > \frac{1}{2}$:*

a) $\bar{F}$ is odd in the future changes,

$$\forall (t,y,x) \in [0,T] \times \mathbb{R} \times \mathbb{R}, \quad \bar{F}(t,y,x) = -\bar{F}(t,y,-x).$$

b) $\bar{F}$ is C^1 in the future changes (10.3.2, p. 283, [122]),

$$\forall (t,y) \in [0,T] \times \mathbb{R}, \quad x \mapsto \bar{F}(t,y,x) \text{ is } C^1.$$

c) $\bar{F}$ has at most polynomial growth in future changes (10.3.3, p. 284, [122]),

$$\forall (t,y,x) \in [0,T] \times \mathbb{R} \times \mathbb{R}, \left|\bar{F}(t,y,x)\right| \le g(x).$$

the derivative of $\bar{F}$ has at most polynomial growth in future changes (10.3.4, p. 284, [122]),

$$\forall (t,y,x) \in [0,T] \times \mathbb{R} \times \mathbb{R}, \left|\frac{\partial \bar{F}(t,y,x)}{\partial x}\right| \le g(x).$$

(e) one has the local estimate (10.3.7, p. 284, [122])

$$\forall (t,s,y,x) \in [0,T] \times [0,T] \times \mathbb{R} \times \mathbb{R}, \left|\bar{F}(t,y,x) - \bar{F}(s,y,x)\right| \leq g(x)|t-s|^{\gamma}.$$

Theorem B.3.3 (Theorem 10.3.2(a), p. 285 [122]). *Under assumptions B.3.1 and B.3.2, the discrete process*

$$\sqrt{\Delta t^N} \sum_{n=1}^{\lceil Nt \rceil} \bar{F}\left(t_{n-1}^N, X_{n-1}^N, \frac{\Delta_n X^N}{\sqrt{\Delta t^N}}\right)$$

converges in law stably to

$$\int_0^t \left(b_s \tilde{\mathbb{E}}\left[\partial_x \bar{F}(s, X_s, \sigma_s Z)\right] + \frac{\tilde{\sigma}_s}{2} \tilde{\mathbb{E}}\left[\partial_x^2 (Z^2 - 1) \bar{F}(s, X_s, \sigma_s Z)\right]\right) ds \\ + \int_0^t \sqrt{\tilde{\mathbb{E}}\left[\bar{F}^2(s, X_s, \sigma_s Z)\right]} dW'_s$$

on a very good filtered extension of the initial probability space. The quadratic covariation between W and W' is

$$[W, W']_t = \int_0^t \frac{\tilde{\mathbb{E}}\left[Z \bar{F}(s, X_s, \sigma_s Z)\right]}{\sqrt{\tilde{\mathbb{E}}\left[\bar{F}^2(s, X_s, \sigma_s Z)\right]}} ds.$$

Proof. Theorem 10.3.2 from Jacod and Protter (2012)[122] applies with hypothesis (a). One summarizes the list of satisfied assumptions: (10.3.2) (10.3.3), (10.3.4), (10.3.7), $q = q'$, (K) and (H) from Jacod and Protter (2012)[122]. In Jacod and Protter's notation (2012)[122], equation (10.3.1 on page 283 yields

$$\bar{V}'^N = \sqrt{\Delta t^N} \sum_{i=1}^{\lceil Nt \rceil} \bar{F}\left(t_{i-1}^N, X_{i-1}^N, \frac{\Delta_i X^N}{\sqrt{\Delta t^N}}\right)$$

as $\bar{F}$ is odd. Therefore,

$$\int \bar{F}(s, X_s, x) \rho_{c_s}(dx) = 0.$$

The limit $\bar{V}'(\bar{F}(X), X)$ from Theorem 10.3.2 decomposes on p. 286 into fo terms:

$$\bar{V}'(\bar{F}(X), X) = \bar{U}'(\bar{F}(X), X) + \bar{A}(\bar{F}(X), X) + \bar{A}'(\bar{F}(X), X) + \bar{U}'(\bar{F}(X), X$$

The four terms satisfy the following equations.[1]

[1] The reader finds the four terms' definitions on page 286. They rely on notation, functions $\hat{\gamma}, \hat{\gamma}', \bar{\gamma}$ that equation (10.3.8) on page 285 introduces.

(a) Define $\bar{U}'$ as a continuous centered Gaussian process with independent increments satisfying

$$\mathbb{E}\left[\left(\bar{U}'(\bar{F}(X),X)_t\right)^2\middle|\mathcal{F}\right] =$$
$$\int_0^t \tilde{E}\left[\left(\bar{F}(s,X_s,\sigma_s Z) - Z\tilde{E}\left[\bar{F}(s,X_s,\sigma_s Z)\right)^2\right]\right] ds$$
$$- \int_0^t \left(\tilde{E}\left[\bar{F}(s,X_s,\sigma_s Z)\right]\right)^2 ds.$$

Algebraic manipulations and the property that $\bar{F}$ is odd yield

$$\mathbb{E}\left[\left(\bar{U}'(\bar{F}(X),X)_t\right)^2\middle|\mathcal{F}\right] =$$
$$\int_0^t \tilde{E}\left[\bar{F}^2(s,X_s,\sigma_s Z)\right] - \tilde{E}\left[Z\bar{F}(s,X_s,\sigma_s Z)\right]^2 ds.$$

This equation is the part of the term

$$\int_0^t \sqrt{\tilde{\mathbb{E}}\left[\bar{F}^2(s,X_s,\sigma_s Z)\right]}dW_s'$$

in the final formula that is independent of W.

b) Define $\bar{U}$ as

$$\bar{U}(\bar{F}(X),X)_t = \int_0^t \tilde{E}\left[Z\bar{F}\left(s,X_s,\sigma_s Z\right)\right] dW_s.$$

$\bar{U}$ leads to the part of the term

$$\int_0^t \sqrt{\tilde{\mathbb{E}}\left[\bar{F}^2(s,X_s,\sigma_s Z)\right]}dW_s'$$

in the final formula that is dependent on W.

) Define $\bar{A}$ as the drift term

$$\int_0^t b_s\tilde{\mathbb{E}}\left[\partial_x\bar{F}(s,X_s,\sigma_s Z)\right] ds.$$

Define $\bar{A}'$ as the drift term

$$\int_0^t \frac{\tilde{\sigma}_s}{2}\tilde{\mathbb{E}}\left[(Z^2-1)\partial_x\bar{F}(s,X_s,\sigma_s Z)\right] ds.$$

□

Remark B.3.4 (Reiterating Remark (10.3.4) from [122]). *Jacod and Protter generalize the central limit theorem to stochastic functionals:*

> *"One could of course suppose that $\bar{F} = \bar{F}(\omega, t, y, x)$ depends on ω as well, in an adapted way. If the conditions (10.3.4)-(10.3.7) are uniform in ω, Theorem 10.3.2 holds, with the same proof" (p. 287).*

From a microstructure perspective, this extends the result's range to include not just static models with local non-linearities but models with both local non-linearities and stochastic parameters, for instance, a time-changed locally concave price impact model.

Example B.3.5 (Application for an exponential time kernel). *Consider the discretized SDE*

$$\Delta_{n+1} I^N = -\beta I_n^N \Delta t^N + \sqrt{\Delta t^N} g\left(\frac{\Delta_n Q^N}{\sqrt{\Delta t^N}}\right).$$

Theorem B.3.3 does not directly apply to an SDE's solution. However, the memory-less property of the exponential moving average leads to a simple transformation:

$$\begin{aligned} I_n^N &= \sum_{i=1}^{n} \left(1 - \beta \Delta t^N\right)^{n-i} \sqrt{\Delta t^N} g\left(\frac{\Delta_i Q^N}{\sqrt{\Delta t^N}}\right) \\ &= \left(1 - \beta \Delta t^N\right)^{n} \sqrt{\Delta t^N} \sum_{i=1}^{n} \left(1 - \beta \Delta t^N\right)^{-i} g\left(\frac{\Delta_i Q^N}{\sqrt{\Delta t^N}}\right). \end{aligned}$$

A two-step solution follows.

(a) Control the gap between the discrete and continuous-time exponential

$$\left(1 - \beta \Delta t^N\right)^{-i} - e^{\beta t_i^N}.$$

(b) Apply Theorem B.3.3 to obtain the convergence

$$\sqrt{\Delta t^N} \sum_{i=1}^{n} e^{\beta t_i^N} g\left(\frac{\Delta_i Q^N}{\sqrt{\Delta t^N}}\right) \to e^{\beta t} I_t$$

where I is the reduced form price impact model from Proposition 2.6.9 [illegible] Section 2.6.3 of Chapter 2.

Example B.3.6 (Smooth pasting near the origin). *Similarly, Theorem B.3 does not* quite *apply to functions like $x \mapsto sign(x)\sqrt{|x|}$ and $x \mapsto sign(x) \log |x$ [illegible] The lack of differentiability around $x = 0$ is a problem. Thankfully, the [illegible]havior around zero is not empirically significant. For example, consider t[illegible] smooth-pasted version g_ϵ of the square root function.*

$$g_\epsilon(x) = \frac{x\sqrt{|x|}}{|x| + \epsilon} + \frac{\epsilon x}{|x| + \epsilon}$$

Two observations follow.

(a) *Empirically, replacing* $g(x) = sign(x)\sqrt{|x|}$ *with* g_ϵ *for small* ϵ *does not change the out-of-sample* R^2 *of the Bouchaud model from Section 7.3.3 in Chapter 7.*

(b) *Conversely, one applies Theorem B.3.3 to* g_ϵ *and takes the limit as* $\epsilon \to 0$ *to obtain the same reduced form model as in Section 7.3.4 from Chapter 7.*

B.4 Further Readings

Functional convergence results are a rich field of Probability Theory with broad financial applications. For example,

(a) Another reference for the reader interested in extending functional convergence theorems is *Limit Theorems for Stochastic Processes*, by Jacod and Shiryaev (2003)[123].

(b) *High-Frequency Financial Econometrics*, by Aït-Sahalia and Jacod (2014)[4] covers applications of functional convergence results to the *statistical estimation* of microstructure models, for example, high-frequency volatility and correlation parameters. The following paragraph captures the core challenges of adapting econometrics to high-frequency trading data:

> "Two final distinguishing characteristics of high-frequency financial data are important, and call for the development of appropriate econometric methods. First, the time interval separating successive observations can be random, or at least time varying. Second, the observations are subject to *market microstructure noise*, especially as the sampling frequency increases. Market microstructure effects can be either information or non-information related, such as the presence of a bid-ask spread and the corresponding bounces, the differences in trade sizes and the corresponding difference in representativeness of the prices, the different informational content of price changes due to informational asymmetries of traders, the gradual response of prices to a block trade, the strategic component of the order flow, inventory control effects, the discreteness of price changes, data errors, etc. The fact that this form of noise interacts with the sampling frequency raises many new questions, and distinguishes this from the classical measurement error problem in statistics" (p. xviii).

The favored modeling approach, writing down a precise microstructure model and deriving a reduced form model in the continuous-time limit, is also at the core of the seminal work by Gatheral, Jaisson, and Rosenbaum, *Volatility is rough* (2018)[105].

A summary of this fascinating topic follows.

- Jaisson and Rosenbaum (2015, 2016)[124, 125] and Gatheral, Jaisson, and Rosenbaum (2018)[105] propose a microstructure model capturing trading's self-exciting nature using nearly unstable Hawkes processes. They *derive* a reduced-form volatility model from this microstructure and find that the reduced-form model captures all known *macroscopic* facts of the implied volatility surface.

  > "Finally, we provide a quantitative market microstructure-based foundation for our findings, relating the roughness of volatility to high-frequency trading and order splitting" (p. 1, [105]).

- Jusselin and Rosenbaum (2020)[198] extend the microstructure model and prove further links between microstructure and rough volatility models.

  > "Furthermore, we prove that this implies that the macroscopic price is diffusive with rough volatility, with a one-to-one correspondence between the exponent of the price impact function and the Hurst parameter of the volatility" (p. 1).

- Rosenbaum and Tomas (2021)[200] further extend the link by studying the multi-dimensional case. Cross-asset microstructure models yield cross-asset macroscopic behaviors.

  > "In this work, we show that some specific multivariate rough volatility models arise naturally from microstructural properties of the joint dynamics of asset prices. To do so, we use Hawkes processes to build microscopic models that reproduce accurately high-frequency cross-asset interactions and investigate their long term scaling limits. We emphasize the relevance of our approach by providing insights on the role of microscopic features such as momentum and mean-reversion on the multidimensional price formation process. We in particular recover classical properties of high-dimensional stock correlation matrices" (p. 1).

C

Solutions to Exercises

C.1 Solutions to Chapter 2

Answer of exercise 1

Closed-form examples I

1. The SDE's general solution is

$$I_t = e^{-\beta t} I_0 + \lambda \int_0^t e^{-\beta(t-s)} dQ_s.$$

Therefore, for a TWAP order, the impact curve is

$$\begin{aligned} I_t &= \int_0^t \lambda e^{-\beta(t-s)} Q ds \\ &= \frac{\lambda}{\beta} Q(1 - e^{-\beta t}) \end{aligned}$$

s $I_0 = 0$.

Figure C.1 plots an example for I.

2. The expected fundamental P&L is

$$\begin{aligned} \mathbb{E}[Y_t] &= -\mathbb{E}\left[\int_0^t I_s Q ds\right] \\ &= -\frac{\lambda}{\beta} Q^2 \int_0^t (1 - e^{-\beta s}) ds \\ &= -\frac{\lambda}{\beta} Q^2 \left(t + \beta^{-1}\left(e^{-\beta t} - 1\right)\right). \end{aligned}$$

The expected accounting P&L is

$$\begin{aligned} \mathbb{E}[X_t] &= \mathbb{E}[Y_t] + \mathbb{E}[I_t Q_t] \\ &= -\frac{\lambda}{\beta} Q^2 \left((t + \beta^{-1}) e^{-\beta t} - \beta^{-1}\right). \end{aligned}$$

Figure C.1 plots an example for X and Y.

I: 10.1201/9781003316923-C

3. For $t \in (1,2)$, one has

$$\begin{aligned} I_t &= e^{-\beta(t-1)} I_{1+} \\ &= \frac{1}{2} e^{-\beta(t-1)} I_{1-} \\ &= \frac{\lambda}{2\beta} e^{-\beta t} Q(e^{\beta} - 1). \end{aligned}$$

Because no trading occurs, the fundamental P&L remains unchanged.

$$\begin{aligned} \mathbb{E}[Y_t] &= \mathbb{E}[Y_1] \\ &= -\frac{\lambda}{\beta} Q^2 \left(1 + \beta^{-1}\left(e^{-\beta} - 1\right)\right). \end{aligned}$$

The accounting P&L changes as the impact state's effect on the position deflates.

$$\begin{aligned} \mathbb{E}[X_t] &= \mathbb{E}[Y_t + I_t Q_t] \\ &= -\frac{\lambda}{\beta} Q^2 \left(1 + \beta^{-1}\left(e^{-\beta} - 1\right)\right) + \frac{\lambda}{2\beta} Q^2 (e^{-\beta(t-1)} - e^{-\beta t}). \end{aligned}$$

The accounting P&L X can become positive temporarily, as seen in Figure C.1. These instances do not lead to price manipulation: the corresponding profits are illusory. Instead, they dissipate when the accounting P&L converges to the fundamental P&L. Section 4.3.1 of Chapter 4 discusses this phenomenon in detail.

Numerical examples II

1. See Figure C.2.
2. The general solution to I for $t \in (0,1)$ is

$$\begin{aligned} I_t &= e^{-\beta t} I_0 + \int_0^t e^{-\beta(t-s)} \frac{\lambda}{v_s} dQ_s \\ &= e^{-\beta t} I_0 + \lambda \int_0^t e^{-\beta(t-s) - 4\cdot(s-0.7)^2} Q ds. \end{aligned}$$

For $t \in (1,2)$, the solution is

$$I_t = \frac{1}{2} e^{-\beta t} I_{1-}.$$

Figure C.3 plots example numerical solutions for I, X, Y.

Answer of exercise 2

1. One has

$$S_T - S_t = \mu(T-t) + \sigma\left(W_T - W_t\right).$$

It follows that $\alpha_t = \mu(T-t)$.

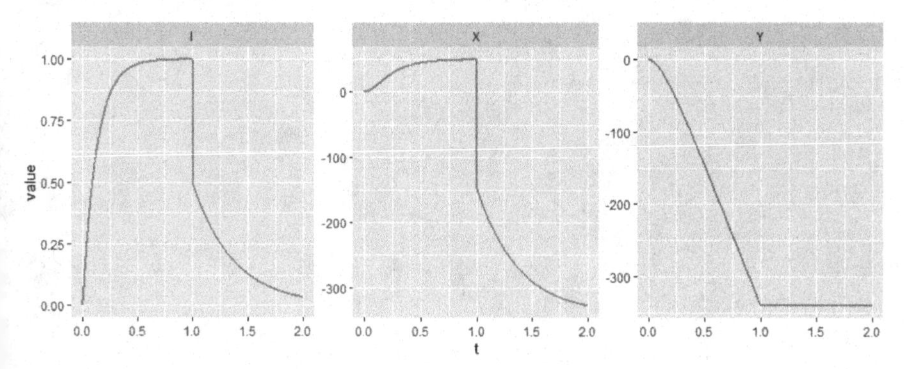

FIGURE C.1
Impact and P&L curves for constant liquidity.

2. Applying Proposition 2.8.3 yields the objective function

$$\mathbb{E}[Y_T] =$$

$$\mathbb{E}\left[\int_0^T \frac{1}{\lambda_t}\left((\beta_t + \gamma_t')\mu(T-t)I_t - (\beta_t + \frac{1}{2}\gamma_t')I_t^2\right)dt + \int_0^T \frac{1}{\lambda_t}I_t\mu dt - \frac{1}{2\lambda_T}I_T^2\right].$$

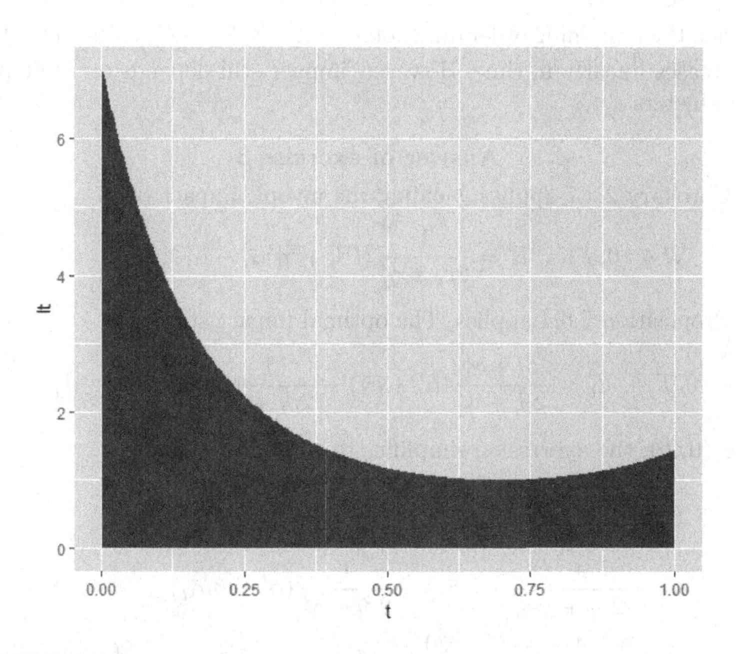

GURE C.2
raday volume profile $v_t = e^{4\cdot(t-0.7)^2}$.

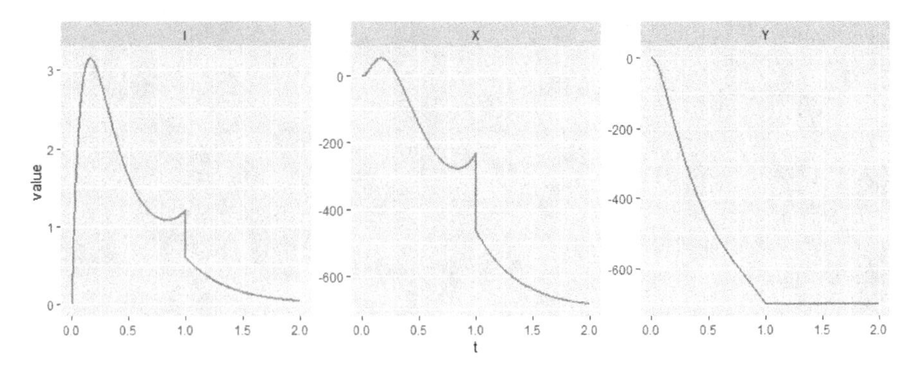

FIGURE C.3
Impact and P&L curves for time-varying liquidity.

3. The objective function in impact space is a pointwise optimization problem. The optimal impact state is

$$\forall t \in (0,T) \quad I_t^* = \frac{\mu}{2\beta_t + \gamma_t'}\left((\beta_t + \gamma_t')(T-t) + 1\right); \quad I_T^* = 0.$$

(a) When the remaining order duration is short, $T - t \ll \frac{1}{\beta_t + \gamma_t'}$, the impact state is proportional to the liquidity time scale $\frac{1}{2\beta_t + \gamma_t'}$.

(b) When the remaining order duration is long, $T - t \gg \frac{1}{\beta_t + \gamma_t'}$, the impact state decreases linearly in time. However, impact still depends on the liquidity parameters β_t, λ_t.

Answer of exercise 3

1. Corollary 2.8.7 applies, yielding the myopic impact state

$$\forall t \in (0,T), \quad I_t^0 = \frac{1}{2\beta_t + \gamma_t'}\left((\beta_t + \gamma_t')\alpha_t - \alpha_t'\right); \quad I_T^0 = 0.$$

2. Proposition 2.6.4 applies. The optimal impact state is

$$\forall t \in (0,T), \quad I_t^1 = \frac{\beta_t + \gamma_t'}{2\beta_t + \gamma_t'}(\alpha_t + \bar{\alpha}_t) - \frac{1}{2\beta_t + \gamma_t'}(\alpha_t' + \bar{\alpha}_t'); \quad I_T^1 = \bar{\alpha}_T.$$

For $t \in (0,T)$, the expression simplifies to

$$\begin{aligned}
I_t^1 &= \frac{\beta_t + \gamma_t'}{2\beta_t + \gamma_t'}(\alpha_t + \bar{\alpha}_t) - \frac{1}{2\beta_t + \gamma_t'}(\alpha_t' + \bar{\alpha}_t') \\
&= \frac{\beta_t + \gamma_t'}{2\beta_t + \gamma_t'}(\alpha_t + \bar{\alpha}_t) - \frac{1}{2\beta_t + \gamma_t'}(\alpha_t' - \beta_t\bar{\alpha}_t) \\
&= \frac{\beta_t + \gamma_t'}{2\beta_t + \gamma_t'}\alpha_t - \frac{1}{2\beta_t + \gamma_t'}\alpha_t' + \frac{\beta_t + \gamma_t'}{2\beta_t + \gamma_t'}\bar{\alpha}_t + \frac{\beta_t}{2\beta_t + \gamma_t'}\bar{\alpha}_t \\
&= I_t^0 + \bar{\alpha}_t.
\end{aligned}$$

3. The equation $I^1 = I^0 + \bar{\alpha}$ yields two implications:

(a) The total impact state is equal in both cases. The chosen point of view shifts the impact of the first order from internal to external.

(b) The corresponding optimal trading processes Q^0 and Q^1 are identical. Indeed,

$$\begin{aligned}
\lambda_t dQ_t^0 &= \beta_t I_t^0 dt + dI_t^0 \\
&= \beta_t(I_t^1 - \bar{\alpha}_t)dt + dI_t^1 - d\bar{\alpha}_t \\
&= \beta_t I_t^1 dt + dI_t^1 \\
&= \lambda_t dQ_t^1
\end{aligned}$$

for $t \in (0, T)$. Manual inspection of the jumps at $t = 0, T$ confirms that they match across both points of view.

Answer of exercise 4

Local square root model under the calendar clock I

1. The liquidity factor λ_t is known, yielding the bound

$$\gamma_t' = -4(t - 0.7) \geq -1.2.$$

Therefore, the lower bound on β guaranteeing no price manipulation is

$$\beta > 0.6.$$

This lower bound corresponds to a maximum half-life for the price impact model of around 0.73 days. Therefore, a price impact model that decays slower than this time scale exhibits price manipulation.

2. One wishes to enter and exit a position during a period where γ' is strongly negative. For example, γ' is negative over the interval $[0.7, 1]$. Consider the following trade pair:

(a) Buy one share at time $t = 0.7$.

(b) Sell one share at time $t = 1$.

The corresponding transaction costs are

$$\begin{aligned}
&I_{0.7-}\Delta_{0.7}Q + I_{1-}\Delta_1 Q + \frac{1}{2}\Delta_{0.7}I\Delta_{0.7}Q + \frac{1}{2}\Delta_1 I\Delta_1 Q \\
&= (0) - \lambda e^{-0.3\cdot 0.01} + \frac{\lambda}{2} + \frac{\lambda}{2}e^{-2(0.3)^2} \\
&= \frac{\lambda}{2}(1 + e^{-0.18} - 2e^{-0.003}) \\
&< 0
\end{aligned}$$

as $1 + e^{-0.18} - 2e^{-0.003} \approx -0.15.$

Local log model under the calendar clock II

1. The liquidity factor λ_t is known, yielding the bound

$$\gamma_t' = -8(t-0.7) \geq -2.4.$$

Therefore, a lower bound on β guaranteeing no price manipulation is

$$\beta > 1.2$$

as $v_t \geq 1$.

2. Consider the following trade pair:

(a) Buy one share at time $t = 0.7$.

(b) Sell one share at time $t = 1$.

For $t \in (0.7, 1)$,

$$\begin{aligned} dI_t &= -\beta e^{4\cdot(t-0.7)^2} I_t dt \\ &> -\beta e^{0.36} I_t dt \\ &> -2\beta I_t dt \end{aligned}$$

and $I_{0.7} = \lambda$. Hence,

$$I_1 > \lambda e^{2\cdot 0.01 \cdot 0.3}$$

The corresponding transaction costs are

$$\begin{aligned} I_{0.7-}\Delta_{0.7}Q &+ I_{1-}\Delta_1 Q + \frac{1}{2}\Delta_{0.7}I\Delta_{0.7}Q + \frac{1}{2}\Delta_1 I\Delta_1 Q \\ &= (0) - I_1 + \frac{\lambda}{2} + \frac{\lambda}{2}e^{-4(0.3)^2} \\ &< \frac{\lambda}{2}(1 + e^{-0.36} - 2e^{-0.006}) \\ &< 0 \end{aligned}$$

as $1 + e^{-0.36} - 2e^{-0.006} \approx -0.29$.

Answer of exercise 5

1. Proposition 2.6.14 applies with $\bar{\alpha} = 0$ and yields

$$\begin{aligned} \mathbb{E}\left[Y_T\right] &= \mathbb{E}\left[-H(J_T) - \int_0^T \beta h(J_t) J_t dt\right] \\ &= \mathbb{E}\left[-\frac{1}{1+c}|J_T|^{1+c} - \int_0^T \beta |J_t|^{1+c} dt\right]. \end{aligned}$$

2. Adding a Lagrange multiplier η leads to the objective function

$$\mathbb{E}\left[\eta J_T - \frac{1}{1+c}|J_T|^{1+c} + \int_0^T \beta\left(\eta J_t - |J_t|^{1+c}\right) dt\right].$$

Therefore, the optimal impact state I_t^* satisfies

$$\forall t \in (0,T), \quad I_t^* = \operatorname{sign}(J_t^*)\,|J_t^*|^c = \frac{\eta}{1+c}; \quad I_T^* = \eta.$$

The constraint yields the equation

$$\begin{aligned}\lambda Q &= J_T + \beta \int_0^T J_t dt \\ &= \eta^{1/c}\left(1 + \frac{\beta T}{(1+c)^{1/c}}\right).\end{aligned}$$

Therefore, one has

$$\eta = (1+c)\left(\frac{\lambda Q}{(1+c)^{1/c} + \beta T}\right)^c.$$

The solution is like the original OW optimal execution problem, which one recovers when $c = 1$. Indeed, the optimal trading strategy targets a constant impact level, exhibits two jumps at times $t = 0, T$, and trades along a TWAP schedule over $(0, T)$.

Global concavity $c < 1$ changes the target impact level

$$\forall t \in (0,T), \quad I_t^* = \left(\frac{\lambda Q}{(1+c)^{1/c} + \beta T}\right)^c \quad I_T^* = (1+c)\left(\frac{\lambda Q}{(1+c)^{1/c} + \beta T}\right)^c.$$

he relationship between I_T and Q is concave.

3. For a TWAP order, $dQ_t = \frac{Q}{T}dt$. Therefore, one has

$$\begin{aligned}J_T &= \int_0^T e^{-\beta(t-s)}\frac{\lambda Q}{T}ds \\ &= \frac{\lambda Q}{\beta T}\left(1 - e^{-\beta T}\right)\end{aligned}$$

d hence,

$$I_T = \left(\frac{\lambda Q}{\beta T}\left(1 - e^{-\beta T}\right)\right)^c.$$

e relationship between I_T and Q is concave.

C.2 Solutions to Chapter 3

Answer of exercise 6

1. The proof is like Proposition 2.6.14 from Chapter 2, with two differences:

(a) There is no external alpha in this model

$$\bar{\alpha} = 0.$$

(b) An additional term

$$\tilde{\alpha} Q_T = \frac{1}{\lambda} \int_0^T \beta \tilde{\alpha} J_t dt + \frac{1}{\lambda} \tilde{\alpha} J_T$$

accounts for the expected P&L contribution of the overnight alpha $\tilde{\alpha}$.

The objective function in impact space is

$$\mathbb{E}[Y_T] = \frac{1}{\lambda} \mathbb{E}\left[\int_0^T \left((\tilde{\alpha} + \alpha_t) \beta J_t - \alpha_t' J_t - \beta |J_t|^{1+c} \right) dt + \tilde{\alpha} J_T - \frac{1}{1+c} |J_T|^{1+c} \right]$$

The problem is myopic in I, and one solves it point-wise:

$$\forall t \in (0,T), \quad I_t^* = \frac{1}{1+c} \left(\tilde{\alpha} + \alpha_t - \beta^{-1} \alpha_t' \right); \quad I_T^* = \tilde{\alpha}.$$

As a sanity check, one recovers the original OW model's solution for $c = 1$.

2. The implied execution problem is a special case of the idealized optima execution problem where the intraday alpha α is absent. The target impac state is

$$\forall t \in (0,T), \quad I_t^* = \frac{1}{1+c} \tilde{\alpha}; \quad I_T^* = \tilde{\alpha}.$$

One relates the overnight alpha $\tilde{\alpha}$ to the terminal position Q_T. Assum that $\tilde{\alpha} > 0$ for simplicity. Then, the order size satisfies

$$\begin{aligned} Q_T &= \frac{1}{\lambda} J_T + \frac{\beta}{\lambda} \int_0^T J_t dt \\ &= \frac{1}{\lambda} \tilde{\alpha}^{1/c} + \frac{\beta T}{\lambda} \left(\frac{\tilde{\alpha}}{1+c} \right)^{1/c} \\ &= \left(1 + \frac{\beta T}{(1+c)^{1/c}} \right) \frac{\tilde{\alpha}^{1/c}}{\lambda}. \end{aligned}$$

Equivalently, one defines the implied alpha $\tilde{\alpha}$ of the target position Q_T as

$$\tilde{\alpha} = (1+c) \left(\frac{\lambda Q_T}{(1+c)^{1/c} + \beta T} \right)^c.$$

3. The simplest prediction stems from the relationship

$$\forall t \in (0,T), \quad I_t^* = \frac{1}{1+c}\tilde{\alpha}; \quad I_T^* = \tilde{\alpha}$$

under the *no intraday alpha* assumptions.

This relationship suggests that the ratio between the target impact state and the overnight or implied alpha should be $\frac{1}{1+c}$. For $c < 1$, this ratio is higher than in the original OW model.

4. By the myopic nature of the control problem in I, the target impact state remains unchanged:

$$\forall t \in (0,T), \quad I_t^* = \frac{1}{1+c}\tilde{\alpha}; \quad I_T^* = \tilde{\alpha}.$$

Therefore, the optimal strategy in trade space is identical for $t \in (0,T]$, and only differs by a jump at time $t = 0$. To determine the jump size, note that

$$J_T - J_0 + \beta \int_0^T J_t dt = \lambda Q.$$

Hence, the modified implied alpha relationship is

$$\alpha = (1+c)\left(\frac{\lambda Q + I_0^{1/c}}{(1+c)^{1/c} + \beta T}\right)^c.$$

Answer of exercise 7

This exercise uses the time-change notation Section 2.5.1 of Chapter 2 ntroduces.

1. An original OW model under the volume clock leads to a model with tochastic decay parameter $\beta_t = \beta v_t$ in the regular clock

$$dI_t = -\beta v_t I_t dt + \lambda dQ_t.$$

he target impact state for the optimal execution problem for $t \in (0,T)$ is

$$I_t^* = \frac{1}{2}\tilde{\alpha}$$

ıere $\tilde{\alpha}$ is the implied alpha of the trade. $dI_t = 0$ yields

$$dQ_t^* = \frac{\beta}{2\lambda}\tilde{\alpha} v_t dt.$$

erefore, the optimal strategy follows a PoV algorithm over $(0,T)$ with a ticipation rate

$$r = \frac{\beta}{2\lambda}\tilde{\alpha}.$$

An alternative proof goes as follows:

(a) The optimal strategy for the original OW model trades at a constant speed for $t \in (0, T)$.

(b) Therefore, the time-changed optimal strategy $\hat{Q}^*$ trades at a constant speed for $t \in (0, \hat{T})$ in the volume clock.

(c) Hence, the optimal strategy Q^* follows a PoV algorithm in the regular clock.

2. First, solve for the optimal time-changed impact state $\bar{I}_t^*$ under the volume clock for $t \in (0, \hat{T})$. The reader finds the proof in exercise 6. Hence, one has

$$\hat{I}_t^* = \frac{1}{1+c}\tilde{\alpha}$$

where $\tilde{\alpha}$ is the trade's implied alpha.

Under the volume clock, the trading process $\hat{Q}^*$ satisfies

$$\begin{aligned} d\hat{Q}_t^* &= \frac{\beta}{\lambda} J_t^* dt + \frac{1}{\lambda} dJ_t^* \\ &= \frac{\beta \tilde{\alpha}^{1/c}}{\lambda (1+c)^{1/c}} dt \end{aligned}$$

for $t \in (0, \hat{T})$.

Under the regular clock, the optimal trading process satisfies

$$dQ_t^* = \frac{\beta \tilde{\alpha}^{1/c}}{\lambda (1+c)^{1/c}} v_t dt$$

for $t \in (0, T)$.

Therefore, the optimal trading strategy follows a PoV algorithm. The participation rate is

$$r = \frac{\beta \tilde{\alpha}^{1/c}}{\lambda (1+c)^{1/c}}.$$

In the globally concave AFS model under the volume clock, the optimal participation rate is a convex function of the trade's implied alpha.

Answer of exercise 8

1. By linearity of the price impact SDE, δI satisfies

$$d\delta I_t = -\beta \delta I_t dt + \lambda d\delta Q_t.$$

By Itô, one has

$$\begin{aligned} d\left(\delta Q_t\right)^2 &= 2\delta Q_t d\delta Q_t + \eta^2 dt \\ &= \left(\eta^2 - 2\gamma (\delta Q_t)^2\right) dt + 2\eta \delta Q_t dW_t. \end{aligned}$$

Hence, the equation

$$\mathbb{E}\left[(\delta Q_t)^2\right] = \frac{\eta^2}{2\gamma}\left(1 - e^{-2\gamma t}\right)$$

holds.

Therefore, in steady state, one has

$$\mathbb{E}\left[(\delta Q_t)^2\right] = \frac{\eta^2}{2\gamma}.$$

Similarly, one has

$$\begin{aligned} d\left(\delta I_t \delta Q_t\right) &= \delta I_t d\delta Q_t + \delta Q_t d\delta I_t + \lambda\eta^2 dt \\ &= -\gamma\delta I_t\delta Q_t dt - \beta\delta I_t\delta Q_t dt - \lambda\gamma\left(\delta Q_t\right)^2 dt \\ &\quad + \lambda\eta^2 dt + (\delta I_t + \lambda\delta Q_t)\eta dW_t. \end{aligned}$$

Hence, in steady state, one has

$$\mathbb{E}\left[\delta I_t \delta Q_t\right] = \frac{\lambda\eta^2}{2(\beta+\gamma)}.$$

2. By Itô, one has

$$\begin{aligned} d\left(\delta I_t\right)^2 &= 2\delta I_t d\delta I_t + \lambda^2\eta^2 dt \\ &= -2\beta\left(\delta I_t\right)^2 dt + 2\lambda\delta I_t d\delta Q_t + \lambda^2\eta^2 dt \\ &= -2\beta\left(\delta I_t\right)^2 dt - 2\lambda\gamma\delta I_t\delta Q_t dt + \lambda^2\eta^2 dt + 2\lambda\eta\delta I_t dW_t. \end{aligned}$$

Therefore, in steady state, one has

$$\begin{aligned} \mathbb{E}\left[(\delta I_t)^2\right] &= \frac{\lambda^2\eta^2}{2\beta}\left(1 - \frac{\gamma}{\beta+\gamma}\right) \\ &= \frac{\lambda^2\eta^2}{2(\beta+\gamma)}. \end{aligned}$$

3. By Corollary 2.5.9 from Chapter 2, the deviation's expected cost is

$$\mathbb{E}\left[Y_T(Q^*)\right] - \mathbb{E}\left[Y_T(Q^* + \delta Q)\right] = \mathbb{E}\left[\int_0^T \frac{\beta}{\lambda}\left(\delta I_t\right)^2 dt + \frac{1}{2\lambda}\left(\delta I_T\right)^2\right]$$

Hence, in steady state, the deviation's expected cost per unit of time is

$$\begin{aligned} \frac{\beta}{\lambda}\mathbb{E}\left[(\delta I_t)^2\right] &= \frac{\beta\lambda\eta^2}{2(\beta+\gamma)} \\ &= \frac{\lambda\eta^2}{2(1+\gamma/\beta)}. \end{aligned}$$

Intuitively, the deviation's expected cost

(a) increases with the size of the push factor λ.

(b) increases with the size of the deviation η.

(c) decreases with the speed of the mean-reversion of the deviation γ relative to β.

Therefore, the tactical deviation must overcome this expected cost to contribute to the execution strategy. The deviation's required profitability depends strongly on its size η and mean-reversion speed γ relative to β.

Answer of exercise 9

1. Assuming no alpha and no trades realize on the second day, the impact state reverts. For a generalized OW model, the reversion occurs according to the equation

$$dI_t = -\beta_t I_t dt.$$

Therefore, the prediction is that the next day's return offsets the previous day's return. The above equation quantifies the second day's expected reversion.

2. The assumption is that there is a positive correlation between the second day's alpha and the first day's overnight alpha. This assumption reduces the second day's expected reversion without trading.

In the OW model, the impact state at the first day's end equals the overnight alpha. Therefore, one computes and empirically compares

(a) the impact state's decrease

$$\begin{aligned} I_T &= e^{-\beta T} I_0 \\ &= e^{-\beta T} \tilde{\alpha}. \end{aligned}$$

(b) the correlated next day alpha

$$\mathbb{E}\left[S_T - S_0\right] = \rho\tilde{\alpha}.$$

3. If a trader submits an order on the second day, the strategy attemp to capture the second day's alpha, considering the elevated impact state a its expected reversion. In that case, instead of exponentially reverting, t impact state settles around the new alpha.

4. If the impact does not revert according to the equation

$$dI_t = -\beta_t I_t dt$$

after the trade ends, then, there are three courses of action:

(a) Confirm that no trading occurs on the second day. One can weaken t assumption to the correlation between the same- and next-day trad being zero across orders.

(b) Confirm there is no autocorrelation in the alpha signal between trading days.

(c) After ruling out (a) and (b), confirm that the decay factor β is correctly calibrated.

Answer of exercise 10

1. For simplicity, assume $Q^i_T - Q^r_T > 0$. Exercise 6 derives the optimal impact state

$$\forall t \in (T, \tau), \quad I^*_t = \frac{1}{1+c}\tilde{\alpha}; \quad I^*_\tau = \tilde{\alpha}$$

where

$$\tilde{\alpha} = (1+c)\left(\frac{\lambda\left(Q^i_T - Q^r_T\right) + I_T^{1/c}}{(1+c)^{1/c} + \beta(\tau - T)}\right)^c.$$

The implied cleanup cost is

$$\begin{aligned}
\int_0^T I_t dQ_t &= \int_T^\tau J^c_t\left(\beta J_t dt + dJ_t\right) \\
&= \int_T^\tau \beta J^{c+1}_t dt + \frac{1}{1+c}J^{1+c}_\tau - \frac{1}{1+c}J^{1+c}_T \\
&= \frac{1+(1+c)^{-1/c}\beta(\tau - T)}{1+c}\left(\frac{\lambda\left(Q^i_T - Q^r_T\right) + I_T^{1/c}}{1+\beta(\tau-T)(1+c)^{-1/c}}\right)^{1+c} \\
&\quad - \frac{1}{1+c}I_T^{1+1/c} \\
&= \frac{\left(\lambda\left(Q^i_T - Q^r_T\right) + I_T^{1/c}\right)^{1+c}}{\left((1+c)^{1/c} + \beta(\tau - T)\right)^c} - \frac{1}{1+c}I_T^{1+1/c}.
\end{aligned}$$

.3 Solutions to Chapter 4

Answer of exercise 11

This simulation is straightforward.

2. For an efficient algorithmic implementation of an exponential moving erage under an arbitrary clock, refer to Section A.3 of Appendix A. The al step is the recursive formula

$$I_{t+1} = \left(1 - \frac{\beta}{N}\right) I_t + \lambda \Delta_{t+1} Q.$$

3. See Figure C.4. Three observations debug a faulty simulation:

(a) For the OW optimal execution strategy, the impact state should be constant while actively trading, with two equal-sized jumps. A deviation from this pattern implies an incorrect implementation of the impact model or the trading strategy.

(b) The accounting P&L should initially increase, due to position inflation, before converging to the fundamental P&L. For the OW model, the position footprint and the transaction costs should cancel while actively trading: accounting P&L is constant in simulation.

(c) The fundamental P&L should always be negative, with the OW model outperforming the slow TWAP and the slow TWAP outperforming the faster TWAP.

4. See Figure C.5. Two observations debug a faulty simulation:

(a) The historical price path should initially be higher than the simulated one, as the faster TWAP pushes the price further.

(b) Once the slower TWAP stops trading, the order should reverse before converging to the same long-term price.

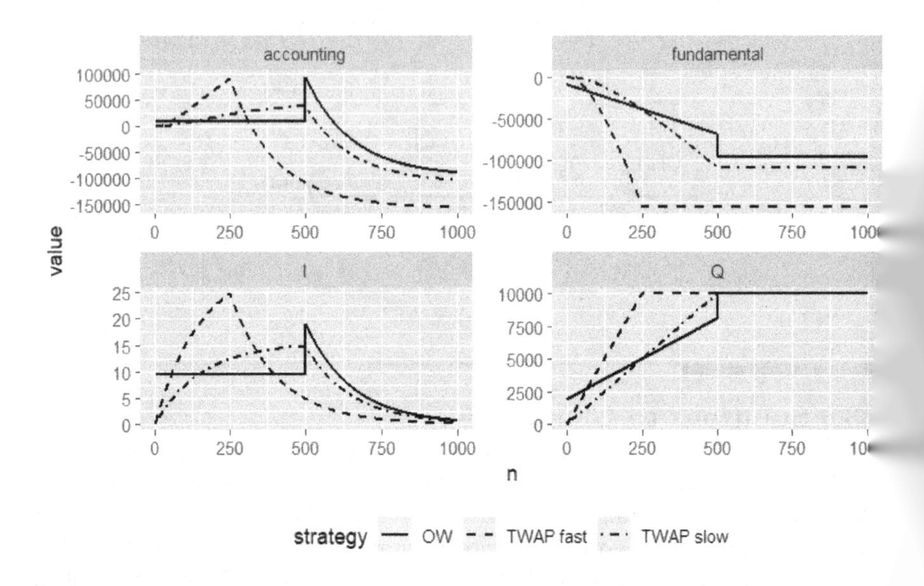

FIGURE C.4
Simulation of trading strategies' price impact and P&L metrics.

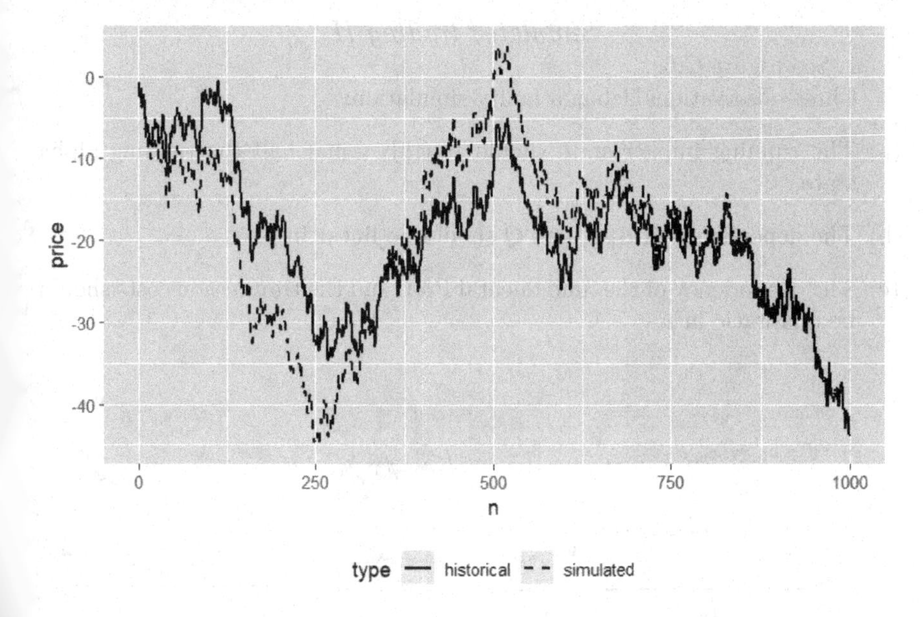

FIGURE C.5
Sample path of an impact-adjusted simulation for the observed price.

Answer of exercise 12

Synthetic alphas I

1. The covariance equals

$$\mathbb{E}\left[(S_T - S_t)\alpha_t\right] = a\mathbb{E}\left[(S_T - S_t)^2\right] = a\sigma^2(T-t).$$

he variances are

$$\mathbb{E}\left[(S_T - S_t)^2\right] = \sigma^2(T-t); \quad \mathbb{E}\left[\alpha_t^2\right] = (a^2+b^2)\sigma^2(T-t).$$

herefore, the correlation is

$$\rho = \frac{a}{\sqrt{a^2+b^2}}$$

2. One wants to satisfy the equations

$$\frac{a}{\sqrt{a^2+b^2}} = 0.05; \quad \frac{a}{a^2+b^2} = 1$$

ere the first equation provides the correlation and the second equation the a between $S_T - S_t$ and α_t. The equations yield

$$a = \rho^2; \quad b = \rho\sqrt{1-\rho^2}.$$

Simulated trading II

1. See Figure C.6.
Three observations debug a faulty simulation:

(a) The running impact state should roughly equal half the running alpha state.

(b) The dependency of α, I, and Q should be linear in ρ.

(c) The dependency of the fundamental P&L and the transaction costs should be quadratic in ρ.

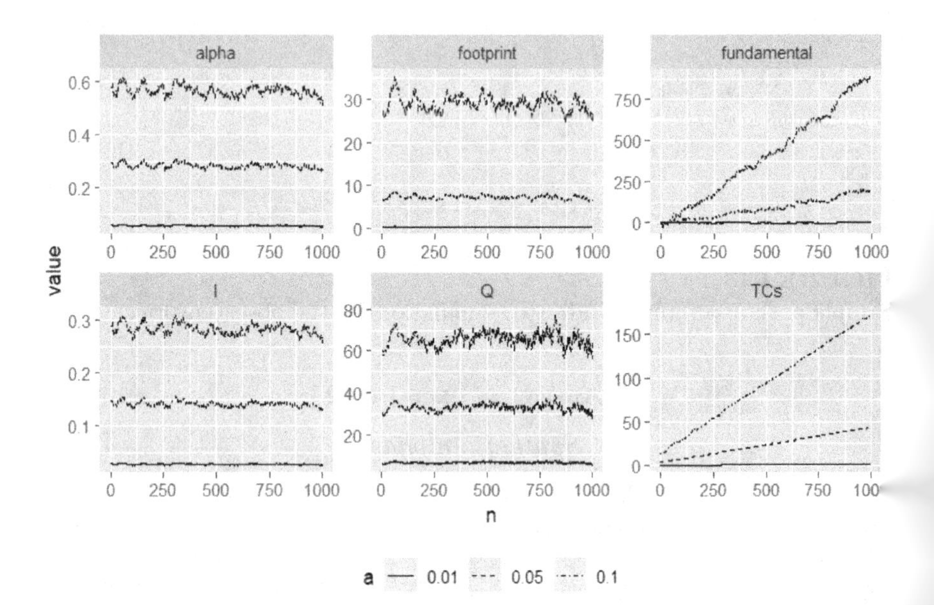

FIGURE C.6
Trading metrics for the optimal trading strategy in the presence of alph signals ranging from $\rho = 0.01$ to $\rho = 0.1$. One estimates the alpha α_t with prediction horizon of $h = 100$ and α'_t with a prediction horizon of $h = 1$. On defines the sign of α, I, Q, and the trading footprint considering the alph signal's sign.

Answer of exercise 13

1. See exercise 12.
2. The reader finds the impact of a TWAP order in exercise 1.

$$I_{t+1} = e^{-\beta} I_t + \frac{\lambda}{\beta}(1 - e^{-\beta})(Q_{t+1} - Q_t).$$

3. The terminal impact state I_{t+1} equals the trade's implied alpha. The implied alpha depends on the trade size $Q_{t+1} - Q_t$ and the initial impact state I_t:

$$\begin{aligned} I_{t+1} &= \Lambda_1\left(\frac{1}{\lambda} I_t + Q_{t+1} - Q_t\right) \\ &= \frac{2}{2+\beta}\left(I_t + \lambda(Q_{t+1} - Q_t)\right). \end{aligned}$$

4. See Figure C.7 for a simulation of α, daily transaction costs, F, and Y.

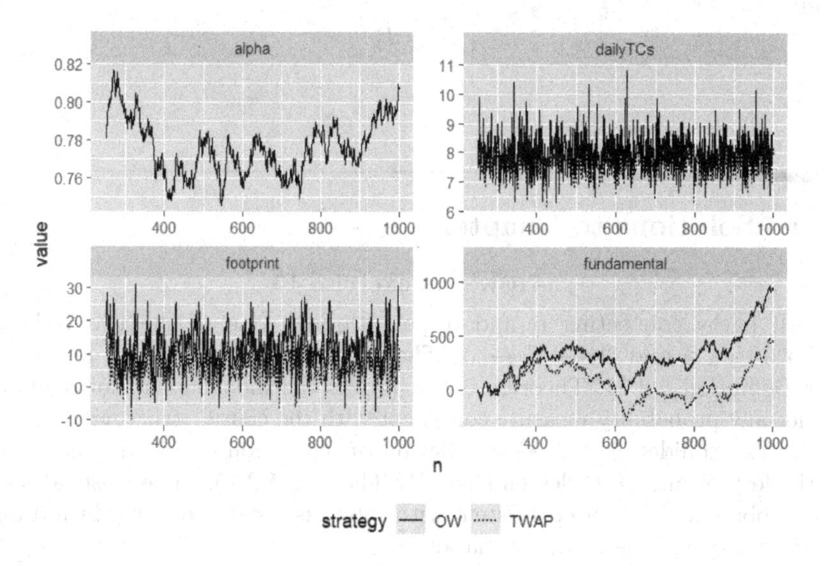

IGURE C.7
imulation of a simplified Markovitz portfolio under two execution schemes: ie Obhizaeva and Wang (OW) execution and TWAP execution.

Answer of exercise 14

1. For a Nash equilibrium, assume the strategy of the $(n-1)$ other players ly jumps at $t = 0$ and is otherwise differentiable. This assumption leads to e partial equilibrium conditions

$$\forall t \in (0,T), \quad I_t^i = \frac{1}{2}\left(\tilde{\alpha} + \alpha_t - \sum_{k \neq i} I_t^k - \beta^{-1}\alpha_t' + \beta^{-1}\sum_{k \neq i}\left(I_t^k\right)'\right)$$

d

$$I_T^i = \tilde{\alpha} - \sum_{k \neq i} I_T^k.$$

Looking for a symmetric equilibrium leads to

$$(n-1)I_t' = \beta\left((1+n)I_t - \tilde{\alpha} - \alpha_t + \frac{1}{\beta}\alpha_t'\right)$$

with terminal condition

$$I_T = \frac{1}{n}\tilde{\alpha}.$$

2. In the mean-field limit, one compares the total impact state $\bar{I}_t = \sum_i I_t^i$ with the single player optimum $I_t^* = \frac{1}{2}\left(\tilde{\alpha} - \alpha_t + \frac{1}{\beta}\alpha_t'\right)$ and $I_T^* = \tilde{\alpha}_T$.

$$\bar{I}_t' = \beta\left(\bar{I}_t - I_t^*\right)$$

and

$$\bar{I}_T = I_T^*.$$

C.4 Solutions to Chapter 5

Answer of exercise 15

1. All paths connecting $\tilde{\alpha}$ and α pass through one of the two colliders $\tilde{\alpha} \to Y \leftarrow \alpha$ and $A \to \bar{S} \leftarrow \alpha$. The empty set d-separates $\tilde{\alpha}$ and α, as per Definition 5.2.14. Theorem 5.2.15 proves the independence of $\tilde{\alpha}$ and α under any probability measure consistent with the causal structure.

2. The collider $A \to \bar{S} \leftarrow \alpha$, rules out d-separation of $\tilde{\alpha}$ and α using the variable $\bar{S}$ or any of its descendants. By Theorem 5.2.15, there exists at least one probability measure consistent with the causal structure for which α and $\tilde{\alpha}$ are not independent conditional on $\bar{S}$.

In this case, the dependence follows from the common consequence $\bar{S}$ c both alpha signals α and $\tilde{\alpha}$. This dependence is Bayesian, and α and $\tilde{\alpha}$ remai causally independent.

3. The chain $S \to A \to \bar{S}$ creates a dependency between S and $\bar{S}$ tha only variable A d-separates. By Theorem 5.2.15, there exists at least or probability measure consistent with the causal structure for which S and are not independent. In this case, the dependence is *causal.*

4. All paths connecting S to $\bar{S}$ either go through A, or go through one the two colliders $S \to Y \leftarrow \bar{S}$ and $\tilde{\alpha} \to Y \leftarrow \alpha$. Therefore, A d-separates and $\bar{S}$. Theorem 5.2.15 proves the independence of S and $\bar{S}$ conditional on under any probability measure consistent with the causal structure.

5. All paths connecting α to A go through the collider $A \to \bar{S} \leftarrow \alpha$ o collider centered on Y. Theorem 5.2.15 proves the independence of α and under any probability measure consistent with the causal structure.

6. This dependence is due to the collider $A \to \bar{S} \leftarrow \alpha$. $\bar{S}$ is a common c sequence of the causally independent variables A and α. By Theorem 5.2. there exists at least one probability measure consistent with the causal str ture for which A and α are not independent conditional on $\bar{S}$.

7. The chain $\tilde{\alpha} \to S \to A$ creates a dependency between the two variables. By Theorem 5.2.15, there exists at least one probability measure consistent with the causal structure for which $\tilde{\alpha}$ and A are not independent.

8. Observing S d-separates the path $\tilde{\alpha} \to S \to A$. All other routes between $\tilde{\alpha}$ and A go through one of the colliders centered around Y. Theorem 5.2.15 proves the independence of $\tilde{\alpha}$ and A conditional on S under any probability measure consistent with the causal structure.

9. Mathematically, the distinction lies between the chain $\tilde{\alpha} \to S \to A$ and the collider $A \to \bar{S} \leftarrow \alpha$.

From a trading perspective, the trade size S and upstream alpha $\tilde{\alpha}$ cause decision A. One the other hand, A causes $\bar{S}$ and, therefore, is causally independent of α. Hence, any dependence between A and α can only be Bayesian.

Answer of exercise 16

1. Consider each counterfactual:

(a) Inspecting the paths from S to Y, the simplest back-door goes through $\tilde{\alpha}$, making it the ideal control variable for identifying do(S) on Y. Applying Theorem 5.2.34, one obtains the identifiability equation

$$p(Y|\,\mathrm{do}(S)) = \int p(Y|\,S, \tilde{\alpha} = x)\, p(\tilde{\alpha} = x)\, dx.$$

Taking expectations yields the identification equation

$$\begin{aligned}\mathbb{E}[Y|\,\mathrm{do}(S)] &= \int y \cdot p(Y = y|\,\mathrm{do}(S))\, dy \\ &= \int\int y \cdot p(Y|\,S, \tilde{\alpha} = x)\, p(\tilde{\alpha} = x)\, dx dy \\ &= \int E[Y|\,S, \tilde{\alpha} = x]\, p(\tilde{\alpha} = x)\, dx.\end{aligned}$$

(b) If not for the link $S \to Y$, the variable α could be used as a control variable to identify the counterfactual. Unfortunately, due to the link, the only viable option is to use Theorem 5.2.34 using the back-door through S. This leads to the identifiability equation

$$\mathbb{E}[Y|\,\mathrm{do}(A)] = \int E[Y|\,A, S = s]\, p(S = s)\, ds.$$

The set $\{S, \alpha\}$ satisfies the back-door criterion from Theorem 5.2.34. This leads to the identifiability equation

$$\mathbb{E}\left[Y|\,\mathrm{do}(A, \bar{S})\right] = \int\int E\left[Y|\,A, \bar{S}, S = s, \alpha = x\right] p(S = s, \alpha = x)\, dsdx.$$

Answer of exercise 17

1. The biased beta on the observational data leads to $\beta \approx 1.1$.
2. This is straightforward.
3. See Figure C.8 and Table C.1. The distribution is broad when the interventional data is small.
4. See Figure C.8 and Table C.1. Causal regularization never produces a beta larger than the one from the observational data. In this case, this truncates the distribution for $\beta > 1.1$. Consequently, the distribution of betas for causal regularization outperforms that using interventional data only, as seen both in Figure C.8 and Table C.1.

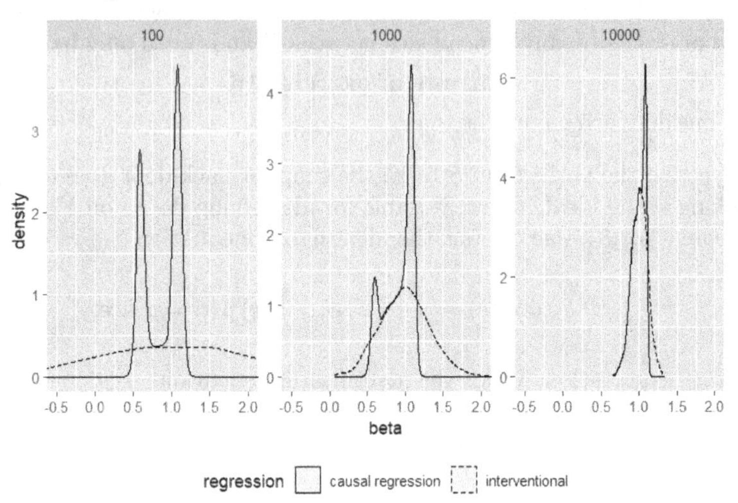

FIGURE C.8
Distribution of estimated β across samples. Each column corresponds to on intervention size. Each data point in the density corresponds to an indeper dent intervention.

TABLE C.1
T-stat of the estimated β across samples.

Interventional data size	interventional regression	causal regularizatior
1e2	0.9	3.7
1e3	3.0	5.1
1e4	10.1	12.1

Answer of exercise 18

1. Consider the matrices

$$X = \begin{pmatrix} \alpha \\ \Delta I \end{pmatrix}$$

and

$$Y = (\Delta P) .$$

The least squares regression coefficients are

$$\begin{pmatrix} \hat{\beta} \\ \hat{\lambda} \end{pmatrix} = \left(X^T X\right)^{-1} X^T Y.$$

The estimator's variance is

$$\begin{aligned} \text{Var}\left[\left(X^T X\right)^{-1} X^T Y \middle| X \right] &= \left(X^T X\right)^{-1} X^T \text{Var}\left[Y|X\right] X \left(X^T X\right)^{-1} \\ &= \sigma^2 \left(X^T X\right)^{-1}. \end{aligned}$$

One has

$$X^T X = \begin{pmatrix} \alpha \cdot \alpha & \alpha \cdot \Delta I \\ \alpha \cdot \Delta I & \Delta I \cdot \Delta I \end{pmatrix}$$

and hence,

$$\left(X^T X\right)^{-1} = \frac{1}{(\Delta I \cdot \Delta I)(\alpha \cdot \alpha) - (\alpha \cdot \Delta I)^2} \begin{pmatrix} \Delta I \cdot \Delta I & -\alpha \cdot \Delta I \\ -\alpha \cdot \Delta I & \alpha \cdot \alpha \end{pmatrix}$$

The variance for $\hat{\beta}$ is

$$\begin{aligned} \text{Var}\left[\hat{\beta}\right] &= \frac{\sigma^2 \Delta I \cdot \Delta I}{(\Delta I \cdot \Delta I)(\alpha \cdot \alpha) - (\alpha \cdot \Delta I)^2} \\ &= \frac{\sigma^2}{n \hat{\sigma}_\alpha^2 (1 - \hat{\rho}^2)} \end{aligned}.$$

where n is the number of data points.

The variance for $\hat{\lambda}$ is

$$\begin{aligned} \text{Var}\left[\hat{\lambda}\right] &= \frac{\sigma^2 \alpha \cdot \alpha}{(\Delta I \cdot \Delta I)(\alpha \cdot \alpha) - (\alpha \cdot \Delta I)^2} \\ &= \frac{\sigma^2}{n \hat{\sigma}_I^2 (1 - \hat{\rho}^2)}. \end{aligned}$$

In both cases, one observes variance inflation by a factor of

$$\frac{1}{1 - \hat{\rho}^2}$$

e to colinearity.

2. Figure C.9 illustrates variance inflation as a function of ρ. The figure presses the equivalent amount of interventional data as a percentage of observational data. One sees that, for low colinearity ρ, the intervention needed is not significantly smaller than the observational data. Even a relation of $\rho = 0.7$ leads to an equivalent intervention size $0.49n$.

For the joint estimation of price impact and alpha, the relevant range of relations is $\rho \in (0.9, 1)$, as typically, one programs the trading strategy to

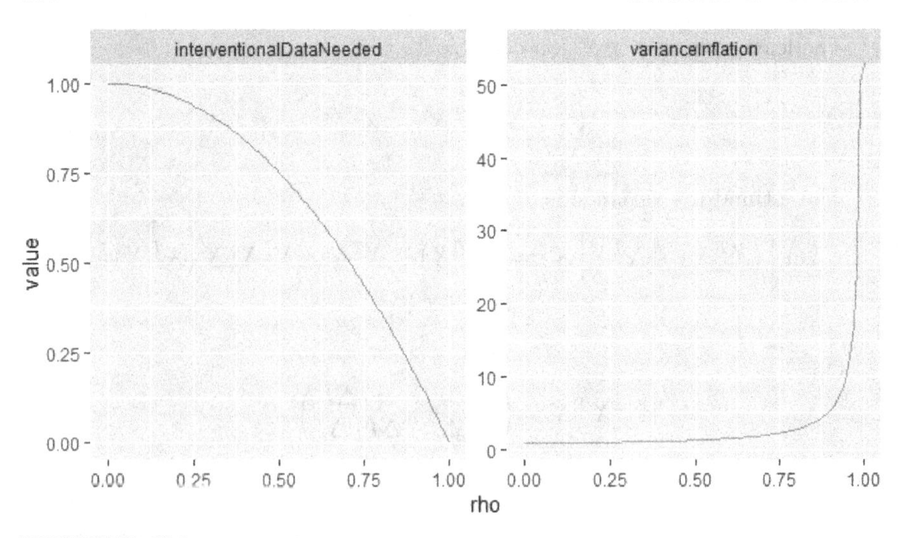

FIGURE C.9
Variance inflation and equivalent intervention size as a function of the colinearity ρ.

follow the signal. Figure C.10 illustrates this regime. For an optimal trading strategy with minimal noise, one could observe $\rho = 0.98$. The correlation leads to an equivalent interventional data of $0.0396n$. Therefore, one halves the confidence interval by setting aside 8% of the trading data for an A-B test compared to using only observational data. Causal regularization further improves the confidence interval by harnessing both data's strengths.

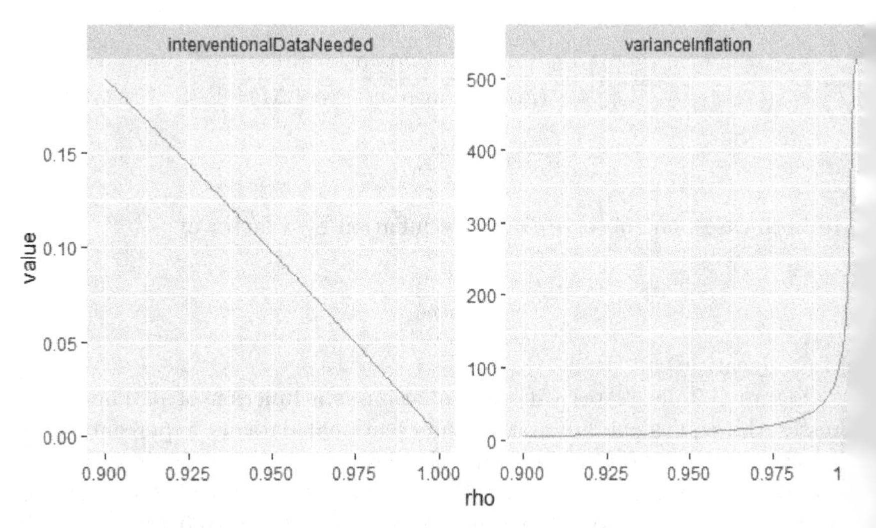

FIGURE C.10
Variance inflation and equivalent intervention size as a function of the co
earity ρ for $\rho \in (0.9, 1)$.

C.5 Solutions to Chapter 6

Answer of exercise 19

1. The link $\alpha_1 \to Q_1$ is the only link into Q_1. α_1 satisfies the back-door criterion from Theorem 5.2.34. This back-door yields the identifiability equation

$$\mathbb{E}\left[\Delta P \middle| \operatorname{do}(Q_1)\right] = \int E\left[\Delta P \middle| Q_1, \alpha_1 = x\right] p\left(\alpha_1 = x\right) dx.$$

2. Adding the link $Q_2 \to Q_1$ creates a second link into Q_1. α_1 and Q_2 cause Q_1 and ΔP. Therefore, the only available back-door is the set $\{\alpha_1, Q_2\}$. Controlling for both yields the identifiability equation

$$\mathbb{E}\left[\Delta P \middle| \operatorname{do}(Q_1)\right] = \int\int E\left[\Delta P \middle| Q_1, \alpha_1 = x, Q_2 = q\right] p\left(\alpha_1 = x, Q_2 = q\right) dxdq.$$

3. The same logic applies to the set $\{\alpha_1, \alpha_2\}$, and one has the identifiability equation

$$\mathbb{E}\left[\Delta P \middle| \operatorname{do}(Q_1)\right] = \int\int E\left[\Delta P \middle| Q_1, \alpha_1 = x, \alpha_2 = y\right] p\left(\alpha_1 = x, \alpha_2 = y\right) dxdy.$$

)ne distinguishes the two previous cases: if Q_1 directly observes Q_2, but not :s upstream alpha α_2, one must control for Q_2. On the other side, if Q_1 has ccess to the two alphas α_1, α_2 but not Q_2, then one must control for α_2 to btain an unbiased estimator of impact.

4. No additional incoming links appear when adding $Q_1 \to Q_2$. Therefore, ie identifiability equation remains the same as in question 1.

$$\mathbb{E}\left[\Delta P \middle| \operatorname{do}(Q_1)\right] = \int E\left[\Delta P \middle| Q_1, \alpha_1 = x\right] p\left(\alpha_1 = x\right) dx.$$

However, the interpretation of the do action differs. In this case, $\operatorname{do}(Q_1)$ easures not only the direct impact $Q_1 \to \Delta P$, but also the impact through e leakage path $Q_1 \to Q_2 \to \Delta P$.

Answer of exercise 20

1. The only path between Q and M that is not a collider is $Q \leftarrow \alpha \to M$. erefore, α *d*-separates Q and M on the causal graph $\mathcal{D}$.
Theorem 5.2.15 proves the independence of Q and M conditional on α.
2. The collider $Q \to D \leftarrow M$ stops *d*-separation conditional on D. By eorem 5.2.15, there exists at least one probability measure consistent with causal structure $\mathcal{D}$ for which Q and M are not independent conditional α, D.

This is an example of a Bayesian dependency. Observing the common consequence D of Q and M makes them dependent. Observing the dark fill provides information about the existence and sign of the order M.

3. α satisfies the back-door criterion, leading to the identification equation

$$p(\Delta P, D|\,\mathrm{do}(Q)) = \int p(\Delta P, D|\,Q, \alpha = x)\, p(\alpha = x)\, dx.$$

Then, one has

$$\begin{aligned}
\mathbb{E}[\Delta P|\,\mathrm{do}(Q), D] &= \int y p(\Delta P = y|\,\mathrm{do}(Q), D)\, dy \\
&= \int y \frac{p(\Delta P = y, D|\,\mathrm{do}(Q))}{p(D|\,\mathrm{do}(Q))} dy \\
&= \frac{\int \int y p(\Delta P = y, D|\,Q, \alpha = x)\, p(\alpha = x)\, dx}{\int p(D|\,Q, \alpha = x)\, p(\alpha = x) dx} \\
&= \frac{\int \mathbb{E}[\Delta P|\,Q, D, \alpha = x]\, p(D|\,Q, \alpha = x)\, p(\alpha = x) dx}{\int p(D|\,Q, \alpha = x)\, p(\alpha = x) dx}.
\end{aligned}$$

In contrast,

$$E[\Delta P|\,\mathrm{do}(Q)] = \int E[\Delta P|\,Q, \alpha = x]\, p(\alpha = x) dx.$$

Answer of exercise 21

1. A causal structure modeling the closing auction should respect the fol- lowing two causal assumptions:

(a) Temporal structure: M_1 and C_1 cannot cause M_0 and C_0.

(b) Revelation times for the closing auction imbalance: C_0 cannot cause M_0 and C_1 cannot cause M_1.

Figure C.11 is an example of such a causal structure. A more complicate causal structure could include alpha signals common to all trading variable

2. No links enter M_0. Therefore, one naively identifies $\mathbb{E}[P_T|\,\mathrm{do}(M_0)]$ un- der the causal graph $\mathcal{C}$. This counterfactual also includes the price impact leakage from M_0 onto other trading variables.

3. M_0 satisfies the back-door criterion, yielding the identifiability equatio

$$\mathbb{E}[P_T|\,\mathrm{do}(C_0)] = \int \mathbb{E}[P_T|\,C_0, M_0 = m]\, p(M_0 = m) dm.$$

4. Unfortunately, under the causal structure $\mathcal{C}$, M_0, M_1, C_0 all direc cause both C_1 and P_T: the identifiability equation requires controlling three variables.

If one makes the Markov-like assumption that the information in M_1 persedes the information in M_0, one erases the link $M_0 \to P_T$. The sa assumption erases $C_0 \to P_T$. Under these additional assumptions, controll for M_1 identifies $\mathbb{E}[P_T|\,\mathrm{do}(C_1)]$.

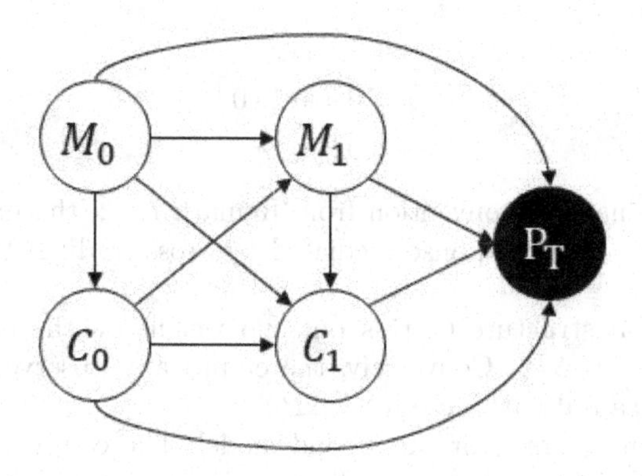

FIGURE C.11
Causal graph $\mathcal{C}$.

C.6 Solutions to Chapter 7

Answer of exercise 22

1. The naive identification formula is the result of a linear regression between ΔP_2 and ΔQ_1 with a confounding variable Q_2:

$$\begin{aligned}\hat{\lambda}_{21}\Delta I(Q_1) &= \mathbb{E}\left[\Delta P_2 | Q_1\right] \\ &= \lambda_{21}\Delta I(Q_1) + \lambda_{22}\mathbb{E}\left[\Delta I(Q_2) | Q_1\right] + \mathbb{E}\left[\epsilon_2 | Q_1\right] \\ &= \left(\lambda_{21} + \rho\lambda_{22}\frac{l_2}{l_1}\right)\Delta I(Q_1).\end{aligned}$$

he corrected identification equation requires the statistician to fit the full ulti-dimensional model. Even though the interaction term the statistician timates only involves the variables ΔQ_1 and ΔP_2, they must cofit the model:

$$\begin{aligned}\hat{\lambda}'_{21}\Delta I(Q_1) &= \int \mathbb{E}\left[\Delta P_2 | Q_1, Q_2 = q\right] p\left(Q_2 = q\right) dq \\ &= \int \left(\lambda_{21}\Delta I(Q_1) + \lambda_{22} q\right) p\left(Q_2 = q\right) dq \\ &= \lambda_{21}\Delta I(Q_1).\end{aligned}$$

2. The naive cross-impact coefficients are

$$\begin{aligned}\hat{\lambda}^0_{21} &= 0.3 + 0 \cdot 0.6 \\ &= 0.3\end{aligned}$$

and

$$\begin{aligned}\hat{\lambda}_{21}^{1} &= 0 + 0.5 \cdot 0.6 \\ &= 0.3.\end{aligned}$$

3. Using the naming convention from Remark 7.5.2, the first model puts no weight on the cross response mediated by cross-trading: terms (b1) and (b3).

On the causal structure $\mathcal{C}$, this puts no weight on the causal pathway $Q_1 \leftarrow \bar{S} \rightarrow Q_2 \rightarrow \Delta P_2$. Conversely, the empirically observed cross-impact stems from the causal pathway $Q_1 \rightarrow \Delta P_2$.

The situation reverses for the second model. The causal pathway $Q_1 \rightarrow \Delta P_2$ carries no weight, and cross-trading via $Q_1 \leftarrow \bar{S} \rightarrow Q_2 \rightarrow \Delta P_2$ mediates the observed cross-impact.

4. While a causal interpretation is preferable, one can interpret the examples within a regression framework. From a statistical perspective, the above results reconcile:

(a) fitting two simple regressions

(b) fitting a multi-dimensional regression

The two methods do not provide the same estimates for the model parameters when the explanatory variables are correlated.

Answer of exercise 23

1. $\bar{\alpha}$ satisfies the back-door criterion from Theorem 5.2.34 in Chapter 5. This control variable yields the identification equation

$$\mathbb{E}\left[\Delta P_2 | \operatorname{do}\left(\bar{Q}\right)\right] = \int \mathbb{E}\left[\Delta P_2 | \bar{Q}, \bar{\alpha} = x\right] p\left(\bar{\alpha} = x\right) dx.$$

2. The scenario removes the node $\bar{\alpha}$ and recovers a naive identification equation for cross-impact.

Answer of exercise 24

1. The solution is identical to the single asset case from Section 2.3.2 of Chapter 2. Applying Corollary 2.3.6, one derives the optimality equation

$$\forall t \in (0,T), \quad \bar{I}_t^* = \frac{\bar{\lambda}}{2 + \beta T}\bar{Q}_T; \quad \bar{I}_T^* = \frac{2\bar{\lambda}}{2 + \beta T}\bar{Q}_T$$

and

$$\forall t \in (0,T), \quad d\bar{Q}_t^* = \frac{\beta}{2 + \beta T}\bar{Q}_T dt.$$

2. The objective function decouples into $n+1$ independent one-dimensional optimal execution problems. Applying Corollary 2.3.6, one derives the factor neutral optimality equations

$$\forall t \in (0,T), \quad I_t^{i*} = \frac{\lambda}{2 + \beta T}\mathring{Q}_T^i; \quad I_T^{i*} = \frac{2\lambda}{2 + \beta T}\mathring{Q}_T^i$$

and therefore,

$$\forall t \in (0, T), \quad d\mathring{Q}_t^{i*} = \frac{\beta}{2 + \beta T} \mathring{Q}_T^i dt.$$

The factor-level solution is identical to the previous question.

3. Again, the objective function decouples into $n + 1$ one-dimensional optimal execution problems. Applying Corollary 3.2.3 from Chapter 3 leads to the factor-neutral optimality equations

$$\forall t \in (0, T), \quad I_t^{i*} = \frac{1}{2}\left(\alpha_t^i - \beta^{-1}\left(\alpha^i\right)_t'\right); \quad I_T^{i*} = 0$$

and

$$\mathring{Q}_T^i = \int_0^T \frac{1}{2\lambda}\left(\beta\alpha_t^i - \left(\alpha^i\right)_t'\right) dt.$$

Similarly, one has the factor-level optimality equations

$$\forall t \in (0, T), \quad \bar{I}_t^* = \bar{\alpha}_t - \beta^{-1}\bar{\alpha}_t'; \quad \bar{I}_T^* = 0$$

and

$$\bar{Q}_T = \int_0^T \frac{1}{2\bar{\lambda}}\left(\beta\bar{\alpha}_t - \bar{\alpha}_t'\right) dt.$$

Answer of exercise 25

1. The simplest causal structure is a two-factor extension of the causal tructure $\mathcal{C}'$, as illustrated in Figure C.12. One may want to empirically verify vhether the factor-neutral impact terms $Q_i \to \Delta P_i$, $Q_j \to \Delta P_j$ are statistiually and economically meaningful.

2. This assumption adds nodes for an alpha signal on the level α_{level} and kew α_{skew} of the implied volatility surface. Figure C.13 illustrates the causal raph.

3. Under the proposed causal structures, the do-actions

$$\mathbb{E}\left[\Delta P_i | \operatorname{do}\left(Q_{\text{level}}\right)\right]; \quad \mathbb{E}\left[\Delta P_i | \operatorname{do}\left(Q_{\text{skew}}\right)\right].$$

pture cross-impact.

Without alpha signals, a naive identification equation holds. With alpha nals, conditioning on the corresponding alpha signal identifies the countertual.

Answer of exercise 26

1. The map is invertible if and only if Λ is invertible. The inverse map is

$$dQ_t = \Lambda^{-1} \cdot \left(dI_t + B \cdot I_t dt\right).$$

2. Lemma 2.3.3 still applies and leads to the risk-neutral objective function

$$J = \mathbb{E}\left[\int_0^T \left(\alpha_t - I_t\right)^T \cdot dQ_t - \frac{1}{2}[I^T, Q]_T + [\alpha^T, Q]_T\right].$$

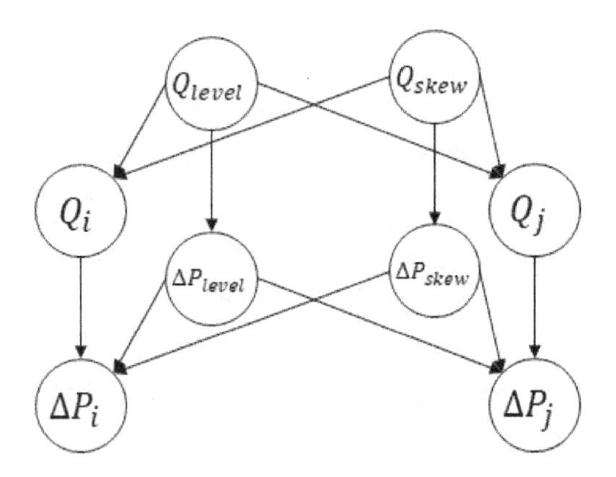

FIGURE C.12
Causal graph for options price impact. One assumes the *level* and *skew* factors of the implied volatility surface carry cross-impact. Individual options also may have an idiosyncratic price impact.

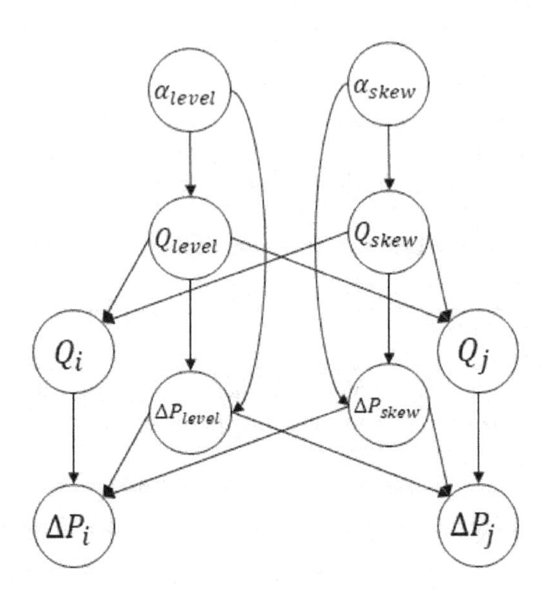

FIGURE C.13
Causal graph for options price impact with alpha signals for the level α_{l} and skew α_{skew} of the implied volatility surface.

Mapping into impact-space and applying integration by parts leads to

$$J = \mathbb{E}\left[\int_0^T (\alpha_t - I_t)^T \cdot \Lambda^{-1} \cdot (dI_t + B \cdot I_t dt) - \frac{1}{2}[I^T, Q]_T + [\alpha^T, Q]_T\right]$$

$$= \mathbb{E}\left[\int_0^T \left(\alpha_t^T \cdot \Lambda^{-1} \cdot B \cdot I_t - I_t^T \cdot \Lambda^{-1} \cdot B \cdot I_t\right) dt\right]$$

$$- \mathbb{E}\left[\int_0^T I_t^T \cdot (\Lambda^T)^{-1} \cdot d\alpha_t + \frac{1}{2} I_T^T \cdot \Lambda^{-1} \cdot I_T\right].$$

There is no price manipulation if Λ^{-1} and $\Lambda^{-1} \cdot B$ are semi-definite positive matrices.

3. The steps are the same as in the previous question. The only difference is that the integration by parts is involved.

$$J = \mathbb{E}\left[\int_0^T (\alpha_t - I_t)^T \cdot e^{-\Gamma_t} \cdot (dI_t + B_t \cdot I_t dt) - \frac{1}{2}[I^T, Q]_T + [\alpha^T, Q]_T\right].$$

One has

$$\int_0^T \alpha_t^T \cdot e^{-\Gamma_t} \cdot dI_t + [\alpha^T, Q]_T = \int_0^T \alpha_t^T \cdot \Gamma_t' \cdot e^{-\Gamma_t} \cdot I_t dt - \int_0^T I_t^T \cdot \left(e^{-\Gamma_t}\right)^T \cdot d\alpha_t$$

nd

$$\int_0^T I_t^T \cdot e^{-\Gamma_t} \cdot dI_t + \frac{1}{2}[I^T, Q]_T = \int_0^T I_t^T \cdot \frac{1}{2}\Gamma_t' \cdot e^{-\Gamma_t} \cdot I_t dt - \frac{1}{2} I_T^T \cdot \Lambda_T^{-1} \cdot I_T.$$

inally, the objective function in impact space is the expectation of

$$\int_0^T \left(\alpha_t^T \cdot \Lambda_t^{-1} \cdot (B_t + \Gamma_t') \cdot I_t - I_t^T \cdot \Lambda_t^{-1} \cdot (B_t + \frac{1}{2}\Gamma_t') \cdot I_t\right) dt$$

$$- \int_0^T I_t^T \cdot \left(\Lambda_t^T\right)^{-1} \cdot d\alpha_t - \frac{1}{2} I_T^T \cdot \Lambda_T^{-1} \cdot I_T.$$

4. $\Lambda_t^{-1} \cdot (B_t + \frac{1}{2}\Gamma_t')$ and Λ_T^{-1} must be semi-definite positive.

Bibliography

[1] E. Abi Jaber and E. Neuman. Optimal liquidation with signals: the general propagator case. *Preprint*, 2022. https://papers.ssrn.com/sol3/papers.cfm?abstract_id=4264823.

[2] J. Ackermann, T. Kruse, and M. Urusov. Reducing obizhaeva-wang type trade execution problems to lq stochastic control problems. *Preprint*, 2022. https://arxiv.org/abs/2206.03772.

[3] ADIA. Adia lab award for causal research in investments. https://www.adialab.ae/call-for-papers, 2022. Accessed 12 January 2023.

[4] Y. Aït-Sahalia and J. Jacod. *High-Frequency Financial Econometrics.* Princeton University Press, 2014.

[5] R. Albuquerque, S. Song, and C. Yao. The price effects of liquidity shocks: A study of the sec's tick size experiment. *Journal of Financial Economics*, 138(3):pp. 700–724, 2020.

[6] I. Aldridge. *High-Frequency Trading.* Wiley Trading, 2010.

[7] A. Alfonsi, A. Fruth, and A. Schied. Optimal execution strategies in limit order books with general shape functions. *Quantitative Finance*, 10(2):pp. 143–157, 2010.

[8] R. Almgren. Using a simulator to develop execution algorithms. http://www.math.ualberta.ca/~cfrei/PIMS/Almgren5.pdf, 2016. Accessed 11 January 2022.

[9] R. Almgren. Real time trading signals. https://www.youtube.com/watch?v=s4IdoWUhRDA, 2018. Accessed 11 January 2022.

[10] R. Almgren and N. Chriss. Optimal execution of portfolio transactions. *Journal of Risk*, 3(2):pp. 5–39, 2001.

[11] R. Almgren et al. Direct estimation of equity market impact. *Risk*, 18:pp. 57–62, 2005.

[12] J. Bacidore. *Algorithmic Trading.* TBG Press, 2020.

[13] J. Bacidore et al. Trading around the close. *The Journal of Trading*, 8(1):pp. 48–57, 2012.

[14] D. Bailey et al. Pseudo-mathematics and financial charlatanism. *American Mathematical Society*, 61(5):pp. 458–471, 2014.

[15] D. Bailey et al. Backtest overfitting in financial markets. *Automated Trader*, issue 39, 2016.

[16] B. Baldacci and I. Manziuk. Adaptive trading strategies across liquidity pools. *Market Microstructure and Liquidity*, 2022. `https://arxiv.org/abs/2008.07807`

[17] G. Banerji. The 30 minutes that can make or break the trading day. *The Wall Street Journal*, 2020. `https://www.wsj.com/articles/the-30-minutes-that-can-make-or-break-the-trading-day-/11583886131`.

[18] P. Bank and D. Baum. Hedging and portfolio optimization in financial markets with a large trader. *Mathematical Finance*, 14(1):pp. 1–18, 2004.

[19] S. Basar. Mifid ii boosts tca. `https://www.marketsmedia.com/mifid-ii-boosts-tca`, 2019. Accessed 1 December 2021.

[20] D. Becherer, T. Bilarev, and P. Fentrup. Optimal liquidation under stochastic liquidity. *Finance and Stochastics*, 22:pp. 39–68, 2018. `https://link.springer.com/article/10.1007/s00780-017-0346-2`.

[21] K. Bechler and M. Ludkovski. Optimal execution with dynamic order flow imbalance. *SIAM Journal on Financial Mathematics*, 6(1):pp. 1123–1151, 2015.

[22] A. Belloni, V. Chernozhukov, and C. Hansen. High-dimensional method and inference on structural and treatment effects. *Journal of Econom Perspectives*, 28(2):pp. 29–50, 2014.

[23] P. Bergault, F. Drissi, and O. Guéant. Multi-asset optimal executic and statistical arbitrage strategies under ornstein-uhlenbeck dynamic *Preprint*, 2021. `https://arxiv.org/abs/2103.13773`.

[24] B. Bernanke. On the implications of the financial crisis f economics. `https://www.federalreserve.gov/newsevents/speec files/bernanke20100924a.pdf`, 2010. Accessed 1 December 2021.

[25] N. Bershova and D. Rakhlin. The non-linear market impact of la trades: Evidence from buy-side order flow. *Quantitative Finan* 13(11):pp. 1759–1778, 2013.

[26] D. Bertsimas and A. Lo. Optimal control of execution costs. *Journa Financial Markets*, 1(1):pp. 1–50, 1998.

[27] F. Black and M. Scholes. The pricing of options and corporate liabilities. *The Journal of Political Economy*, 81(3):pp. 637–654, 1973.

[28] Bloomberg. Transaction cost analysis solutions for mifid ii. `https://www.bloomberg.com/professional/solution/regulation/mifid-ii/mifid-btca`. Accessed 1 December 2021.

[29] G. Bordigoni et al. Strategic execution trajectories. *Econometrics: Mathematical Methods & Programming eJournal*, 2021.

[30] G. Bosilca et al. Dague: A generic distributed dag engine for high performance computing. *Parallel Computing*, 2012. Volume 38, issues 1-2 pages 37-51 (see `https://www.sciencedirect.com/science/article/abs/pii/S0167819111001347`)

[31] J.P. Bouchaud. The inelastic market hypothesis: A microstructural interpretation. *Quantitative Finance*, 22(10):pp. 1785–1795, 2022.

[32] J.P. Bouchaud, J. Farmer, and F. Lillo. How markets slowly digest changes in supply and demand. *Handbook of Financial Markets: Dynamics and Evolution*, pp. 57–160, 2009. See `https://www.sciencedirect.com/science/article/pii/B978012374258250006 3#:~:text=The%20most%20important%20of%20which%20is%20long%20memory,correlations%20in%20the%20initiation%20of%20buying%20versus%20selling`.

[33] J.P. Bouchaud, J. Kockelkoren, and M. Potters. Random walks, liquidity molasses and critical response in financial markets. *Quantitative Finance*, 6(2):pp. 115–123, 2006.

[34] J.P. Bouchaud et al. Fluctuations and response in financial markets: The subtle nature of random price changes. *Quantitative Finance*, 4(2):pp. 176–190, 2004.

[35] J.P. Bouchaud et al. Slow decay of impact in equity markets. *Market Microstructure and Liquidity*, 1(2), 2015. pp. 1-15 (see `https://www.worldscientific.com/doi/10.1142/S2382626615500070`)

[36] J.P. Bouchaud et al. Dissecting cross-impact on stock markets: An empirical analysis. *CFM*, 2016. `https://www.cfm.fr/insights/dissecting-cross-impact-on-stock-markets-an-empirical-/analysis`.

[37] J.P. Bouchaud et al. *Trades, Quotes and Prices.* Cambridge University Press, 2018.

[38] Quantitative Brokers. A brief history of implementation shortfall. `https://quantitativebrokers.com/blog/a-brief-history-of-implementation-shortfall`, 2018. Accessed 30 November 2021.

[39] Quantitative Brokers. The paradox of the pre-trade cost model. https://quantitativebrokers.com/blog/the-paradox-of-the-pre-trade-cost-model, 2019. Accessed 13 January 2022.

[40] B. Brookfield. Delta-based simulations systems, US Patent Application US20160224995A1 2015.

[41] M. Brunnermeier and L.H. Pedersen. Predatory trading. *The Journal of Finance*, 60(4):pp. 1825–1863, 2005.

[42] S. Bubeck and N. Cesa-Bianchi. Regret analysis of stochastic and non-stochastic multi-armed bandit problems. *Foundations and Trends in Machine Learning*, 5(1):pp. 1–122, 2012.

[43] F. Bucci et al. Slow decay of impact in equity markets: Insights from the ancerno database. *Market Microstructure and Liquidity*, 4(3), 2018. pages 1-6 (see https://www.worldscientific.com/doi/epdf/10.1142/S2382626619500060)

[44] F. Bucci et al. Co-impact: Crowding effects in institutional trading activity. *Quantitative Finance*, 20(2):pp. 193–205, 2020.

[45] B. Bulthuis et al. Optimal execution of limit and market orders with trade director, speed limiter, and fill uncertainty. *International Journal of Financial Engineering*, 4(2), 2017. p 1-28 (see https://www.worldscientific.com/doi/abs/10.1142/S2424786317500207#:~:text=We%20study%20the%20optimal%20execution%20of%20market%20and,to%20provide%20better%20control%20on%20the%20trading%20rates.)

[46] B. Burns and M. Wieth. Causality and reasoning: The monty ha[illegible] dilemma. *CogSci 2003: 25th Annual Conference of the Cognitive Scien[illegible] Society*, 2004.

[47] E. Busseti and F. Lillo. Calibration of optimal execution of financi[illegible] transactions in the presence of transient market impact. *Journal [illegible] Statistical Mechanics Theory and Experiment*, 9, 2012. Volume 201[illegible] issue 9, pages 2-32 (see https://iopscience.iop.org/article/1[illegible]1088/1742-5468/2012/09/P09010/meta)

[48] S. Butcher. Is this the best trading job in a bank? [illegible] the worst? https://www.efinancialcareers.co.uk/news/2018/C[illegible]central-risk-desks-banking, 2018. Accessed 10 November 2021.

[49] S. Butcher. Why so few people learn the hottest coding language[illegible] finance. https://www.efinancialcareers.com/news/2020/10/k[illegible]finance-jobs, 2020. Accessed 22 October 2022.

[50] S. Butcher. The coding language you can learn in months for top finance jobs. `https://www.efinancialcareers.com/news/2021/07/learn-kdb`, 2021. Accessed 22 October 2022.

[51] U. Çetin, M. Soner, and N. Touzi. Option hedging for small investors under liquidity costs. *Finance and Stochastic*, 14:pp. 317–341, 2010.

[52] E. Çinlar. *Probability and Stochastic.* Springer, 2011.

[53] F. Caccioli, J.P. Bouchaud, and J.D. Farmer. A proposal for impact-adjusted valuation: Critical leverage and execution risk. *Preprint*, 2012. `https://arxiv.org/abs/1204.0922`.

[54] J. Campbell, A. Lo, and C. MacKinlay. *The Econometrics of Financial Markets.* Princeton University Press, 1997.

[55] Capital.com. Market impact cost. `https://capital.com/market-impact-cost-definition`. Accessed 9 November 2021.

[56] F. Capponi and R. Cont. Multi-asset market impact and order flow commonality. *Preprint*, 2020.

[57] B.I. Carlin, M.S. Lobo, and S. Viswanatha. Episodic liquidity crises: Cooperative and predatory trading. *The Journal of Finance*, 62(5):pp. 2235–2274, 2005.

[58] R. Carmona and F. Delarue. *Probabilistic Theory of Mean Field Games with Applications I.* Springer, 2016.

[59] R. Carmona and F. Delarue. *Probabilistic Theory of Mean Field Games with Applications II.* Springer, 2016.

[60] R. Carmona and L. Leal. Optimal execution with quadratic variation inventories. *Preprint*, 2021. `https://arxiv.org/abs/2104.14615`.

[61] R. Carmona and K. Webster. The self-financing equation in limit order book markets. *Finance and Stochastics*, 23(3):pp. 729–759, 2019.

[62] R. Carmona and J. Yang. Predatory trading: a game on volatility and liquidity. Preprint 2008, `https://carmona.princeton.edu/download/fe/PredatoryTradingGameQF.pdf`

[63] A. Cartea and S. Jaimungal. Incorporating order-flow into optimal execution. *Mathematics and Financial Economics*, 10(3):pp. 339–364, 2016.

[64] A. Cartea, S. Jaimungal, and J. Penalva. *Algorithmic and High Frequency Trading.* Cambridge University Press, 2015.

[65] P. Casgrain and S. Jaimungal. Trading algorithms with learning in latent alpha models. *Mathematical Finance*, 29(3):pp. 735–772, 2019.

[66] P. Casgrain and S. Jaimungal. Mean field games with partial information for algorithmic trading. *Preprint*, 2020. https://arxiv.org/abs/1803.04094.

[67] R. Cesari, M. Marzo, and P. Zagaglia. Effective trade execution. *Preprint*, 2012. https://arxiv.org/abs/1206.5324.

[68] G.P. Chaudhuri. Flow & tell with ishares| jan 2021. https://www.ishares.com/us/insights/flow-and-tell-january-2022#etfs-at-the-intersection-of-trading, 2021. Accessed 18 April 2022.

[69] G.P. Chaudhuri. Flow & tell with ishares| nov 2021. https://www.ishares.com/us/insights/flow-and-tell-november-2021#sector-rotations, 2021. Accessed 18 April 2022.

[70] Y. Chen, U. Horst, and H. H. Tran. Portfolio liquidation under transient price impact – theoretical solution and implementation with 100 nasdaq stocks. *Preprint*, 2019. https://arxiv.org/abs/1912.06426.

[71] European Commission. Directive 2014/65/eu of the european parliament and of the council. https://eur-lex.europa.eu/legal-content/EN/TXT/PDF/?uri=CELEX:02014L0065-20160701&from=EN, 2014. Accessed 1 December 2021.

[72] R. Cont and F. Capponi. Cross-impact in equity markets. https //www.youtube.com/watch?v=MI42T7IxFbY&t=4s, 2020. Accessed 1 April 2022.

[73] R. Cont, M. Cucuringu, and C. Zhang. Price impact of order flo imbalance: Multi-level, cross-sectional and forecasting. *Preprint*, 202 https://arxiv.org/abs/2112.13213.

[74] R Cont, A. Kukanov, and S. Stoikov. The price impact of order boo events. *Journal of Financial Econometrics*, 12(1):pp. 47–88, 2013.

[75] R. Cont and E. Schaanning. Fire sales, indirect contagion a systemic stress testing. *Norges Bank*, 2017. Volume 2017, iss 2, pages 1-50 (https://www.norges-bank.no/contentasset bb47f56979fe4adf9249d1c0ab55c7d1/working_paper_2_17.pdf?v 03/17/2017132952&ft=.pdf)

[76] R. Cont and L. Wagalath. Running for the exit: Distressed sell and endogenous correlation in financial markets. *Mathematical Finar* 23(4):pp. 718–741, 2013.

[77] R. Cont and L. Wagalath. Fire sales forensics: Measuring endogen risk. *Mathematical Finance*, 26(4):pp. 835–866, 2016.

[78] R. Cont and L. Wagalath. Institutional investors and the dependence structure of asset returns. *International Journal of Theoretical and Applied Finance*, 19(2):pp. 835–866, 2016.

[79] A. Criscuolo and H. Waelbroeck. Optimal execution and alpha capture. *The Journal of Trading*, 7(2):pp. 48–56, 2012.

[80] G. Curato, J. Gatheral, and F. Lillo. Optimal execution with non-linear transient market impact. *Quantitative Finance*, 17(1):pp. 41–54, 2017.

[81] Z. Da and S. Shive. Exchange traded funds and asset return correlations. *European Financial Management*, 33(1):pp. 136–168, 2017.

[82] N.M. Dang. Optimal execution with transient impact. *Market Microstructure and Liquidity*, 3(1), 2017. pp. 1-41 (`https://www.worldscientific.com/doi/10.1142/S2382626617500083`)

[83] Lopez de Prado. Causal factor investing: Can factor investing become scientific? *ADIA Lab Research Paper Series*, 2022. `https://papers.ssrn.com/sol3/papers.cfm?abstract_id=4205613`.

[84] E. Derman. *My Life as a Quant: Reflections on Physics and Finance.* Wiley, 2004.

[85] J. Donier and J. Bonart. A million metaorder analysis of market impact on the bitcoin. *Market Microstructure and Liquidity*, 1(2), 2015. pages 1-18 (`https://www.worldscientific.com/doi/abs/10.1142/S2382626615500082`)

[86] R. Donnelly and M. Lorig. Optimal trading with differing trade signals. *Applied Mathematical Finance*, 27(4):pp. 317–344, 2020.

[87] Z. Eisler, J.P. Bouchaud, and J. Kockelkoren. The price impact of order book events: Market orders, limit orders and cancellations. *Preprint*, 2018. `https://arxiv.org/abs/0904.0900`.

[88] I. Ekren and J. Muhle-Karbe. Portfolio choice with small temporary and transient price impact. *Mathematical Finance*, 29(4):pp. 1066–1115, 2019.

[89] Cboe Exchange. U.S. equities market volume summary. `https://www.cboe.com/us/equities/market_share/`, 2022. Accessed 21 April 2022.

[90] New York Stock Exchange. Daily taq. `https://www.nyse.com/market-data/historical/daily-taq`. Accessed 9 November 2021.

[91] A. Farahani et al. A concise review of transfer learning. *2020 International Conference on Computational Science and Computational Intelligence (CSCI)*, 2020.

[92] J.D. Farmer et al. The market impact of large trading orders: Correlated order flow, asymmetric liquidity and efficient prices. *Preprint*, 2008.

[93] Finextra. Central risk books - the new black for capital markets. `https://www.finextra.com/blogposting/13121/central-risk-books---the-new-black-for-capital-markets`, 2016. Accessed 10 November 2021.

[94] FINRA. Algorithmic trading: Notices. `https://www.finra.org/rules-guidance/key-topics/algorithmic-trading#notices`. Accessed 13 December 2021.

[95] FINRA. Guidance on effective supervision and control practices for firms engaging in algorithmic trading strategies. `https://www.finra.org/rules-guidance/notices/15-09`. Accessed 13 December 2021.

[96] A. Frazzini, R. Israel, and T. Moskowitz. Trading costs. *Preprint*, 2018. `https://papers.ssrn.com/sol3/papers.cfm?abstract_id=3229719`.

[97] A. Fruth, T. Schöneborn, and M. Urusov. Optimal trade execution and price manipulation in order books with time-varying liquidity. *Mathematical Finance*, 24(4):pp. 651–695, 2013.

[98] A. Fruth, T. Schöneborn, and M. Urusov. Optimal trade execution an[illegible] price manipulation in order books with stochastic liquidity. *Mathemat[illegible] ical Finance*, 29(2):pp. 507–541, 2019.

[99] A. Fruth, T. Schöneborn, and M. Urusov. Optimal trade execution i[illegible] order books with stochastic liquidity parameters. *SIAM Journal o[illegible] Financial Mathematics*, 12(2):pp. 788–822, 2021.

[100] G. Fu, U. Horst, and X. Xia. Portfolio liquidation games with sel[illegible] exciting order flow. *Mathematical Finance*, 32(4):pp. 1020–1065, 202[illegible]

[101] N. Gârleanu and L.H. Pedersen. Dynamic trading with predictable r[illegible] turns and transaction costs. *Journal of Finance*, 68(6):pp. 2309–234[illegible] 2013.

[102] N. Gârleanu and L.H. Pedersen. Dynamic portfolio choice with frictio[illegible] *Journal of Economic Theory*, 165:pp. 487–516, 2016.

[103] J. Gatheral. *The Volatility Surface, A Practitioner's Guide.* Wiley, 20[illegible]

[104] J. Gatheral. No-dynamic-arbitrage and market impact. *Quantita[illegible] Finance*, 10(7):pp. 749–759, 2010.

[105] J. Gatheral, T. Jaisson, and M. Rosenbaum. Volatility is rough. *Qu[illegible] titative Finance*, 18(6):pp. 933–949, 2018.

[106] J. Gatheral, A. Schied, and A. Slynko. Exponential resilience and decay of market impact. *Econophysics of Order-Driven Markets*, 2011. pp 225–236, publisher: Springer (see `https://link.springer.com/chapter/10.1007/978-88-470-1766-5_15`)

[107] B. Ghojogh and M. Crowley. The theory behind overfitting, cross validation, regularization, bagging, and boosting:tutorial. *Preprint*, 2019. `https://arxiv.org/abs/1905.12787`.

[108] S. Gitlin et al. Supercharging a/b testing at uber. `https://www.uber.com/blog/supercharging-a-b-testing-at-uber/`, 2022. Accessed 30 August 2022.

[109] P. Graewe and U. Horst. Optimal trade execution with instantaneous price impact and stochastic resilience. *SIAM Journal on Control and Optimization*, 55(6):pp. 3707–3725, 2017.

[110] D. Griffin and M. Alastair. Deutsche bank said to lose money on risk-management trades. `https://www.bloomberg.com/news/articles/2018-11-21/deutsche-bank-said-to-lose-money-on-trades-meant-to-/improve-risk`, 2018. Accessed 10 November 2021.

[111] O. Guéant. *The Financial Mathematics of Market Liquidity*. CRC Press, 2016.

[112] F. Guilbaud, M. Mnif, and H. Pham. Numerical methods for an optimal order execution problem. *Journal of Computational Finance*, 16(3):pp. 1460–1559, 2013.

[113] X. Guo, C.A. Lehalle, and R. Xu. Transaction cost analytics for corporate bonds. *Quantitative Finance*, 22(7):pp. 1295–1319, 2022.

[114] X. Guo et al. *Quantitative Trading*. CRC Press, 2017.

[115] W. Hadley et al. Dplyr overview. `https://dplyr.tidyverse.org/`. Accessed 10 December 2021.

[116] C. Harvey et al. Quantifying long-term market impact. *The Journal of Portfolio Management*, 48(3):pp. 25–46, 2022.

[117] C. Holden and A. Subrahmanyam. Long-lived private information and imperfect competition. *The Journal of Finance*, 47(1):pp. 247–270, 1992.

[118] Y. Huang and M. Valtorta. Pearl's calculus of intervention is complete. *Proceedings of the Twenty-Second Conference on Uncertainty in Artificial Intelligence*, 2006.

[119] Institutional Investor. A strategy for all seasons (including inflation). *Institutional Investor*, 2022. `https://www.institutionalinvestor.com/article/b1yvyrlq7716dc/a-strategy-for-all-seasons-including-inflation`.

[120] iShares. ishares factor etfs. `https://www.ishares.com/us/strategies/smart-beta-investing`, 2022. Accessed 19 April 2022.

[121] M. Isichenko. *Quantitative Portfolio Management.* Wiley, 2021.

[122] J. Jacod and P. Protter. *Discretization of Processes.* Springer, 2012.

[123] J. Jacod and A. Shiryaev. *Limit Theorems for Stochastic Processes.* Springer, 2003.

[124] T. Jaisson and M. Rosenbaum. Limit theorems for nearly unstable hawkes processes. *The Annals of Applied Probability*, 25(2):pp. 600–631, 2015.

[125] T. Jaisson and M. Rosenbaum. Rough fractional diffusions as scaling limits of nearly unstable heavy tailed hawkes processes. *The Annals of Applied Probability*, 26(5):pp. 2860–2882, 2016.

[126] D. Janzing. Causal regularization. *Advances in Neural Information Processing Systems*, 2019.

[127] R. Johnson. The state of transaction cost analysis - 2019. `https://www.greenwich.com/market-structure-technology/state-of-transaction-cost-analysis-2019`, 2019. Accessed 1 December 2021.

[128] A. Kaeck, V. van Kervel, and N. Seeger. Price impact versus bid–ask spreads in the index option market. *Journal of Financial Markets*, 59 2021. 59(A), 2022, pages 1-22 (see `https://www.sciencedirect.com science/article/pii/S1386418121000550?via%3Dihub`)

[129] R. Kissell. *Algorithmic Trading Methods.* Academic Press, 2021.

[130] W. Knight. An ai pioneer wants his algorithms to understand the 'why'. `https://www.wired.com/story/ai-pioneer-algorithms-understand-why/`, 2019. Accessed 11 March 2022.

[131] P. Kolm, J. Turiel, and N. Westray. Deep order flow imbalance: Extracting alpha at multiple horizons from the limit order book. *Econometric Modeling: Capital Markets*, 2021. `https://papers.ssrn.com/sol3/papers.cfm?abstract_id=3900141`

[132] P. Kolm and N. Westray. What happened to the rest? a principled approach to clean-up costs in algorithmic trading. *Risk* 2021. `https://www.risk.net/cutting-edge/investments/78658[illegible]a-principled-approach-to-clean-up-costs-in-algo-trading`.

[133] P. Kolm and N. Westray. Mean-variance optimization for simulation order flow. *The Journal of Portfolio Management*, 2022.

[134] KX. Changes in 4.0. `https://code.kx.com/q/releases/ChangesIn4.0/`. Accessed 11 December 2021.

[135] KX. Developing with kdb+ and the q language. `https://code.kx.com/q/`. Accessed 10 December 2021.

[136] KX. Kx developer tools. `https://code.kx.com/q/devtools/`. Accessed 10 December 2021.

[137] KX. Kx download & licenses. `https://kx.com/developers/download-licenses/`. Accessed 10 December 2021.

[138] KX. Kx jupyterq. `https://code.kx.com/q/ml/jupyterq/`. Accessed 10 December 2021.

[139] KX. Qsql query templates. `https://code.kx.com/q/basics/qsql/`. Accessed 10 December 2021.

[140] KX. Realtime databases. `https://code.kx.com/q/learn/startingkdb/tick/`. Accessed 10 December 2021.

[141] KX. Starting kdb+. `https://code.kx.com/q/learn/startingkdb/`. Accessed 10 December 2021.

[142] KX. kdb microservices for ultra-high velocities on a postage stamp footprint. `https://www.brighttalk.com/webcast/19020/570694`, 2023. Accessed 12 January 2023.

143] A.S. Kyle. Continuous auctions and insider trading. *Econometrica: Journal of the Econometric Society*, 53(6):pp. 1315–1335, 1985.

44] F. Lartimore and D. Rohde. Replacing the do-calculus with bayes rule. *Preprint*, 2019. `https://arxiv.org/abs/1906.07125.pdf`.

45] J. Lataillade and A. Chaouki. Equations and shape of the optimal band strategy. *Preprint*, 2020. `https://arxiv.org/abs/2003.04646`.

46] Y.T. Lee, R. Fok, and Y.J. Liu. Explaining intraday pattern of trading volume from the order flow data. *Journal of Business Finance and Accounting*, 28(1-2):pp. 199–230, 2001.

7] C.-A. Lehalle and S. Laruelle. *Market Microstructure in Practice.* World Scientific, 2018.

8] H. Leland. Option pricing and replication with transactions costs. *The Journal of Finance*, 40(5):pp. 1283–1301, 1985.

9] W. Leontief. Academic economics. *Science*, 1982.

] K.C. Li. Asymptotic optimality of cl and generalized cross-validation in ridge regression with application to spline smoothing. *The Annals of Statistics*, 14(3):pp. 1101–1112, 1986.

[151] F. Lillo and J.D. Farmer. The long memory of the efficient market. *Studies in Nonlinear Dynamics & Econometrics*, 8(3), 2004. pp. 1-35 (see `https://econpapers.repec.org/article/bpjsndecm/default8.htm`)

[152] F. Lillo, J.D. Farmer, and R.N. Mantegna. Econophysics: Master curve for price-impact function. *Nature*, 421:pp. 129–130, 2003.

[153] Linkedin. Linkedin jobs search. `https://www.linkedin.com/jobs/search/?keywords=kdb`. Accessed 13 December 2021.

[154] Lobster. Lobster. `https://lobsterdata.com/index.php`. Accessed 10 December 2021.

[155] Lobster. Lobster sample files. `https://lobsterdata.com/info/DataSamples.php`. Accessed 10 December 2021.

[156] C. Lorenz and A. Schied. Drift dependence of optimal trade execution strategies under transient price impact. *Finance and Stochastics*, 17:pp. 743–770, 2013.

[157] K. Lott. Ein verfahren zur replikation von optionen unter transaktionkosten in stetiger zeit. *Dissertation*, 1993. working paper/ dissertation (see `https://www.econbiz.de/Record/ein-verfahren-zur-replikation-von-optionen-unter-transaktionskosten-in-stetiger-zeit-lott-klaus/10000374092`)

[158] H. Lovell. Quantitative brokers. `https://thehedgefundjournal.com quantitative-brokers-10-years-of-optimizing-execution`, 202(Accessed 12 January 2023.

[159] P. Mackintosh. The 2022 intern's guide to trading. `https://www nasdaq.com/articles/the-2022-interns-guide-to-trading`, 202 Accessed 5 July 2022.

[160] A. Madhavan and D. Morillo. The impact of flows into exchange-trad funds: Volumes and correlations. *The Journal of Portfolio Manageme* 44(7):pp. 96–107, 2018.

[161] Capital Fund Management. Packed in like sardines. *CFM*, 2019.

[162] V. Markov. Bayesian trading cost analysis and ranking of broker al rithms. *Preprint*, 2019. `https://arxiv.org/abs/1904.01566`.

[163] H.M. Markowitz. Portfolio selection. *The Journal of Finance*, 7(1): 77–91, 1952.

[164] I. Mastromatteo, B. Tóth, and J.P. Bouchaud. Agent-based models latent liquidity and concave price impact. *Physical Review E*, 89 2014. pp. 1-36 (see `https://www.researchgate.net/publicati`

262338197_Agent-based_models_for_latent_liquidity_and_concave_price_impact)

[165] I. Mastromatteo et al. Trading lightly: Cross-impact and optimal portfolio execution. *Risk*, 2017. July 2017 issue, pp. 78–83.

[166] L.P. Mertens et al. Liquidity fluctuations and the latent dynamics of price impact. *Quantitative Finance*, 22(1):pp. 149–169, 2022.

[167] A. Micheli, J. Muhle-Karbe, and E. Neuman. Closed-loop nash competition for liquidity. *Preprint*, 2021. https://arxiv.org/abs/2112.02961.

[168] Microsoft. Visual studio marketplace. https://marketplace.visualstudio.com/search?term=kdb&target=VSCode&category=All%20categories&sortBy=Relevance. Accessed 10 December 2021.

[169] J. Muhle-Karbe, Z. Wang, and K. Webster. A leland model for delta hedging in central risk books. *Preprint*, 2022. https://papers.ssrn.com/sol3/papers.cfm?abstract_id=4049864.

[170] J. Muhle-Karbe, Z. Wang, and K. Webster. Stochastic liquidity as a proxy for nonlinear price impact. *Preprint*, 2022. https://papers.ssrn.com/sol3/papers.cfm?abstract_id=4286108.

171] J. Muhle-Karbe and K. Webster. Information and inventories in high-frequency trading. *Market Microstructure and Liquidity*, 3(2), 2017. pp. 1–15

172] Nasdaq. Nasdaq equities market data. https://www.nasdaq.com/solutions/nasdaq-equities-market-data-solution. Accessed 9 November 2021.

73] R. Neate and K. Makortoff. Regulators around the world monitor collapse of us hedge fund. https://www.theguardian.com/business/2021/mar/29/credit-suisse-nomura-archegos-sell-off-hedge-fund, 2021. Accessed 21 June 2022.

74] Netflix. Experimentation and causal inference. https://research.netflix.com/research-area/experimentation-and-causal-inference, 2018. Accessed 24 June 2022.

5] Netflix. A survey of causal inference applications at netflix. https://netflixtechblog.com/a-survey-of-causal-inference-applications-at-netflix-/b62d25175e6f, 2022. Accessed 24 June 2022.

] E. Neuman and M. Voß. Trading with the crowd. *Preprint*, 2021. https://arxiv.org/abs/2106.09267.

[177] E. Neuman and M. Voss. Optimal signal-adaptive trading with temporary and transient price impact. *SIAM Journal on Financial Mathematics*, 13:pp. 551–575, 2022.

[178] J. Novotny et al. *Machine Learning and Big Data with Kdb+/q*. Wiley, 2019.

[179] NYSE. An insider's guide to the nyse closing auction. `https://www.nyse.com/article/nyse-closing-auction-insiders-guide`. Accessed 24 March 2022.

[180] A. Obizhaeva and J. Wang. Optimal trading strategy and supply/demand dynamics. *Journal of Financial Markets*, 16(1):pp. 1–32, 2013.

[181] Basel Committee on Banking Supervision. Stress testing principles. `https://www.bis.org/bcbs/publ/d450.pdf`, 2018. Accessed 12 November 2021.

[182] Basel Committee on Banking Supervision. Explanatory note on the minimum capital requirements for market risk. `https://www.bis.org/bcbs/publ/d457_note.pdf`, 2019. Accessed 11 January 2022.

[183] Committee on the Global Financial System. Market-making and proprietary trading: Industry trends, drivers and policy implications. *CGFS* (52), 2014. `https://www.bis.org/publ/cgfs52.pdf`.

[184] Committee on the Global Financial System. Fixed income market liquidity. *CGFS*, (55), 2016. `https://www.bis.org/publ/cgfs55.pdf`.

[185] The pandas development team. Pandas comparison with sq `https://pandas.pydata.org/docs/getting_started/comparison/comparison_with_sql.html`. Accessed 10 December 2021.

[186] J. Pearl. Robustness of causal claims. *20th Conference on Uncertain in Artificial Intelligence*, 2004.

[187] J. Pearl. *Causality*. Cambridge University Press, 2009.

[188] L.H. Pedersen. *Efficiently Inefficient: How Smart Money Invests a Market Prices Are Determined*. Princeton University Press, 2015.

[189] J. Powrie, Z. Zhang, and S. Zohren. Simulating financial m kets at the atomic level. `https://www.man.com/maninstitut simulating-financial-markets`, 2021. Accessed 11 January 2022.

[190] M. Prosperi et al. Causal inference and counterfactual prediction in chine learning for actionable healthcare. *Nature Machine Intellige* 2020.

[191] P. Protter. *Stochastic Integration and Differential Equations*. Sprin 2005.

[192] N. Psaris. *Fun Q: a Functional Introduction to Machine Learning in Q.* Vector Sigma, 2020.

[193] C. Ramey, S. Pulliam, and J. Chung. Archegos founder bill hwang, former cfo charged with securities fraud. `https://www.wsj.com/articles/archegos-founder-and-cfo-charged-with-securities-fraud-/11651059901`. Accessed 21 June 2022.

[194] Microsoft Research. Microsoft research summit 2021. `https://www.microsoft.com/en-us/research/event/microsoft-research-summit-2021/`, 2021. Accessed 11 March 2022.

[195] D.J. Rich. Causal forecasting at lyft. `https://eng.lyft.com/causal-forecasting-at-lyft-part-1-14cca6ff3d6d`, 2022. Accessed 7 September 2022.

[196] T. Roncalli. *Handbook of Financial Risk Management.* CRC Press, 2020.

[197] T. Roncalli et al. Liquidity stress testing in asset management part 2. modeling the asset liquidity risk. *Preprint*, 2021. `https://arxiv.org/abs/2105.08377`.

[198] M. Rosenbaum and P. Jusselin. No-arbitrage implies power-law market impact and rough volatility. *Mathematical Finance*, 30(4):pp. 1309–1336, 2020.

[199] M Rosenbaum and M. Tomas. A characterisation of cross-impact kernels. *Preprint*, 2021. `https://arxiv.org/abs/2107.08684`.

[200] M. Rosenbaum and M. Tomas. From microscopic price dynamics to multidimensional rough volatility models. *Advances in Applied Probability*, 53(2):pp. 425–462, 2021.

[201] D. Rothenhäusler et al. Anchor regression: Heterogeneous data meet causality. *Journal of the Royal Statistical Society Series B*, 83(2):pp. 215–246, 2021.

[202] D. Rubin. Direct and indirect causal effects via potential outcomes. *Scandinavian Journal of Statistics*, 31(2):pp. 161–170, 2004.

[203] E. Said et al. Market impact: A systematic study of the high frequency options market. *Quantitative Finance*, 21(1):pp. 69–84, 2021.

[204] A. Schied and T. Schöneborn. Liquidation in the face of adversity: Stealth vs. sunshine trading. *EFA 2008 Athens Meetings Paper*, 2008.

[205] A. Schied, E. Strehle, and T. Zhang. High-frequency limit of nash equilibria in a market impact game with transient price impact. *SIAM Journal on Financial Mathematics*, 8(1):pp. 589–634, 2017.

[206] M. Schneider and F. Lillo. Cross-impact and no-dynamic-arbitrage. *Quantitative Finance*, 19(1):pp. 137–154, 2019.

[207] B. Schölkopf et al. Towards causal representation learning. *Proceedings of the IEEE*, 109(5):pp. 612–634, 2021.

[208] E. Sciulli. The cio agenda: Alpha opportunities and hidden costs in trading. `https://www.man.com/maninstitute/cio-agenda-alpha-opportunities`, 2021. Accessed 13 January 2022.

[209] European Securities and Markets Authority. Guidelines on liquidity stress testing in ucits and aifs. `https://www.esma.europa.eu/sites/default/files/library/esma34-39-897_guidelines_on_liquidity_stress_testing_in_ucits_and_aifs_en.pdf`, 2020. Accessed 1 December 2021.

[210] US Securities and Exchange Commission. Tick size pilot program. `https://www.sec.gov/ticksizepilot`, 2016. Accessed 10 November 2021.

[211] US Securities and Exchange Commission. Assessment of the plan to implement a tick size pilot program. `https://www.sec.gov/files/TICK%20PILOT%20ASSESSMENT%20FINAL%20Aug%202.pdf`, 2018. Accessed 10 November 2021.

[212] US Securities and Exchange Commission. Market manipulation. `https://www.sec.gov/files/Market%20Manipulations%20and%20Case%20Studies.pdf`, 2019. Accessed 9 November 2021.

[213] US Securities and Exchange Commission. Staff report on algorithmic trading in u.s. capital markets. `www.sec.gov/files/Algo_Trading_Report_2020.pdf`, 2020. Accessed 9 November 2021.

[214] I. Shipster and J. Pearl. Identification of conditional interventional distributions. *Proceedings of the Twenty-Second Conference on Uncertainty in Artificial Intelligence*, 2006.

[215] M. Sotiropoulos and A. Battle. Extended transaction cost analysis (tca). *Deutsche Bank*, 2017. Working paper (See `https://static.autobahn.db.com/microSite/docs/Tca_Extensions.pdf`)

[216] TheStreet Staff. What are etfs and how do they work? `https://www.thestreet.com/dictionary/e/exchange-traded-funds-etf`, 2021. Accessed 19 April 2022.

[217] E. Strehle. Optimal execution in a multiplayer model of transient price impact. *Market Microstructure and Liquidity*, 3(3), 2018. pp. 1–17.

[218] Y. Su et al. The price impact of generalized order flow imbalance. *Preprint*, 2021. `https://arxiv.org/abs/2112.02947`.

[219] A. Subrahmanyam. Transaction taxes and financial market equilibrium. *The Journal of Business*, 71(1):pp. 81–118, 1998.

[220] SWFI. Fund rankings - sovereign wealth fund. `https://www.swfinstitute.org/fund-rankings/sovereign-wealth-fund`, 2022. Accessed 12 January 2023.

[221] thewallstreetlab.com. Episode #77 robert almgren – a deep dive into optimal trade execution in fixed income and futures using quantitative methods. `https://thewallstreetlab.com/77-robert-almgren-a-deep-dive-into-optimal-trade-execution-in-fixed-income-and-futures-using-quantitative-methods`. Accessed 10 January 2022.

[222] R. Tibshirani. Regression shrinkage and selection via the lasso. *Journal of the Royal Statistical Society*, 58(1):pp. 267–288, 1996.

[223] M. Tomas, I. Mastromatteo, and M. Benzaquen. Cross impact in derivative markets. *Preprint*, 2022. `https://arxiv.org/abs/2102.02834`.

[224] M. Tomas, I. Mastromatteo, and M. Benzaquen. How to build a cross-impact model from first principles: Theoretical requirements and empirical results. *Quantitative Finance*, 22(6):pp. 1017–1036, 2022.

225] B. Toth, Z. Eisler, and J.P. Bouchaud. The short-term price impact of trades is universal. *Market Microstructure and Liquidity*, 3(2), 2017.

226] I. Tulchinsky et al. *Finding Alphas.* Wiley, 2015.

227] R. Velu, M. Hardy, and D. Nehren. *Algorithmic Trading and Quantitative Strategies.* CRC Press, 2020.

28] M. Voß. A two-player price impact game. *Preprint*, 2019. `https://arxiv.org/abs/1911.05122`.

29] S. Vyetrenko and S. Xu. Risk-sensitive compact decision trees for autonomous execution in presence of simulated market response. *Preprint*, 2021. `https://arxiv.org/abs/1906.02312`.

30] H. Waelbroeck and C. Gomes. The role of trading in portfolio performance attribution. *Journal of Performance Measurement*, 22(1): pp. 52–67, 2017.

1] H. Waelbroeck et al. Methods and systems related to securities trading, US Patent 8,301,548 2012.

2] S. Wang, R. Schäfer, and T. Guhr. Price response in correlated financial markets: Empirical results. *Preprint*, 2015. `https://arxiv.org/abs/1510.03205`.

[233] S. Wang et al. Signal processing: Systematic m&a arbitrage. *Deutsche Bank Markets Research*, 2016.

[234] K. Webster and N. Westray. Getting more for less - better a/b testing via causal regularization. *Preprint*, 2022.

[235] K. Webster et al. A portfolio manager's guidebook to trade execution. *Deutsche Bank Markets Research*, 2015.

[236] D. Weisberger. Central risk books (crbs) are coming of age. https://www.marketsmedia.com/thoughts-central-risk-books-market-access-rules, 2017. Accessed 10 November 2021.

[237] Wikipedia. Alpha profiling. https://en.wikipedia.org/w/index.php?title=Alpha_Profiling, 2017. Accessed 14 January 2022.

[238] A. Woodie. Kx welcomes new languages to speedy analytics database. https://www.datanami.com/2021/11/02/kx-welcomes-new-languages-to-speedy-analytics-database/. Accessed 20 December 2021.

[239] J. Wu et al. Signal processing: The logistics of supply chain alpha. *Deutsche Bank Markets Research*, 2015.

[240] O. Zakharov. Q insight pad. http://www.qinsightpad.com/download.html. Accessed 10 December 2021.

[241] E. Zarinelli et al. Beyond the square root: Evidence for logarithmi dependence of market impact on size and participation rate. *Marke Microstructure and Liquidity*, 1(2):pp. 1–38, 2015.

[242] W.X. Zhou. Universal price impact functions of individual trades in a order-driven market. *Quantitative Finance*, 12(8):pp. 1253–1263, 201

[243] I. Zovko. Navigating dark liquidity. *Preprint*, 2017. https://arxiv.org/abs/1710.06350.

Index

Printed in the United States
by Baker & Taylor Publisher Services